**Pesticide Selectivity, Health and the Environment**

Since the middle of the twentieth century, pesticides have played a vital role in preventing depradations by weeds, pest and diseases on crops, as well as combating insect carriers of human and animal diseases. However, the non-target effects of pesticides in the environment, and their potential to cause harm to the health of human beings has led to public mistrust of these compounds. This text considers mechanisms of pesticide selectivity and evaluates risks to the environment and public health that arise from pesticide use. Key features include comprehensive accounts of modern pesticides, current regulatory protocols, considerations of dose and exposure in relation to risk, with a final chapter on comparative risk analysis.

BILL CARLILE lectured on pesticide selectivity and toxicology for many years at Nottingham Trent University, where he acted as an instructor and examiner in pesticide application, as well as carrying out research and contractual work on the mode of action of pesticides and resistance. He now lives and works in Ireland.

# Pesticide Selectivity, Health and the Environment

BILL CARLILE
*Nottingham Trent University*

CAMBRIDGE UNIVERSITY PRESS
Cambridge, New York, Melbourne, Madrid, Cape Town, Singapore, São Paulo

Cambridge University Press
The Edinburgh Building, Cambridge CB2 2RU, UK

Published in the United States of America by Cambridge University Press, New York

www.cambridge.org
Information on this title: www.cambridge.org/9780521811941



First published 2006

Printed in the United Kingdom at the University Press, Cambridge

*A catalogue record for this publication is available from the British Library*

ISBN-13 978-0-521-81194-1 hardback
ISBN-10 0-521-81194-5 hardback

ISBN-13 978-0-521-01081-8 paperback
ISBN-10 0-521-01081-0 paperback

---



Every effort has been made to correctly describe the use of pesticides mentioned in this book. Users of pesticides in the UK, and many other countries, must be aware of approved uses of these compounds and adhere to label instructions. Neither the author nor the publisher will accept any responsibility for any damage, accidental or otherwise, arising from the use of any of the pesticides mentioned in this book.

---

# Contents

# Preface

This text considers the risks to human health and other biota in the environment arising from pesticide use. The basis of pesticide use lies in selectivity of compounds: the ability at the dose applied to control target weeds, pests and diseases without causing unacceptable effects to human health or other non-target species. The text considers the mechanisms of selectivity that explain why many pesticides are specific in their action – some to an almost extraordinary degree. Considerations of dose form an integral part of the evaluation, and risks to human health and the environment are compared with those arising from other hazards.

The stimulus to produce this text originated in the late 1990s when, in the same week, the author experienced great frustration in questioning speakers with ingrained, almost blinkered opinions: the first from a major pesticide company keen to emphasise the very low risks to health from pesticide residues, but who would not acknowledge the problems arising from the use of his company's products in their concentrated form in developing countries; and secondly from an organic cooperative who claimed that major medical problems were likely to arise from pesticide residues in food, but would not recognise the much greater problem that might arise from naturally occurring carcinogens, particularly mycotoxins. Such polarised attitudes to pesticides are common. Indeed, the non-target effects of pesticides have been much debated since the publication of *Silent Spring* by Rachel Carson in 1961. The adverse effects of pesticides continue to be prominently aired in the popular media, and there is consequently an extensive mistrust of pesticides among the public at large.

The author is grateful to those publishers and other organisations who have allowed reproduction of many figures and tables. The availability of much information, especially from governmental organisations, in the public domain via the internet is now very welcome; with the usual proviso that, as with other contentious issues, there are many less-than-credible websites that refer to pesticides. The author wishes to acknowledge the critical comments on the text from Ken Pallett, Head of Ecotoxicology at Bayer; Phil Russell, formerly of Schering; Graham Partington, co-founder of Agrisearch UK and now at Risley Crop Consultants; colleagues at Nottingham Trent University especially Mark Darlison and Chris Terrell-Nield, who along with Jo Martin from Agrisearch UK and Tom Robinson from Syngenta kindly provided some of the illustrations; and particularly Sarah Price, who copy-read the text, and provided many excellent suggestions for improvement. Having said that, any errors, omissions or inherent bias in the text are strictly the author's responsibility.

The author wishes to especially express his appreciation of the knowledge gained from students and their supervisors via the superb sandwich training schemes associated with undergraduate courses at Nottingham Trent University, as well as at Harper-Adams University College.

Finally, heartfelt thanks go to Sylvia for her encouragement and patience during the preparation of this text.

# Acknowledgements and permissions

The author is grateful to the following who kindly allowed reproduction or adaptation of Figures and Tables as indicated.

British Crop Protection Council Enterprises, Alton, Hampshire for Figures 2.2, 8.3; 8.5 and Tables 2.1; 8.5; 8.6; 8.7; 8.8, 9.6; 10.12, 10.13 and 10.18.

The UK Department of Environment, Food and Rural Affairs, in particular the Pesticides Safety Directorate, York and Central Science Laboratory, Sand Hutton for Figures 1.8; 7.6; 7.7; and Tables 1.2; 6.1; 6;2; 6.5; 7.8; 7.9; 7.10; 8.2; 9.2; 9.4; 9.13; 10.11. These figures and tables are reproduced through Core Licence No C02W0007764: and Value Added Licence V2005000611.

The Food and Agriculture Organisation of the United Nations for Figures 1.1; 1.5; 1.7; 10.1

Taylor and Francis, Andover, Hampshire, UK for Tables 3.1 and 9.8

Robert Timm at the University of Kansas for Tables 5.1 and 5.2

John Wiley and Sons for Figures 6.4 and 6.6

Tom Robinson for Figure 7.1; Joe Martin for Figure 7;3; Chris Terrell-Nield for the images in Figure 4.1

Bayer CropScience for Figures 7.2 and 8.7

CABI Publishing, Wallingford, Oxfordshire for Figure 7.5 and Table 7.12

The UK Committee on Carcinogenicity for Figure 7.8

The UK Department of Health for Figure 7.9 and Table 7.2

Oxford University Publishing for Table 7.3

The Scandinavian Journal of Work and Environmental Health for Table 7.4

The UK Drinking Water Inspectorate for Table 7.14

The Royal Society of Chemistry for Figures 8.4 and 8.6

The Mammal Society for Figure 10.2

Richard Tren for Figure 10.4

The United States National Center for Policy Analysis for Tables 10.4 and 10.7

The UK Environment Agency for Figure 8.8, and Tables 8.11 and 9.2 (in return for a fee)

# Trademarks and registered trademarks

The following products are trademarks or registered trademarks of the following companies/corporations

**BASF plc**: Basagran; Bavistin; Butisan; Caramba; Comet; Corbel; Dagger; Ensign; Invader; Katamaran; Laser; Masai; Nemolt; Opus; Ronilan: Rovral; Stomp; Terpal; Torque

**Bayer CropScience Limited**: Admire; Artist; Atlantis; Bacara; Bayleton; Baytan; Betanal; Calypso; Cerone; Cheetah; Decis; Draza; Eagle; Elvaron; Flamenco; Folicur; Harvest; Hoegrass; Javelin; Mimic; Mitac; Merlin; Oxytril; Proline; Regent; Ronstar; Scala; Sencorex; Sonata; Sportak; Tattoo; Teldor; Temik; Torch; Totril; Twist

**Belchim Crop Protection**: Centium; Platform; Ranman

**Certis Europe BV**: Applaud; Casoron; Cercobin; Frupica; Sequel

**Crompton (Uniroyal) Registration**: Dimilin

**DowAgrosciences Limited**: Boxer; Dithane; Dursban; Electis; Fazor; Flexidor; Fortress; Garlon; Karathane; Kerb; Profume; Starane; Systhane; Telone; Tordon; Treflan

**Du Pont (UK) Limited**: Ally; Benlate; Curzate; Debut; Lexus; Sanction; Tanos; Vydate

**Headland Agrochemicals Limited**: Agroxone; Pointer

**Makhteshim-Agan (UK) Limited**: Falcon; Goltix

**Margarita Internacional**: Matador; Rubigan

**Monsanto UK Limited**: Avadex; Latitude; Ramrod; Roundup

**Novartis Animal Health**: Neporex

**Nufarm UK Limited**: Arelon

**Sipcam UK Limited**: Rogor; Tairel

**Sumitomo Chemical Agro Europe SA**: Admiral; Regulex

**Syngenta Crop Protection UK Limited**: Acanto; Actellic; Alto; Amistar; Aphox; Axial; Bravo; Beret; Callisto; Cruiser; Cultar; Dicurane; Fubol; Fusilade; Gesaprim; Gesatop; Gramoxone; Grasp; Hallmark; Insegar; Moddus; Nimrod; Patrol; Plover; Plenum; Reglone; Shirlan; Storite; Tilt; Topic; Unix

**Universal Crop Protection Limited**: Cuprokylt

# Abbreviations and acronyms

ACP: Advisory Committee on Pesticides
ADI: Acceptable Daily Intake
ARfD: Acute Reference Dose
BASIS: British Agrochemical Standards Inspection Scheme
EQS: Environmental Quality Standard
FOCUS: Forum for coordination of pesticide fate models and their use
FRAC: Fungicide Resistance Action Committee
HRAC: Herbicide Resistance Action Committee
IEDI: International Estimate of Dietary Intake
IRAC: Insecticide Resistance Action Committee
LERAP: Local Environmental Risk Assessment Plan for Pesticides
MAC: Maximum Admissible Concentration
MCL: Maximum contaminant level
MCLG: Maximum contaminant level goal
MEL: Maximum exposure limit
NEDI: National estimate of dietary intake
NOAEL: No Observed Adverse Effect Level
OECD: Organisation for Economic Cooperation and Development
OES: Occupational Exposure Standard
PIC: Prior Informed Consent
RRAG: Rodenticide Resistance Action Group
TMDI: Theoretical Maximum Daily Intakes
WIGRAMP: Working Group on the Risk Assessment of Mixtures of Pesticides
UNEP: United Nations Environmental Programme
USEPA: United States Environmental Protection Agency

# Units

Concentrations of pesticides are often expressed in terms of mass per unit weight as milligrams per kilogram (mg/kg) or as mass per unit volume as milligrams per litre (mg/l). These values are often referred to as parts per million, respectively as wt/wt or wt/vol.

One thousand times lower in concentration are the identical units of micrograms per kilogram (μg/kg); micrograms per litre (μg/l); or parts per billion.

A further thousand times lower are nanograms per kilogram (ng/kg); nanograms per litre (ng/l); or parts per trillion.

Finally picogram values are one thousand times lower than nanogram concentrations.

# 1
# The rationale, principles and regulation of pesticide use

## 1.1 Introduction

Major improvements in the living standards of human beings have occurred in many parts of the world over the last century. Life expectancies have risen markedly in many countries throughout this period, and much of this increase is due to improvement in health (Figure 1.1). The principal reasons for better health include establishment of comfortable living conditions, improved sanitation, development of drugs, vaccines, antibiotics and other medical advances as well as the provision of good quality drinking water and food.

A plentiful supply of good quality food available throughout the year is taken for granted by the many inhabitants of prosperous countries in the European Union (EU), North America and parts of the Pacific rim. Food shortages in these areas no longer occur and the effects of malnutrition and indeed starvation, manifest in the earlier years of the last century, are rarely seen. Food production in excess of national requirements is common. The United States is a major exporter of maize, soya and wheat. In the UK the production of wheat rose from less than 80% of national requirements in the 1970s to over 120% in the 1990s leading to a considerable export trade by the middle of that decade. Indeed current food production may well be sufficient to support the nutritional needs of the entire population of the globe. The problem, particularly for the less wealthy developing countries, is the cost of production allied to their often-poor infrastructure, internal disputes within these countries and in some cases the political/economic willingness of developed countries to redistribute food.

The self-sufficiency in foodstuffs of many developed countries has involved the establishment and maintenance of sophisticated intensive agricultural systems. Consistent improvements in quality and yield of crops which have occurred over the last century, and particularly the last 50 years, have been achieved by the breeding of superior crop cultivars or varieties, extensive use

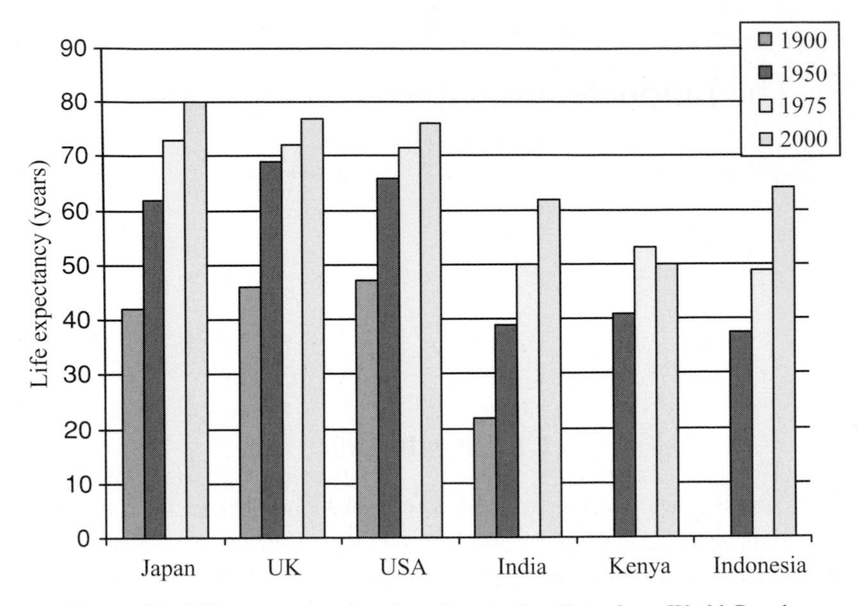


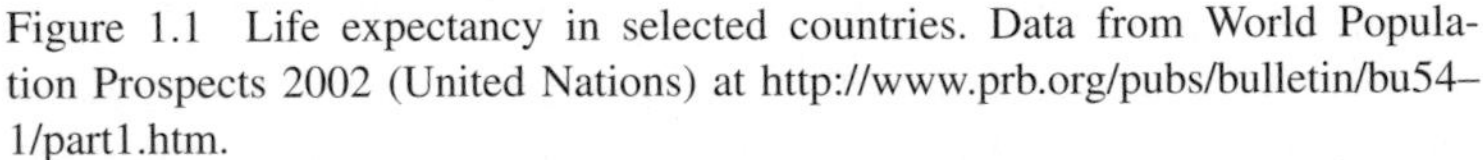
Figure 1.1 Life expectancy in selected countries. Data from World Population Prospects 2002 (United Nations) at http://www.prb.org/pubs/bulletin/bu54–1/part1.htm.

of inorganic fertilisers and implementation of measures to prevent or halt the effects of weeds, pests and diseases. In Western Europe during the first decade of the twenty-first century, yields of over 10 tonnes of wheat can be obtained from an area of land about the size of an average soccer pitch. From the same area under glass, growers can achieve yields of about 250 tonnes of tomatoes in a year. Furthermore the quality of both grain and fruit may be very good, with little damage from attacks by insects and fungi in the field or after harvest.

In such intensive systems of agriculture and horticulture, the competitive effects of weeds, the rotting and nutrient depletion of crops by pathogens and destructive attacks by invertebrate species have been dramatically reduced. Cultivation practices, choice of cultivar and techniques of biocontrol have helped to reduce the incidence of weeds, pests and diseases, but the principal agents used for control of these organisms since the middle of the twentieth century have been chemicals, more commonly known as pesticides.

Pesticides are chemical substances designed to kill or inhibit the growth of unwanted organisms, usually in a selective manner. Under the Control of Pesticides Regulations in the UK (1.5), pesticides are considered as 'any substance, preparation or micro-organism prepared or used for any of the following purposes:

1. protecting plants or wood or other plant products from harmful organisms;
2. regulating the growth of plants;
3. giving protection against harmful creatures;
4. rendering such creatures harmless;
5. controlling organisms with harmful or unwanted effects on water systems (including sewage treatment works), buildings or other structures, or on manufactured products;
6. protecting animals against ectoparasites' (Anon, 2000a).

Undesirable species may compete with, parasitise or consume crops and food intended for humans or livestock. Others may directly infect livestock and domestic pets. Further examples of situations which may lead to pesticide application include protection of industrial premises, restaurants and other food outlets; dwellings and furniture from attack by rodents, insects and fungi; clearance of weeds from roads, pavements, railway tracks and other hard surfaces; removal of unwanted species from waterways, and even selective removal of pest or alien species from sites of particular environmental significance.

Pesticides include insecticides, fungicides and herbicides, which are primarily used to kill insects, fungi and weeds respectively. Most regulatory authorities include plant growth regulators as pesticides, although these are used to modify crop or plant growth rather than kill unwanted species. Other compounds include molluscicides and rodenticides. Agricultural pesticides include those used on crops, in ornamental horticulture, forestry and the home garden as well as herbicides for use on non-crop land and in watercourses. Non-agricultural uses include wood preservation, pesticides used in public hygiene and as anti-fouling products and masonry treatments. The use of pesticides became common in most areas of the world after 1945, and in the 50 or so years since their widespread adoption and use, life expectancies have continued to increase across the globe (Figure 1.1), with the exception of parts of Africa where human immunodeficiency virus/acquired immunodeficiency syndrome (HIV/AIDS) has reduced life expectancies since the late 1980s.

## 1.2 Selectivity, toxicity and dose response

The aggregation together of large hectarages of genetically identical plants as crops has provided almost ideal conditions for the development of pathogens, pests and weed species. Given suitable environmental conditions, it is not surprising that microorganisms, insects and weed species with efficient dispersal mechanisms have on occasion caused severe losses in the yield of crops. Efforts to control the problems caused by pests, diseases and weeds have continued throughout human existence. Measures have involved attempts to control the target species without harming crops, human beings or other organisms. The selective nature of control measures is evident within all procedures used to prevent attacks by pests and diseases, as well as the methods employed in weed control. For example, cultivation techniques used in weed control aim to remove competing species whilst leaving the crop unharmed. Successful biocontrol of insect and other pest species in crops is reliant on the selectivity of predators. Breeding plants resistant to plant diseases relies on the incorporation of genes that will specifically prevent invasion by pathogenic fungi, bacteria and viruses.

This underlying concept of selectivity is also evident with the use of chemicals for crop protection and other areas of pest control, such as insect and rodent control in homes, offices and restaurants as well as vector-borne diseases of human beings and other animals. The control of most weeds, and many pests and diseases of crops, of rodents and insects in public health, as well as vectors of human diseases is achieved through use of naturally occurring or synthetic chemicals that are toxic to the target organism but which in the prescribed circumstances of their use may present few risks to those applying the product, consumers or the environment. The basis of this selectivity, and evaluation of the effects that pesticides may have on human health and other components of the environment is the underlying theme of this text.

Albert in his seminal work '*Selective Toxicity*' (Albert, 1985) defines a remedy as having selectivity if it can influence one kind of living cell without affecting others. This concept may be extended to an influence on one organism in the presence of others. It must also be qualified in that selectivity may involve pronounced effects on one organism, with minor effects on others. The extent to which these latter effects are acceptable is a major consideration in the development of drugs and other medicinal products as well as pesticides.

The basis of selective toxicity lies in exploiting the differences that occur within the metabolic and biosynthetic pathways of different organisms. A substance may inhibit a reaction or process that has evolved in one species, but not another. Photosynthesis only occurs in plants and a few bacteria, and some herbicides inhibit reactions that occur solely in the light reactions of this process. Other herbicides specifically block the synthesis of aromatic and branched-chain amino acids that are produced only by plants, and these compounds, some of which are effective at very low dose rates, have no acute effects on organisms other than plant species.

In some cases the site of action of a therapeutic agent may be almost identical in many organisms, and it is only subtle biochemical differences at the molecular level that enable selectivity through, for example, binding of an inhibitor to an enzyme from one species but not another. Sometimes a difference in only a few amino acids may determine whether a molecule is safe or toxic to an organism.

Another mechanism of selectivity relies on detoxification of a therapeutic agent in non-target organisms. Some insecticides, such as the pyrethroids (4.2.1), are rapidly broken down in warm-blooded animals and this allows their use in treatment of insect pests such as lice and fleas. Strobilurin fungicides (3.7.3) are also rapidly degraded in mammals. Differential metabolism of some herbicides allows selective weed control in crops.

The selective toxicity of both naturally occurring and synthetic chemicals has been widely exploited by human beings for a variety of purposes. Crude extracts from plant and animal tissues, as well as elemental substances (such as arsenic) have been used since prehistoric times as poisons to deliberately kill human beings and other animals. Selectivity here has been conferred by application to the individual or victim! Such methods of applying or positioning of toxic substances such that they may only be experienced by target organisms is used in control of rodents by rodenticides, and in the use of some general biocides.

The beneficial properties of many chemical substances have been identified and the active ingredients of many have been characterised and used in the treatment of illness and disease. The fundamental concepts of selectivity, often allied to dose, became evident during the development and use of medicines. Therapy essentially involved the use of a substance or dose that was sufficient to cure the ailment without killing the patient. This principle of selective toxicity is inherent in the use of all medicines today, although safety margins for many modern products are much greater than before. The therapeutic index is the ratio of a toxic dose to a therapeutic dose, and is used as an indicator of safety of medicines. The larger the index, the better the therapeutic value of the compound.

Many direct benefits to human health have resulted from the use of selectively toxic agents. The concept of selectivity is very well illustrated by the development and use of antibiotics, the principal antimicrobial agents used for control of bacterial (and some fungal) diseases of humans and domestic animals. Many antibiotics are natural products produced by fungi and bacteria that live saprophytically on dead material in soil. Some recently developed pesticidal molecules such as the strobilurin fungicides also have their origins in such organisms. Other antibiotics such as the fluoroquinolone group are, like many pesticides, purely synthetic in origin. Antibiotics inhibit processes unique to their targets (Table 1.1), in this case disease-inciting bacteria in humans and other mammals including pets and farm animals. Some such as the penicillins and cephalosporins block a biosynthetic pathway that only occurs in bacteria. Others may block processes, such as protein synthesis or even specific enzymes

Table 1.1 *Modes of action and basis of selectivity of antibiotics*

| Antibiotic group | Basis of selectivity |
|---|---|
| Penicillins; cephalosporins | Block incorporation of acetylmuramic acid into the peptidoglycans of the bacterial cell wall: a process that only occurs in bacteria |
| Erythromycin; streptomycin; gentamicin; chloramphenicol | Block protein synthesis in ribosomes of bacteria, but not in mammals |
| Fluoroquinolones | Inhibit the activity of DNA gyrase, but only in bacteria |

such as DNA gyrase, that occur both in bacteria and other organisms. Differences at the molecular level explain the selective effects of such antibiotics. For example, protein synthesis in bacteria is carried out in ribosomes that are very different from those in human beings. Several antibiotics block protein synthesis in bacterial (70S) ribosomes, but not in the ribosomes (80S) of eukaryotic organisms. Human beings possess DNA gyrase but it is structurally distinct from bacterial DNA gyrase and unaffected by fluoroquinolone antibiotics. Some antibiotics may have adverse or side-effects in their use, and where these occur they may be transitory and greatly outweighed by the benefits conferred to the health of the individual.

In some cases the dose of compounds required to achieve control of disease in human beings is close to the dose that will cause adverse effects on human health. Indeed, some antibiotics, such as gentamicin, may have adverse effects on some recipients at doses only slightly higher than that required for elimination of bacterial diseases. This also occurs with many other compounds used to treat medical, sometimes life-threatening, conditions. For example, the cardiac glycosides are valuable medicinal products obtained directly from plants and are used in the treatment of congestive heart failure. However, the margin between therapeutic and toxic doses is small, and indeed cardiac glycosides have been used at higher doses than in cardiotherapy as deliberate poisons.

These and other examples serve to show the importance of the dose of a compound in relation to its toxic properties. Chemical substances may have physiological effects on a given organism at a particular dose, and no effects on others at the same dose. Furthermore molecules may have adverse effects on the same organism at a certain dose but none or even beneficial effects at a lower dose. Such dose–responses are fundamental to the evaluation of risks posed by all chemical substances – both naturally occurring and synthesised in the laboratory. Perhaps the classic instance is oxygen, essential for almost all forms of life and beneficial if given in slightly higher concentrations to human beings suffering from respiratory ailments. However, if given in very high concentrations respiratory failure may occur, as well as blindness in young babies. Lack of oxygen in human beings, as with most organisms, will lead to death. Other examples are quoted by Berry (1990). Many substances normally present in the human diet are essential in small quantities, but may be severely toxic in large amounts. Tryptophan is an essential amino acid which occurs naturally in many foods. In searches for so-called natural remedies for insomnia and stress, tryptophan has been consumed at doses as high as 17 g per day, and this has resulted in cases, sometimes fatal, of eosinophilia-myalgia syndrome. L-Tryptophan has now been withdrawn in the USA as a dietary supplement. In the UK the use of vitamin $B_6$ as a dietary supplement at much higher doses than that required for maintenance of good health has prompted warnings from government health authorities. Deaths from botulism have occurred from ingestion of foodstuffs containing the botulinum toxin produced by *Clostridium botulinum* at doses as low as a milligram of toxin per kilogram of body weight. However, at even lower doses, application of the toxin to facial muscles has become very popular in cosmetic surgery in Europe and the USA to enhance the appearance of ageing skin.

In plants, the principles of dose and response are demonstrated well in the biphasic action of the auxin plant hormones. Auxins can exert almost diametrically opposite effects depending on their concentration in plant tissues and the physiological/developmental stage of the plant. In most plants, growth by cell elongation and division is promoted at low concentrations. However, at increasing concentrations, growth abnormalities such as twisting and bending of stems, tissue proliferation in roots and shoots and subsequent chlorosis/necrosis occur. This activity at higher concentrations has been exploited in the development of hormone-based herbicides (2.2).

Toxicity allied to dose and response may be further demonstrated by reference to two of the longest used (and naturally occurring) pesticides employed for disease and pest control in crops. Homer recorded the application of sulphur in Greece for control of insect pests of vine, and the use of this element for control of pests and diseases such as powdery mildews has continued over the last two to three thousand years. Sulphur is applied to crops at relatively high doses (over 5 kg per hectare) and at this rate gives good control of powdery mildews as well as some control of mites. At these (and higher concentrations) sulphur has little effect on most of the plants upon which it is sprayed, or on human beings and most other animals. Sulphur is selectively absorbed by powdery mildews, readily penetrates the exoskeleton of mites, and exerts its effects by inhibiting respiration in these organisms. However, a few plants including some apple cultivars are sensitive to sulphur at the concentrations applied to achieve control of powdery mildews, and so of course sulphur should not be used on these cultivars.

Copper salts control certain fungi, being selectively accumulated to toxic concentrations by pathogens such as *Phytophthora infestans*, the causal agent of potato blight. Copper compounds were the first commercial fungicides to be marketed. Copper is less benign than sulphur to many plants, and may cause phytotoxicity if applied at higher than the recommended dose for control of pathogens. The margin of selectivity achieved through dose is far less with copper than with sulphur in disease control. Copper is also toxic to human beings if ingested in high amounts. It is not, however, readily absorbed through the skin, illustrating the importance of routes of exposure when considering the toxic effects of molecules.

However, both copper and sulphur are essential to all living organisms by virtue of their role in cellular processes. Sulphur is a component of the amino acids cysteine and methionine, and copper is a cofactor necessary for the activity of certain enzymes such as superoxide dismutase, tyrosinase and other amine oxidases. These elements are needed in small quantities for the metabolism and growth of all organisms – including fungi. It is only when applied at high dose that their toxic effects override the essential metabolic role.

## 1.3 Concepts of hazard and risk

Organisms are exposed to hazards throughout their lives. A hazard is an event or phenomenon that has the potential to cause harm. The sea is a hazard to human beings and some other vertebrate species since swimming or falling into it may lead to drowning. Similarly driving a car makes the car a hazard since cars are involved in accidents. Hazards are the source of risks. Risk itself is the likelihood of an adverse outcome resulting from a hazard. For example, insurance companies state that cars are more likely to be involved in accidents when driven by inexperienced drivers. This type of risk may be estimated by reference to appropriate data, here accident statistics, and these estimations used in decision making. Insurance companies appreciate the higher risks of damage to vehicles (and human beings) from inexperienced drivers, and adjust their premiums accordingly.

The risks to human health from chemical substances of both synthetic and natural origin are less easily assessed than physical hazards. Risk assessment from chemical substances to both human beings and the environment as a whole is becoming increasingly sophisticated and generally involves:

1. Hazard identification – determination of the adverse effects that a chemical may cause.
2. Dose – response assessment – quantification of the relationship between dose and any adverse effects.
3. Exposure assessment – the intensity, frequency and duration of exposures through contact, inhalation or ingestion under different conditions.
4. Risk characterisation – integration of the information and uncertainties in the previous three steps to predict the adverse effects of exposure to human and other non-target populations.

A major problem in assessing the risks of chemicals to human beings is the paucity of data linking use of products with health effects. People are exposed to very many chemicals during their lifetime and with a few notable exceptions, such as asbestos, nicotine and the thalidomide drugs, it has proved difficult to directly link adverse health effects with exposure to individual chemicals.

Furthermore, problems arise where risk assessment is affected by judgements of value (Berry, 1990). The analgesic paracetamol is often sold in packets that contain a lethal dose to human beings, but there is an acceptance that most of the population would not consume such a dose. However, there are about 800 fatalities a year in the UK due to consumption of paracetamol. Household cleaning products usually include sodium hypochlorite or sodium hydroxide at concentrations sufficient to cause severe skin irritation. Such concentrations are necessary to kill unwanted organisms and clean stains from surfaces and the availability of products that are able to do this is widely accepted. Nevertheless, skin lesions and burns resulting from contact with cleaning fluids are commonly seen in the casualty departments of hospitals in the UK.

Value judgements have become more common, paradoxically linked in the case of chemicals such as pesticides with the ability to detect these at very low concentrations in air, soils, water and foodstuffs. Increasingly sophisticated methods of detection – such as gas-chromatography linked to mass spectrometry, which can detect molecules at very low concentrations (at one part in several millions) – have enabled the detection and identification of all manner of compounds, both natural and synthetic, in a wide range of situations. Problems often arise through a failure to appreciate the link between dose and effect. The fact that the compound is there and can be detected is often perceived as a health or environmental threat, notwithstanding its concentration or the extent of exposure. There is also an innate fear among many members of the public of the invisible – something which pesticides share with radioactivity and gaseous pollutants.

## 1.4 Pesticide discovery and development

Pesticides are hazards due to their ability to kill organisms. They may present great risks to certain, normally target species, but owing to the selectively toxic nature of many pesticides, may present much lower risks to non-target species. However, the fact that most pesticides are deliberately introduced into the environment means that the assessment of risks to human health, non-target organisms and the ecosystem as a whole is a fundamental consideration in the use of these compounds.

This was not always the case. Early efforts to control weeds, pests and diseases in crops involved a search for measures that would prevent or cure the weed, disease or pest at a dose that would not harm the plant. Absence of toxicity to human beings was obviously an additional desirable property, but until relatively recent times this feature was not always considered to be of prime importance. The principal concern, particularly in the not uncommon times of famine and food shortage, was to protect plants, especially crops, from the effects of pests, diseases and weeds.

Some early compounds were inherently toxic to human beings and many other forms of life (Martin, 1964). In the absence of protective clothing, it is not surprising that many chemicals employed in early efforts to control pests, disease and weeds in crops caused ill effects, and even deaths among those using these treatments. The toxic nature of substances such as arsenic and cyanide used in efforts to combat attacks by insects, as well as compounds such as dinitroorthocresol (DNOC) led to fatalities among workers handling and applying these substances. In the case of DNOC, deaths occurred as late as 1950, and the use of this compound was not prohibited in the UK until 1990. Value judgements of risks to human health from pesticide use were very different from those existing today in the developed world.

Although the toxic effects to human beings of chemicals used in crop protection were at least acknowledged in historic times, the impact of chemical treatments on the environment only became appreciated following their widespread use for the control of pests, microorganisms and weeds from the middle of the twentieth century. The principal difference between the use of chemicals in the alleviation of human disease and illness and their use in the control of microorganisms, pests and weeds is that the former are targeted at the treatment of the individual whereas the latter usually involves application techniques that inevitably result in the release of the chemical into the atmosphere and soil, from where it may be washed or leached into water courses. This deliberate introduction of pesticides into the environment has led to the presence of some compounds, albeit in very low concentrations, in

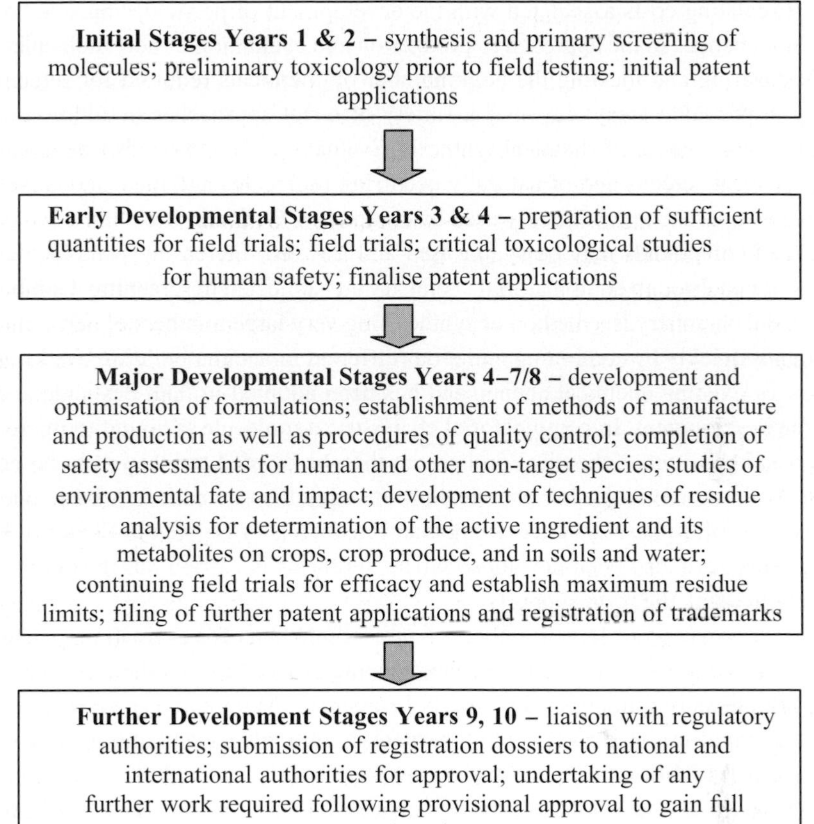


Figure 1.2 Timetable of development of a new pesticide.

almost all areas of the globe, such as polar regions that are remote from sites of application.

In the early twenty-first century, evaluation of the safety to human beings and other non-target species forms the major part of development costs of chemicals used in the control of weeds, pests and microorganisms. Extensive testing for their toxicological and ecotoxicological properties follows the initial selection of molecules that show promise as agents of control of weeds, microorganisms or pest species. The costs of bringing a crop protection chemical to market are estimated at 100–150 million Euros in the early years of the twenty-first century, and most of these costs are associated with toxicological and ecotoxicological studies. A typical programme of pesticide discovery is shown in Figure 1.2.

Escalating costs associated with the development of pesticides have led to major changes in the approach to production and evaluation of new molecules. Discovering and locating the huge number of chemicals required for screening in pesticide assays involves a variety of novel approaches in addition to traditional sources of chemical synthesis. Evaluation of compounds from pharmaceutical screens and of naturally occurring molecules and their derivatives are two approaches commonly used by companies. Collections or 'libraries' of related compounds may be synthesised and assayed. Increasingly, molecules developed through combinatorial chemistry are subjected to screening. Combinatorial chemistry is a method of synthesising very large numbers of new compounds quickly by combining numerous different molecular building blocks of new or existing chemical agents, and has been adopted in both pesticide and drug development. In combinatorial chemistry, a molecule is bound to microscopic beads; other chemical groups can then be bonded to these resin-based beads; in turn even further chemical groups may then be added. A basic outline of a solid-phase system of combinatorial chemistry is given in Figure 1.3. Systems have also been developed where synthesis is carried out in solution, without using the resin support.

The provision of large numbers of new compounds for evaluation as new pesticides has led to a demand for rapid testing procedures and the consequent development of miniaturised systems of screening. Many of these systems are based on the use of 96-well microtitre plates commonly used in immunoassays (Figure 1.4). Microtitre plate wells can accommodate unicellular organisms such as algae and yeasts as well as some small invertebrates. Even isolated protoplasts and specific target enzymes can be used in these microscreens. Growth of unicellular algae, small organisms and protoplasts can be monitored spectroscopically, as can reactions with target enzymes. Robotic techniques are commonly used to operate these high throughput-screening (HTS) systems. Prior to the establishment of high throughput screens, companies would usually synthesise about 0.5–1 g of a new compound for studies with whole organisms, and most major agrochemical companies screened between 5000 and 20 000 compounds a year in their search for new pesticides. With high throughput technology, companies may screen over 100 000 compounds in a year, and in the miniaturised screens as little as 2 mg of the new molecule may be sufficient for tests with a range of organisms and target enzymes (Evans, 1999).

Developments in increasing the numbers of molecules available for evaluation in pesticide screens have been accompanied by efforts to discover new target sites in organisms that might be exploited in a selective manner. The science of genomics has enabled the identification of gene sequences and characterisation of the specific biochemical functions directed by these sequences. Proteomics involves the identification of protein structures and functions associated with

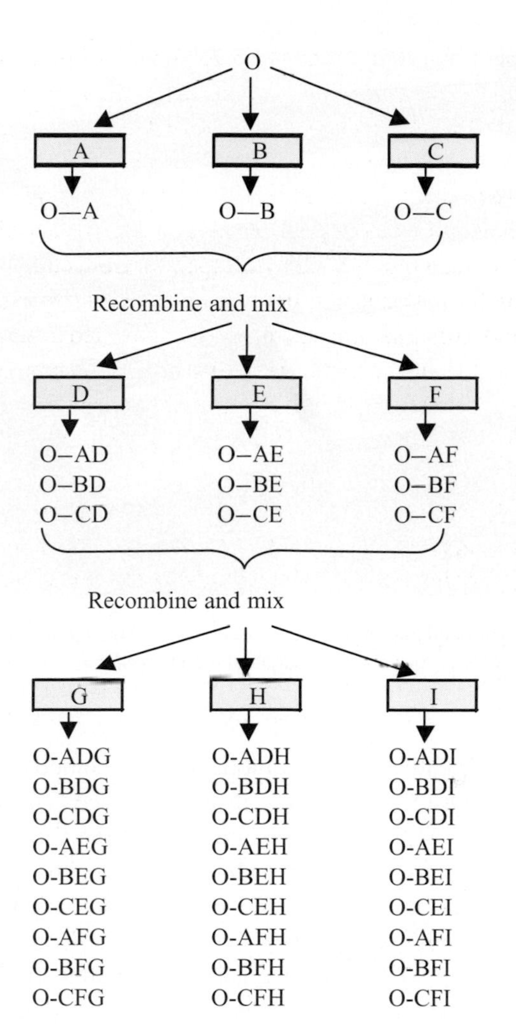


Figure 1.3 A simple example of combinatorial chemistry. Three basic molecules, A, B and C, are linked to a polymer bead, O. The linked beads are mixed, split and added in turn to further molecules (D, E and F). These bind to the beads producing nine different combinations. Again, these can be mixed, split and added to further molecules: in the diagram, 27 combinations are shown. Further splitting and recombining can give very large numbers of chemical combinations.

gene sequences. Genomics and proteomics may enable the identification of new biochemical target sites in weeds, pests and diseases and these target sites may then be exposed to the vast array of compounds produced through combinatorial chemistry in a search for molecules with unique, selective modes of action.

Although techniques of combinatorial chemistry, genomics and proteomics may improve the chances of finding molecules with potential as pesticides

Figure 1.4 A 96-well microtitre plate, commonly used in high throughput screens for pesticide activity, along with an automatic plate reader.

(as well as pharmaceuticals and antibiotics), these compounds must still be shown to work against the target pest, disease or weed. This has proved one of the disadvantages with these procedures, in that the compounds identified may have shown promise in screens with isolated enzyme systems and even unicellular organisms, but when exposed to large plants, target fungi and invertebrates in conventional screens they may prove ineffective. The molecule may not be taken up by the target organisms, or after uptake may not be distributed or translocated to the biochemical site of action. The compound may also be degraded on the surface of the target organisms, or internally before it gets to the biochemical site of action. Consequently, the uptake, distribution and metabolism of compounds must be considered along with candidate molecules and target sites in processes of development of new pesticidal compounds.

## 1.5 Regulation of pesticide use

The potential of pesticides to kill, albeit selectively, and their deliberate and widespread introduction into the environment during use has led to strict control of the sale and application of these compounds in many countries. Regulations governing marketing and use have become more complex and stringent over the last 20 years, and the pesticide industry is regarded as one of the most regulated of all, with a large number of national and international directives, codes and protocols administered and advised by a vast (and increasing) number of committees and other bodies.

In the UK, the Agricultural Chemicals Approval Scheme (ACAS) and Pesticides Safety Precautions Scheme (PSPS) that operated for many years in a voluntary manner were replaced by legislation under the Food and Environmental Protection Act (FEPA) of 1985. The Control of Pesticides Regulations (1986), Plant Protection Products Regulations (1995) and Plant Protection Products (Basic Conditions) Regulations (1997) as part of the act form the principal legislation relating to pesticide use in the UK. In the USA, the Federal Insecticide, Fungicide and Rodenticide Act (FIFRA) of 1951 has been superseded by the Food Quality Protection Act (FQPA) of 1996 (Wilkinson and Barolo, 1999).

Before 1991, member states of the EU operated individual schemes for pesticide registration and use. The introduction of the EU Directive 91/414/EEC aimed to coordinate the regulation of pesticides throughout the EU (Flynn, 1999: Pesticides Safety Directorate, 2003a), and will ultimately replace national legislation such as the Control of Pesticides Regulations in the UK. The Directive defines the principles and procedures to be used for authorisation of plant protection products, and its annexes outline the basis for coordination or harmonisation of data requirements and regulatory decisions. The Uniform Principles developed within the Directive established common criteria for evaluation of products at a national level, and were introduced in 1997. The implementation of Directive 91/414/EEC has led to an EU-wide regulatory process for evaluating the safety of pesticides to human beings and the environment, whilst leaving the responsibility for approval of plant protection products in individual countries to member states. The Directive has established a positive list – Annex 1 – that lists those pesticides that have been judged to be 'without unacceptable risk' to people or the environment.

All new active molecules proposed for use as pesticides within the EU must be deemed acceptable under the Directive. Article 5 of the Directive requires that the use of plant protection products and their residues should not have any harmful effects on human or animal health or on ground water, or have any other unacceptable influence on the environment. In this respect, all existing active ingredients (about 800) introduced to the market prior to 2000 had transitional approval, pending their re-evaluation using modern toxicological and environmental protocols with a view to inclusion in Annex 1.

In the EU three phases of re-evaluation and re-registration have been established. Initially those compounds judged to present the greatest risks to human health and the environment were reviewed, followed by a second group considered to present lower risks, and a third group deemed to present very minor risks. The EU intended that the entire process of re-registration and review for all groups be completed by 2003: because of the immense amount of work required on the part of both companies and regulatory authorities, the deadline for review of the third group of substances has been extended to 2008. A further group, principally disinfectants with some uses in crop protection, will also be reviewed by 2008. Additionally, pesticides may be further reviewed under the Biocides Directive (98/8/EC) that considers their use for pest control in public hygiene, some aspects of veterinary use and in homes and industrial premises. The Directive has similar aims and objectives to 91/414/EEC, which deals with pesticides in crop protection. Identification of products to be supported for re-registration was sought by 2002, and evaluation of these using modern toxicological and environmental protocols is required by 2012. Approved substances will be listed in Annex 1 of the Directive.

In the EU re-registration is leading to a considerable reduction in the numbers of pesticides available for the control of weeds, pests and diseases. Re-registration demands have led to the revocation of approval of many older compounds. Most of these withdrawals have resulted because older compounds are off-patent, with many generic products, and companies are not willing to sponsor the high costs of efficacy and toxicological testing to modern standards. This has led to pressure groups and others referring to these compounds as banned pesticides with the inference that they may present undue risks to human health and the environment.

Regulatory authorities in many other parts of the world have also demanded re-registration of older compounds. In the USA, re-registration demands have been linked to the accelerated approval of so-called reduced-risk compounds. The Environmental Protection Agency gives priority in its registration programme to those pesticides that meet its reduced-risk criteria, defined as a low impact on public health; low toxicity to non-target organisms; low potential for groundwater contamination; lower use rates; low pest resistance potential and compatibility with Integrated Pest Management systems (United States Environmental Protection Agency, 2003). The Australian Pesticides and Veterinary Medicines Authority has reviewed the use of many pesticides since the late 1990s.

In the UK, the Pesticides Safety Directorate (PSD) is responsible for product approval as part of the EU Directives. Within PSD, the Advisory Committee on Pesticides (ACP) was established in the UK under the Food and Environmental Protection Act of 1985 and is responsible for pesticide regulation and approval in the UK. It is assisted in its decisions by various subcommittees such as the Pesticides Residues Committee.

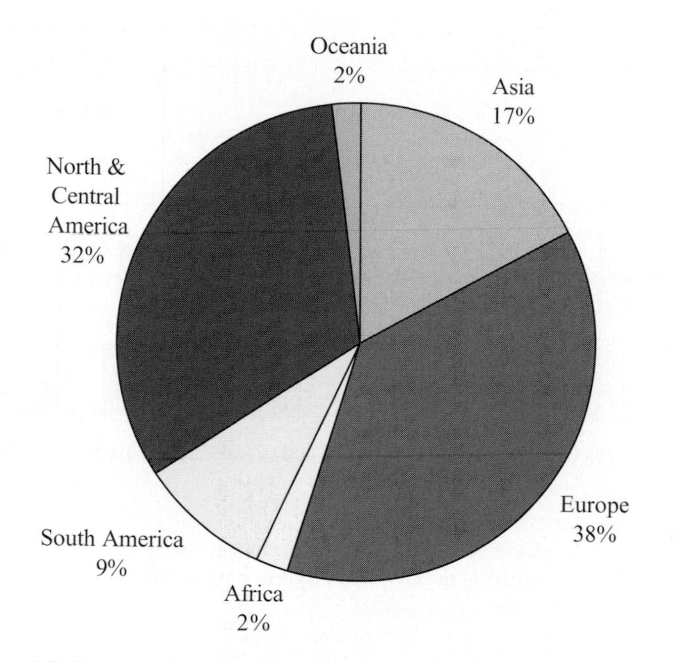


Figure 1.5 Global pesticide use (from the *Yearbook of the Food and Agricultural Organisation of the United Nations*, 1998, published by Food and Agricultural Organisation).

## 1.6 Extent of the pesticide industry

The products of major manufacturers of pesticides are used throughout the world. The pesticide industry is characterised by large multi-national companies, and mergers to form such corporations became common during the 1990s. Spiralling costs of research and development, allied to the increasingly stringent requirements of regulatory authorities have led to mergers between long-established companies. In 1998, the pesticide interests of AgrEvo (itself a product of an earlier merger between Schering and Hoechst) and Rhone–Poulenc joined to form Aventis; subsequently in 2001, the Bayer Corporation acquired Aventis. In 1999, the pesticide arms of two of the world's major crop protection corporations – Novartis (again, formed by an earlier merger of Ciba-Agriculture and Sandoz) and Zeneca were amalgamated to form Syngenta, the world's largest agrochemical corporation. In 2001, the BASF Corporation acquired the pesticide division of the Cyanamid Corporation.

Globally, about 2.5 million tonnes of active ingredients is used annually in efforts to control target weed, disease and pest species, with worldwide sales of pesticides estimated at US$27 000 million per annum in 2001. Data in Figure 1.5 show the principal areas of pesticide use to be in North America and Europe.

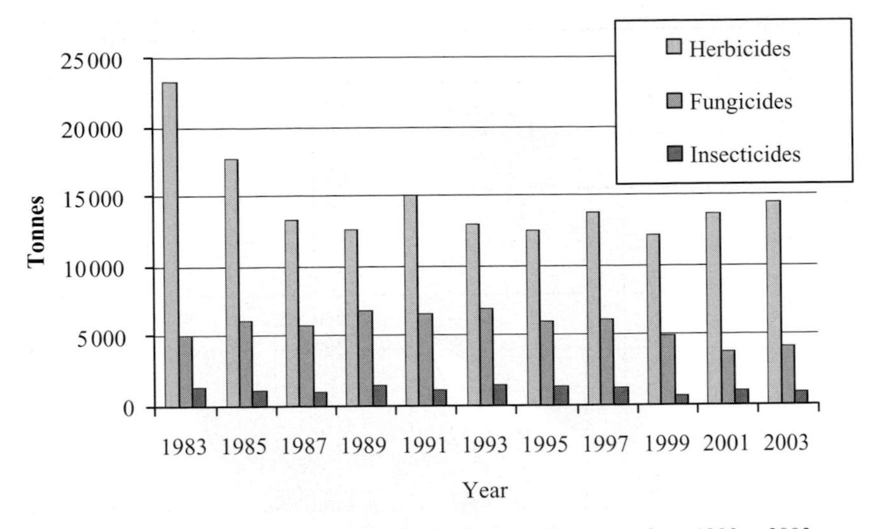


Figure 1.6 UK sales of pesticide active ingredients (in tonnes) from 1983 to 2003 (from the annual reports of the UK Crop Protection Association).

About 45% of all pesticides used worldwide are herbicides, followed by insecticides with around 32% of the market and fungicides at 17%. The USA is the world's major user of pesticides with about 213 000 tonnes of herbicides, 112 000 tonnes of insecticide and 24 000 tonnes of fungicide being deployed in 1997. In the UK and some other Northern European countries, herbicides account for about half of all pesticides used, followed by fungicides and then insecticides (Figure 1.6). Weed control is a priority in the USA and Europe. The situation is very different in other parts of the world, and especially tropical and sub-tropical countries where insecticides are the principal pesticides used and reflect the problems that arthropod pests pose in warmer climates (Figure 1.7).

In some countries such as the Netherlands the amount of pesticide used has declined markedly (Figure 1.8). This is due to several factors including development of compounds effective at low doses, improvements in application technology, deployment of chemicals according to need rather than on a routine basis, national reduced use programmes for pesticides and schemes to encourage organic farming.

Data from the pesticide use survey in the UK have been used to determine the principal active ingredients applied in agricultural situations (Table 1.2). Table 1.2, although including to some extent data derived from mixtures of pesticides, infers that the dose of some compounds needed to control weeds, pests and diseases is much less than that of others that in many cases can do

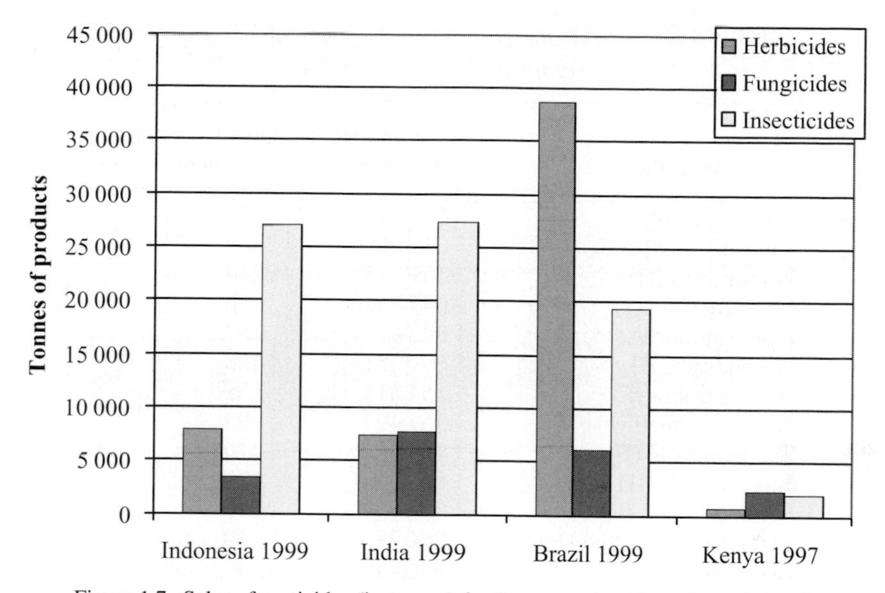


Figure 1.7 Sales of pesticides (in tonnes) for four countries. From the website of the Food and Agricultural Organisation of the United Nations at http://apps.fao.org/lim500/nph-wrap.pl?Pesticides&Domain=LUI&servlet=1.

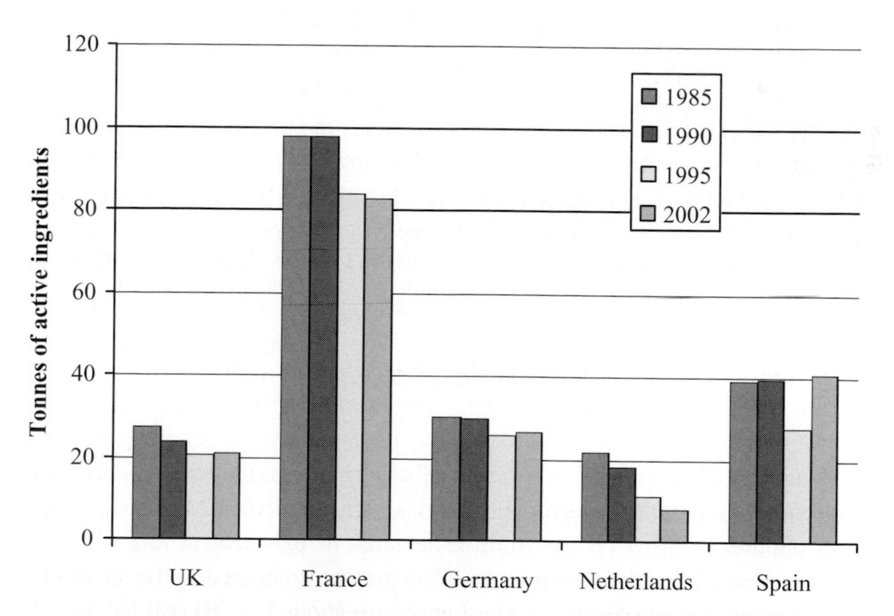


Figure 1.8 Volume of active ingredients used for crop protection in selected European countries (Data from UK Department of the Environment, Food and Rural Affairs (2000). *Design of a Tax or Charge Scheme for Pesticides: Annex C3 Overview of the Pesticide Industry*; available at http://www.defra.gov.uk/environment/pesticidestax/19.htm and the European Crop Protection Association website at http://www.ecpa.be/library/reports/ 12560ECPAStaReview2002.pdf).

Table 1.2 *Pesticide use (excluding sulphuric acid) on arable crops in the UK in 2002*

| Active ingredient(s) | Area treated (hectares) | Quantity applied (tonnes) |
|---|---|---|
| Epoxiconazole (F) | 3 665 844 | 178 |
| Chlormequat (PGR) | 3 017 068 | 2885 |
| Isoproturon (H) | 2 239 215 | 2254 |
| Cypermethrin (I) | 2 047 883 | 50 |
| Fenpropimorph (F) | 1 823 617 | 267 |
| Glyphosate (H) | 1 699 251 | 1488 |
| Azoxystrobin (F) | 1 654 411 | 158 |
| Kresoxim-methyl (F) | 1 557 497 | 93 |
| Tebuconazole (F) | 1 134 382 | 116 |
| Mecoprop-P (H) | 1 079 646 | 527 |
| Pendimethalin (H) | 1 078 753 | 1057 |
| Metsulfuron-methyl (H) | 960 286 | 10 |
| Diflufenican (H) | 936 747 | c. 40 (much used in mixtures) |
| Mancozeb (F) | 889 253 | 1146 |
| Metaldehyde (M) | 866 502 | 334 |
| Fluroxypyr (H) | 853 493 | 97 |
| Flusilazole (F) | 848 570 | 80 |
| Trifluralin (H) | 822 491 | 682 |
| Clodinafop-propargyl (H) | 806 980 | 41 |
| Chlorothalonil (F) | 774 434 | 424 |

F indicates a fungicide; H a herbicide; PGR a plant growth regulator; I an insecticide and M a molluscicide.
Data with permission from Pesticide Usage Survey Report 187: Arable Crops in Great Britain (2002). Garthwaite, D. G., Thomas, M. R., Dawson, A. and Stoddart, H. at the Central Science Laboratory, York. Available at http://www.csl.gov.uk/science/organ/pvm/puskm/arable2002.pdf.

the same job. For example, 55 853 kg of cypermethrin, the most widely used insecticide (and indeed pesticide) in UK agriculture in 1997 (Anon, 2000a), was applied to just over 2.3 million hectares in that year, a rate of about 24 g of active ingredient per hectare. This may be compared to the 99 000 kg of the organophosphate insecticide dimethoate applied to 291 000 ha – a rate of 334 g/ha. Even more striking are the comparisons between herbicides. Over 2.3 million kilograms of isoproturon was applied to about 1.5 million hectares of UK crops in 1997 (an average of 1600 g/ha), compared to just over 3000 kg of metsulfuron-methyl on 765 118 ha (about 25 g/ha). The dose of modern fungicides such as epoxiconazole and cyperaconazole at around 66 and 42 g/ha respectively may be compared to that of chlorothalonil at 424 g/ha, and sulphur, which, in order to exert its fungitoxic effects, must be applied at nearly 5000 g/ha (Anon, 2000a).

This increased potency of pesticides reflects modern developments of selective compounds that are effective at much lower doses than herbicides, fungicides and insecticides brought to the market 40–50 years ago. The basis of selectivity of both old and modern pesticides is the underlying theme of the next four chapters.

# PART I

# Pesticides and their mode of action

The following four chapters consider the selectivity of pesticides in relation to their mechanisms of action. Compounds are generally grouped according to their mode of action, and in some cases this coincides with the chemical family to which they belong, such as the organophosphate and carbamate insecticides, benzimidazole fungicides and auxin-type herbicides. Both the chemical structure and a representative trade name are given. *This does not constitute a recommendation* but hopefully serves to link product names known by farmers, growers and agronomists to the active ingredients that they contain. In general, trade names current in the UK at the time of publication of this book are used, but many of these product names have been adopted more widely by manufacturers. The dose rates given are derived from appropriate product manuals, and refer to the range of maximum doses permitted for the product, again at the time of publication of this book

It is important to note that many pesticides are used in combination to extend the spectrum of activity of the product, and in some cases to cope with problems of pest, weed or pathogen resistance. However, the amount of active ingredient in mixtures rarely, if ever, exceeds the dose recommended for compounds used on their own. In figures of pesticide structure, numbers in parentheses after the dose rate indicate whether more than one application of a pesticide is permitted, and give the maximum number of applications allowed in a year or growing season. Such multiple applications apply mainly to fungicides and insecticides.

# 2

# Herbicides and plant growth regulators

## 2.1 Introduction

The removal of unwanted plant species from crops is mainly accomplished in intensive systems of agriculture by the use of herbicides. The losses in yield due to the competitive effects of weeds have long been recognised, and even today, if weeds are not controlled, very high reductions in yield may occur as shown by the data in Table 2.1. Such high yield losses can easily be prevented by use of appropriate herbicides in the agricultural systems of developed countries, but reductions in yield of 50% or more due to weed infestation are still common in poorer areas of the world. Losses in yield may also be accompanied by reductions in crop quality. For example, cereal and legume grains may be smaller in weed-infested than in weed-free crops, and the grain itself may be contaminated with weed seeds. Unwanted plant species in crops may carry plant pests and diseases, and weeds such as field bindweed (*Convolvulus arvensis*) and cleavers (*Galium aparine*) may also hinder harvesting operations.

Vegetation may also be controlled in amenity and other non-crop areas by herbicidal means, although non-chemical procedures of weed control are often employed. Weeds give an unsightly appearance to flower beds and shrub borders, both in domestic gardens and municipal displays. Colonisation by weeds may also result in the deterioration of hard surfaces such as roads, other tarmac areas, pavements and railway tracks as well as presenting a physical hazard to people and modes of transport. Weeds may even prove problematic in waterways, with species as diverse as the water hyacinth that blocks canals and clogs decorative ponds and other standing bodies of water, to the microscopic blue-green algae whose toxins have caused the death of domesticated stock and pets after drinking water from lakes polluted with these organisms.

Weeds may also pose problems of skin irritation and in some cases their seeds or fruits may prove toxic on ingestion, with examples including, respectively,

Table 2.1 *Single and combined effects of weed species on wheat biomass*

| | Crop biomass in May (grams per square metre) | Grain yield (tonnes/hectare) |
|---|---|---|
| Weed free | 678.5 | 5.01 |
| *Galium aparine* (*G.a.*) (cleavers) | 513.2 | 1.33 |
| *Matricaria perforata* (*M.p.*) (scentless mayweed) | 500.7 | 1.07 |
| *Papaver rhoeas* (*P.r.*) (common poppy) | 458.1 | 0.81 |
| *G.a.+ M.p.* | 444.2 | 0.72 |
| *G.a.+ P.r.* | 338.5 | 0.40 |
| *M.p. + P.r.* | 334.0 | 0.55 |

Data from Wright, K. J., Seavers, G. P. and Wilson, B. J. (1997). Competitive effects of multiple weed species on wheat biomass and wheat yield. Proceedings of the Brighton Crop Protection Conference, Weeds, 497–502.

stinging nettles (*Urtica dioica*) and the berries of the so-called deadly nightshade (*Atropa belladonna*) that have on occasion proved fatal. Seeds of weeds such as corncockle (*Agrostemma githago*) and black nightshade (*Solanum nigrum*) that can contaminate cereals and legume grains respectively during harvesting may prove toxic to consumers. Other weeds may be toxic to animals: poisoning from pyrollizidine alkaloids present in the Oxford ragwort (*Senecio jacobaea*) is a common cause of death and debilitation in ponies and horses in the UK. The problems which weeds cause in crops and other situations are outlined in Figure 2.1, and some of these illustrated in Figure 2.2.

The extensive use of herbicides is linked to their versatility. When compared to labour-intensive methods of weed control such as hand pulling, hoeing, ploughing and/or harrowing, herbicides are more efficient and cost-effective, and the tedium of these mechanical methods is eliminated. In some cases, a single spray of herbicide may prevent weed infestation for many months or, in some non-crop situations, for years.

Particular success has been achieved in the last century with the use of herbicides for weed control in non-row crops such as cereals. Mechanisms of selectivity have been exploited to the extent that unwanted plant species may be eliminated without harm to the crop in question. Broad-leaved weeds may be removed from cereal crops. Molecules also exist which at the appropriate dose will control grass weeds in broad-leaved crops. Furthermore, herbicides

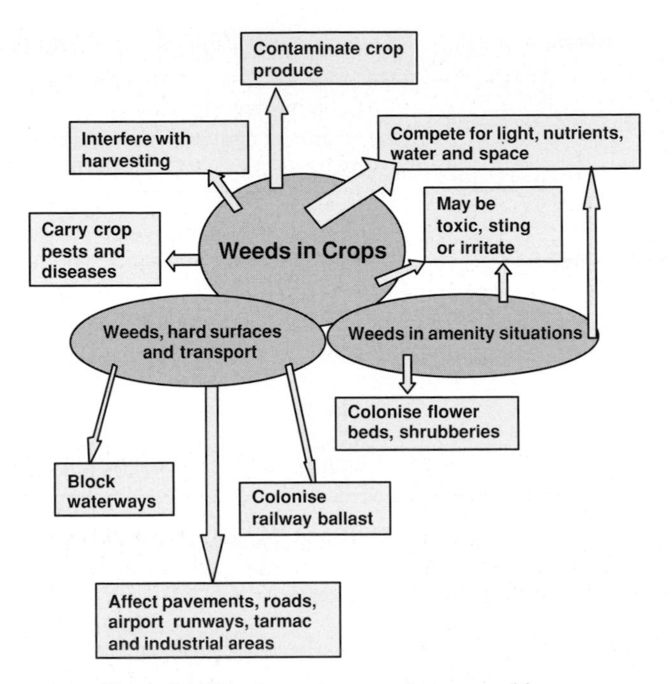


Figure 2.1 Situations where weeds cause problems.

are available that will kill grass weeds, some of which are very closely related to cereals, in cereal crops – a remarkable example of selectivity.

Selectivity between plant species can be achieved in some cases by using different rates of herbicides, and by manipulating the methods and timing of application, as well as adding safeners (2.9) to formulations. However, the basis of selectivity in most cases is explained by events at the biochemical site of action of the herbicide. Primary inhibition at specific biochemical sites may lead to symptoms such as chlorosis, necrosis and stunting, and ultimately death of susceptible plants occurs.

Inhibition of processes that occur only in plants is the principal basis of herbicide selectivity. Some herbicides exert their effects by interfering with the hormonal regulation of plant growth. Many inhibit photosynthesis. Others exploit specific enzyme systems in plants such as those involved with amino acid and carotenoid biosynthesis. Inhibition of these processes means that most herbicides are not acutely toxic to species other than plants. However, the primary site of action of some herbicides occurs not only in plants but also in other organisms and in theory such compounds may be expected to be acutely toxic: in practice they may have little or no effect on the health of the non-target species. Selectivity in such cases may be linked to subtle differences at the

Poppies (*Papaver rhoeas*) in cereals.
Photograph by kind permission of Joe Martin

Weed competition in young wheat

Cleavers (*Galium aparine*) in an ornamental shrub (*Berberis* spp)

Blackgrass (*Alopecurus myosuroides*) in wheat

Weeds on a mainline railway in the UK

Figure 2.2 The damaging effects of weeds.

biochemical level between the target site in plants and other organisms, to the failure of the herbicide to reach the biochemical site of action in non-target species, and/or detoxification in the latter.

The degree of selectivity needed for control of weeds in crops is not, of course, required for non-crop situations and so herbicides capable of killing all plant species are often used. Such so-called total herbicides may also be used before planting or germination of crops to clear weeds. Some of these compounds have been associated with the development of herbicide-resistant crops: here selectivity is conferred by the incorporation of genes whose products allow the crop to break down or avoid the action of the applied herbicide. Other species growing in the crop may not be able to degrade the herbicide, which thus kills them.

Early efforts to control weeds in crops by selective use of chemicals involved the use of substances such as copper sulphate and sulphuric acid. Doses were applied which killed broad-leaved weeds but (hopefully) did not cause too much damage to the growing crop. Selectivity was principally due to differences in crop morphology. After application, the chemicals tended to roll down the narrow erect cereal leaves whilst being retained by broad-leaved plants and indeed being channelled towards the growing points in these species. Nevertheless a certain amount of crop damage was almost inevitable with these corrosive chemicals applied at high dose, and risks to the health of those applying compounds such as sulphuric acid were obviously considerable.

In the early years of the twentieth century, efforts were made to discover chemicals that would offer better weed control and which had fewer adverse effects than materials such as sulphuric acid. Dinitrophenols and dinitrocresols such as dinitroorthocresol (DNOC – Figure 4.15) were developed in the 1930s for selective weed control in cereals and as insecticides (4.4) in other crops. These compounds were taken up more rapidly by broad-leaved species but, as general respiratory inhibitors, they sometimes proved acutely toxic to those applying them, and indeed fatalities among spray operators using DNOC were reported as late as the mid 1950s. Clearly, more selective compounds of low toxicity to human beings were desirable. These arrived in the form of the (plant) hormone-based herbicides discovered and developed during the period 1935–1945.

## 2.2 The auxin hormone herbicides

The discovery and development of these compounds is recounted in detail by Kirby (1980), and represents one of the major agricultural advances of the twentieth century. They provided a means of selectively controlling broad-leaved weeds in the world's major food crops – cereals. Their selectivity in controlling broad-leaved weeds is marked and they can be effective at dose rates of less than 1 kg/ha.

Initial research during the late 1930s at the Jealotts Hill Research Station in Berkshire, England involved spraying mixtures of oat and the broad-leaved weed charlock (*Sinapis arvensis*) with the chemical rooting stimulant 1-naphthylacetic acid. The weeds were killed and the cereals remained unaffected. The dose of around 10 kg/ha was uneconomic, but the results led to a search for other compounds effective at lower doses whilst retaining the selectivity. One of the compounds subsequently discovered was 4-chloro-2-methylphenoxyacetic acid (4-chloro-*o*-tolyloxyacetic acid) or MCPA, which was active at a rate of 1–1.5 kg/ha against many broad-leaved weeds in cereals.

The actual discovery of the growth regulatory properties of 1-naphthylacetic acid and other phenoxyacetic acids was made at the Boyce Thompson Institute in New York. Research there in the late 1930s showed that phenoxyacetic acids were particularly good growth regulators, with one of the most potent chemicals for producing seedless tomatoes being 2,4-dichlorophenoxyacetic acid (2,4-D). At the same time, workers at Rothamsted Experimental Station in Hertfordshire, England were working with derivatives of indolyl acetic acid (now of course known to be one of the principal endogenous plant growth regulatory substances) during studies into nodulation by *Rhizobium*. It quickly became clear that one of these, again 2,4-D, provided outstanding selective control of broad-leaved plants.

Little of the research carried out by both British and American workers was published at the time of their discovery because these molecules were considered for use as biological war agents. Fortunately they were never used for this purpose during the Second World War, although sadly the hormone weedkiller 2,4,5-trichlorophenoxyacetic acid (2,4,5-T) brought immense notoriety to the group, and herbicides in general, following its use as a defoliant during the Vietnam War. Ironically, it seemed that it was the selectivity of the chemicals – they would not kill cereals – that may have led to the decision not to use them during the Second World War.

In 1945 the first systemic hormone herbicide produced on a commercial scale – 2,4-D – was introduced in the USA, quickly followed in 1946 in the UK by the sodium salt of MCPA (Figure 2.3). These compounds were translocated and very small quantities were needed to kill plants. They affected broad-leaved plants systemically, causing stems to twist and bend, the roots to develop abnormal swellings and the leaves to turn yellow and die. In trials on turf and golf greens, broad-leaved weeds were killed without detriment to the grasses.

Wherever the new hormone-based weedkillers were used, the results were spectacular in terms of weed control. In the eastern counties of England, the

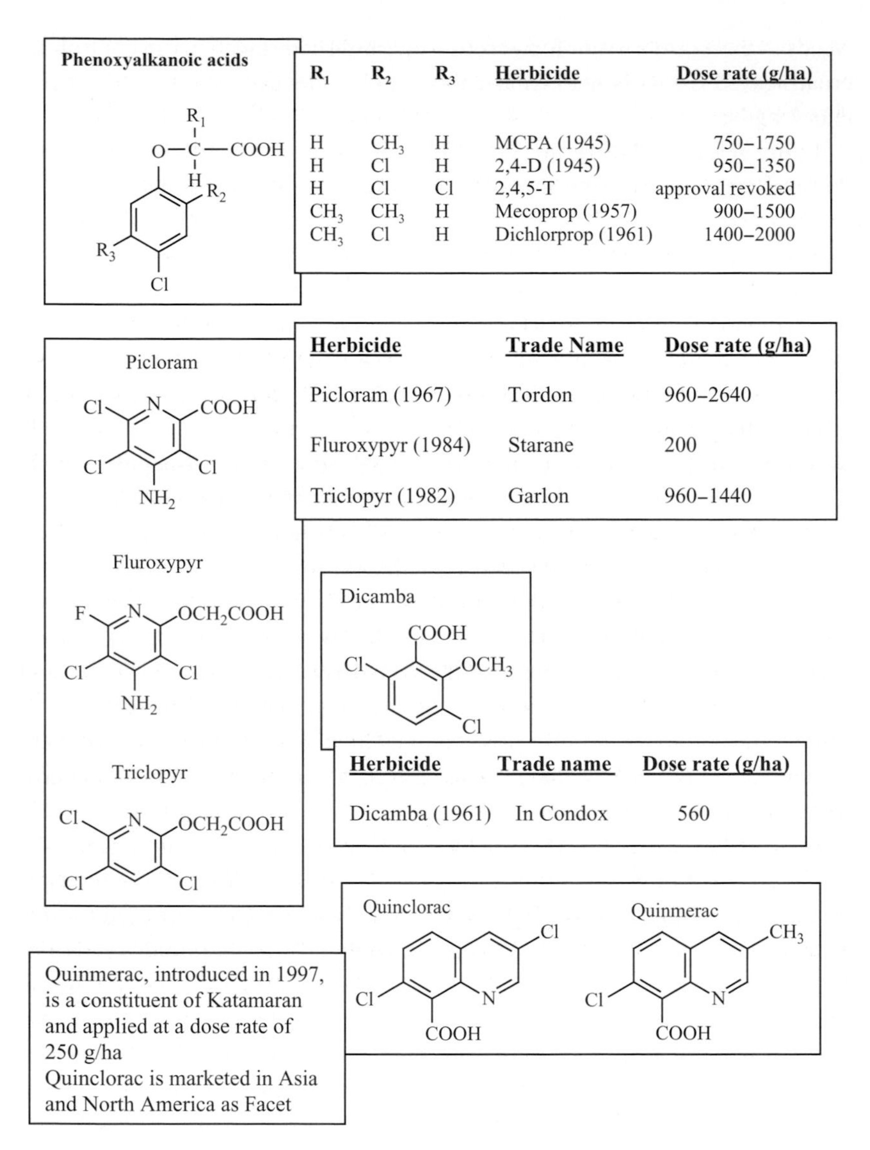


| $R_1$ | $R_2$ | $R_3$ | Herbicide | Dose rate (g/ha) |
|---|---|---|---|---|
| H | $CH_3$ | H | MCPA (1945) | 750–1750 |
| H | Cl | H | 2,4-D (1945) | 950–1350 |
| H | Cl | Cl | 2,4,5-T | approval revoked |
| $CH_3$ | $CH_3$ | H | Mecoprop (1957) | 900–1500 |
| $CH_3$ | Cl | H | Dichlorprop (1961) | 1400–2000 |

| Herbicide | Trade Name | Dose rate (g/ha) |
|---|---|---|
| Picloram (1967) | Tordon | 960–2640 |
| Fluroxypyr (1984) | Starane | 200 |
| Triclopyr (1982) | Garlon | 960–1440 |

| Herbicide | Trade name | Dose rate (g/ha) |
|---|---|---|
| Dicamba (1961) | In Condox | 560 |

Figure 2.3 Auxin hormone herbicides.

red and yellow fields of wheat and barley, due to the high incidence of poppies (*Papaver rhoeas*– Figure 2.2) and charlock (*Sinapis arvensis*), became a thing of the past. The superb control of weeds with attendant increases in yield during a post-War era of food shortage led to further intensive research and development of hormone-based herbicides (Figure 2.3).

Further developments included MCPB, the butyric acid analogue of MCPA. This compound can be used on a number of crops, not just monocotyledons, for weed control; allowing, for example, selective weed control in legume crops. In the 1950s, the 2-phenoxypropionic acid herbicides were developed with mecoprop – 2-(4-chloro-*o*-tolyoxy) propionic acid, also known as CMPP – having good selective activity against cleavers (*Galium aparine*), which is poorly controlled by phenoxyacetic acids, and chickweed (*Stellaria media*) in cereals. Further research in the 1960s and 1970s led to the introduction of benzoic acids such as dicamba, pyridine derivatives including clopyralid and triclopyr, the extremely persistent compound picloram, and fluroxypyr, which was used very extensively for weed control in cereals in the UK during the 1980s and 1990s. In the late 1980s a new generation of hormone herbicides, the quinoline carboxylic acids, was developed, including quinmerac and quinclorac (Grossman, 1998), the former being widely used for weed control in broad-leaved crops such as oilseed rape. Uniquely among auxin herbicides, the quinoline carboxylic acids have the capacity to control some grass weeds in cereal crops.

The morphological changes (Figure 2.4) that occur in susceptible species following application of these herbicides are characteristic of an overdose of auxins. Under normal conditions, the synthesis of auxin and its role in plant growth and development is under strict control. Applications of hormone herbicides may result in the uptake of around 100 μg of synthetic hormone per plant – about 1000 times more auxin than is already present in the plant. Control systems within the plant cannot cope with such a massive auxin overdose, and death effectively occurs due to uncontrolled growth (Cobb, 1991).

Alterations in the permeability of membranes, particularly to cations, occur within a few minutes of application of these herbicides to susceptible plants. Ethylene is released and this results in pronounced epinastic effects characterised by severe twisting of petioles and leaves. This growth distortion can occur within a few hours of application (Figure 2.4). Stomatal function is also affected and carbohydrate reserves may be mobilised. Cambial activity may be stimulated and, as a consequence, in the week following application stems generally thicken and elongate with the formation of adventitious roots. Leaf chlorosis, root disintegration and plant death soon follow.

Despite their extensive use for over half a century, the molecular basis of action of auxin hormone herbicides is still not clear, largely due to the lack of understanding of the precise mode of action of natural auxins. The activity of natural auxins may follow their binding to an auxin-binding protein on the plant cell membrane: changes in plasma membrane potential may then quickly follow. Expression of certain genes is stimulated, leading to synthesis of key regulatory enzymes such as 1-aminocyclopropane-1-carboxylic

Figure 2.4 Selectivity of the hormone herbicide MCPA.

acid (ACC) synthase. ACC synthase is involved in the production of ethylene, and stress metabolites such as abscisic acid. Large doses of auxins may lead to massive changes in membrane permeability and over-expression of ACC synthase leading ultimately to the gross morphological changes noted above.

In terms of acute toxicity, hormone weedkillers only affect plant tissues. Receptors for these synthetic auxins may be present only on the plasma membrane of susceptible plant species, and not in other organisms. Further selectivity for individual plant groups such as MCPB in legume crops is due to metabolism. Oxidation of the molecule to MCPA occurs in broad-leaved species other than legumes, the latter thus remaining unaffected. However, the reasons why most of the hormone-based herbicides affect dicotyledonous rather than monocotyledonous plant species are, as with the mode of action, not yet clear. Selectivity has been linked to the morphological nature of the leaf surface in the two groups, with monocotyledons retaining far less herbicide in their upright, narrow leaves than broad-leaved dicotyledonous plants. Differences in uptake of hormone herbicides have been reported between species but these differences are not sufficient to explain selectivity. It is possible that selectivity may be due to a difference in sensitivity or structure of auxin receptors in dicotyledons and monocotyledons, but confirmation of this hypothesis requires the full characterisation of auxin-binding proteins from plant cell membranes. The situation has been further complicated by the development of the quinoline carboxylic acids such as quinclorac, which is able to control some monocotyledonous weeds in rice: whether this specificity is due to slight differences in the receptors of plant species or metabolism/detoxification of quinclorac in rice remains to be resolved.

## 2.3 Inhibitors of photosynthesis

The transformation of light energy to chemical energy in the form of adenosine triphosphate (ATP), and the subsequent use of this energy to fix carbon from the atmosphere into organic compounds occurs only in plants and a few bacterial species. The exploitation of this fundamental process of photosynthesis as a herbicidal target began in the early 1950s with the development of the substituted urea diuron and the triazine herbicides atrazine and simazine. These compounds are soil-applied, persistent and have been used extensively in crops and non-crop situations for long-term control of weeds and unwanted vegetation. Foliar applied photosynthetic inhibitor herbicides were also developed that act at the same site as diuron and the triazines. The bipyridylium compounds such as paraquat act at a different site to all other photosynthetic inhibitor herbicides, and were initially marketed in the late 1950s.

Most of the herbicides that inhibit photosynthesis have no acute effects on other biological processes and thus present few risks to the health of human beings and other animals. However, the mechanism of action of the bipyridylium compounds in plants is mimicked in the lungs of human beings and other animals, and so these compounds in their concentrated form pose considerable risks to health. It is unlikely that bipyridylium compounds would receive approval under the regulatory schemes currently in force for pesticide registration.

All herbicides that inhibit photosynthesis directly do so by interfering with the light reactions of the process. No herbicides appear to directly inhibit the fixation of carbon dioxide, although this process is, of course, profoundly affected following inhibition of photochemical reactions.

Herbicides that interfere directly with photosynthesis do so by either blocking the flow of electrons in photosystem II or by diverting the flow of electrons in photosystem I of the photosynthetic light reactions. Photosynthetic pigments within the thylakoid membranes of chloroplasts absorb light, and the energy thus trapped is used in the process of photosynthesis. Photolysis of water occurs in the thylakoid membrane resulting in the production of protons ($H^+$), electrons ($e^-$) and oxygen. The initial light reactions of photosynthesis may be summarised:

$$2H_2O \longrightarrow 4H^+ + 4e^- + O_2$$

The electrons and protons thus produced pass through the thylakoid membrane. Electrons are transferred along a series of carrier molecules within the membrane and during their passage energy is generated. Two photosystems (I and II) are involved in the transformation of light energy into chemical energy in the form of reduced nicotinamide adenine dinucleotide phosphate (NADP) and ATP. These latter high-energy compounds may then be employed in the fixation of $CO_2$ and subsequent provision of organic molecules for plant growth.

The principal toxic effects resulting from the action of all photosynthetic inhibitor herbicides, including those that prevent synthesis of the photosynthetic apparatus (Hess, 2000), are expressed through disruption of thylakoid, chloroplast and then plant cell membranes. This disruption follows the oxidation, due to an excess of activated oxygen species, of unsaturated fatty acids in these membranes. Peroxidation results in the degradation of unsaturated fatty acids such as linolenic acid that constitute a major part of cell membranes in chloroplasts and plant cells. The structural alterations caused by this lipid peroxidation lead to destabilisation of membranes, and their function in maintaining ionic balance is impaired, resulting in leakage of solutes from both chloroplasts and plant cells. Substantial tissue damage ensues leading ultimately to plant death.

### 2.3.1 Inhibition of photosynthesis at photosystem II

Symptoms in plants treated with herbicides that interfere with photosystem II (PS II) develop gradually over several days. Treated plants initially become chlorotic, largely because of chlorophyll destruction due to photo-oxidation reactions leading to membrane damage in the chloroplast. Necrosis then proceeds, linked to membrane destruction resulting from lipid peroxidation.

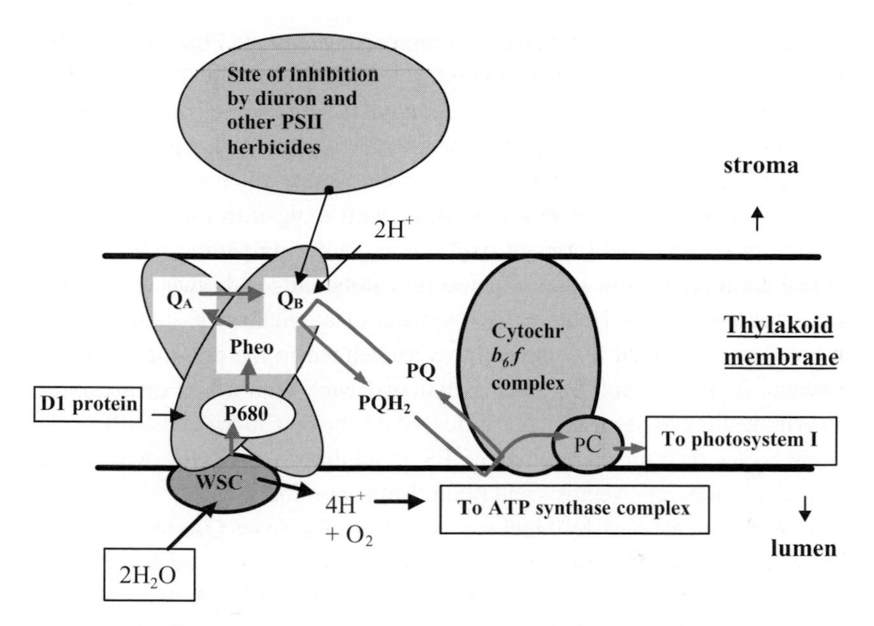


Figure 2.5 Electron flow in photosystem II and the site of inhibition by herbicides. WSC is the water-splitting complex from which electrons originate; they then pass to, respectively, p680, which is the chlorophyll reaction centre of photosystem II; Pheo (pheophytin); $Q_A$ and $Q_B$, protein-bound quinones; PQ, plastoquinone, and $PQH_2$, which is plastohydroquinone; and finally to PC, plastocyanine, and onwards to photosystem I.

The precise biochemical site of action of the PSII inhibitor herbicides has been identified as the plastoquinone $Q_B$ that is bound to a protein (the D1 protein) present in the chloroplast thylakoid membranes. Herbicides that inhibit photosynthesis at PS II bind to the D1 protein and in doing so compete for the $Q_B$ binding site. During photosynthesis, $Q_B$ accepts electrons from another protein-bound plastoquinone, $Q_A$ (Figure 2.5). In the presence of PS II inhibitor herbicides, $Q_B$ is displaced from its binding site on the D1 protein, cannot accept electrons from $Q_A$, and an activated form of $Q_A$ (denoted $Q_A{}^-$) results. The generation of energy for reduction of $CO_2$ is impaired and the process of carbon fixation disrupted. However, the blockage of electron flow from $Q_A$ has much more serious consequences for photosynthetic activity through effects on chloroplast structure and function.

The energy associated with $Q_A{}^-$ that cannot be dissipated by flow through $Q_B$ may activate oxygen, which is of course present in abundance in the chloroplast, leading to the production of toxic forms of this molecule in the thylakoid. The principal toxic form produced is singlet oxygen ($^1O_2$). When chlorophyll accepts light energy, it becomes activated to the so-called singlet energy state, the light energy being passed to the reaction centre denoted P680 and initiating electron transfer from P680 to pheophytin. When electron flow is blocked from $Q_A$ to $Q_B$, singlet chlorophyll energy accumulates and some is transformed to a triplet energy state. Some triplet chlorophyll is formed during photosynthesis under normal conditions, but its effects are ameliorated by dissipation through carotenoids. Triplet chlorophyll may react with oxygen to form singlet oxygen and other activated forms of oxygen: again during photosynthesis under normal conditions these are dissipated by carotenoids and other antioxidants such as α-tocopherol. However, this system of quenching of singlet oxygen and triplet chlorophyll is completely overwhelmed in the presence of photosynthetic inhibitors, and lipid peroxidation occurs, with a loss of membrane integrity leading to tissue damage as described above. Chlorosis usually occurs within a few days of application of PS II inhibitor herbicides, with tissues gradually yellowing. Necrosis and plant death soon follow.

Many herbicides act by blocking, electron flow from $Q_A$, to $Q_B$ including, substituted ureas, triazines, uracils, phenylcarbamates and hydroxybenzonitriles (Figure 2.6). Studies using partially degraded D1 protein show that sensitivity to substituted ureas such as diuron may be removed, whilst at the same time sensitivity to some other herbicides which bind to the $Q_B$ site on the D1 protein (such as the hydroxybenzonitriles) may be enhanced. This suggests that there is more than one binding site for herbicides on the D1 protein.

Several PS II inhibitors are persistent and soil applications may allow weed control for many months. Plants readily absorb ureas and triazines from soil, and the herbicides are rapidly translocated to photosynthetic tissues. Atrazine and simazine have been widely used for long-term control of weeds in tree and bush crops such as apples and raspberries in temperate zones, and citrus, cocoa, tea and coffee in warmer parts of the world. Triazines leach very slowly through the soil profile, and thus have little effect on deep-rooted species.

Diuron and the triazines were also used extensively in the UK and elsewhere until the 1990s for control of vegetation in non-crop situations, often being applied at high doses for this purpose. Although recommended dose rates for use in crops such as maize are between 1 and 2 kg/ha (Figure 2.6) application rates of 20 kg/ha were not uncommon on roads, railway track beds and hard surfaces in industrial areas. Although these compounds may leach only slowly in soils, high application rates and run-off from hard surfaces have led to the widespread occurrence of these compounds in natural waters and subsequent restrictions on their use, particularly in non-crop situations.

Some of the foliar-applied compounds developed in the 1960s and 1970s have allowed selective weed control in crops. Chlorotoluron and the very widely used substituted urea herbicide isoproturon offer selective control of

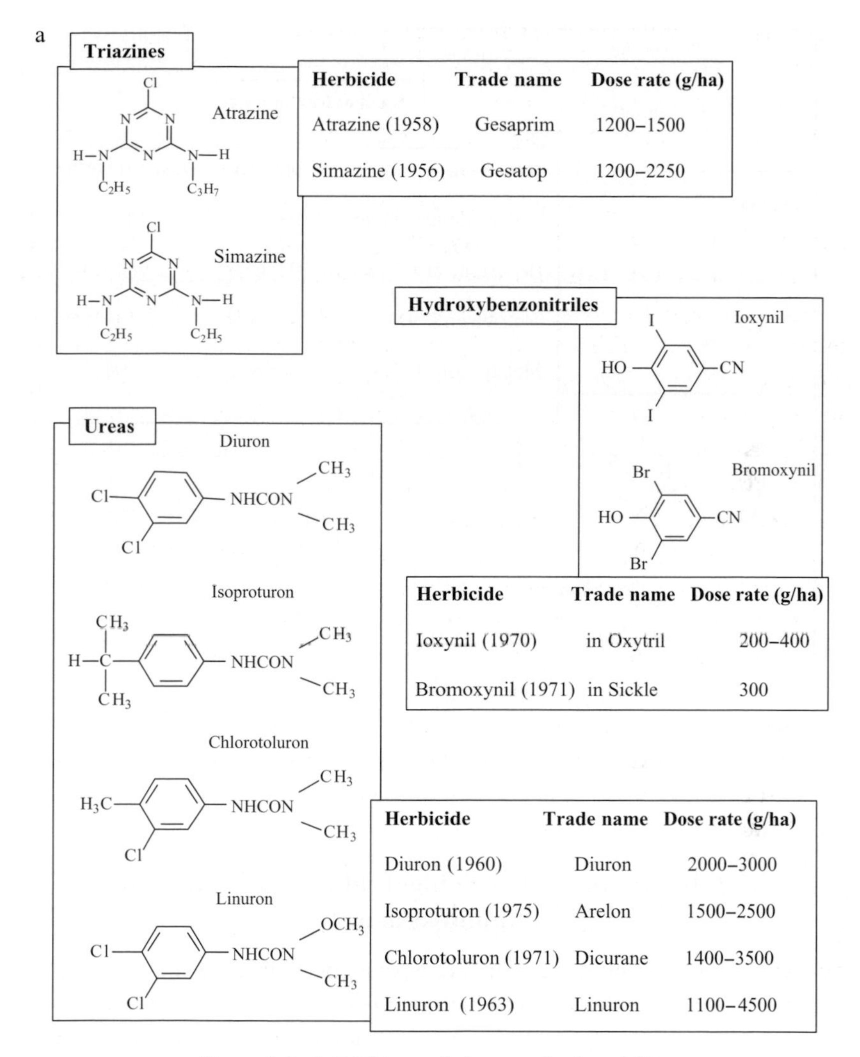


| Herbicide | Trade name | Dose rate (g/ha) |
|---|---|---|
| Atrazine (1958) | Gesaprim | 1200–1500 |
| Simazine (1956) | Gesatop | 1200–2250 |

| Herbicide | Trade name | Dose rate (g/ha) |
|---|---|---|
| Ioxynil (1970) | in Oxytril | 200–400 |
| Bromoxynil (1971) | in Sickle | 300 |

| Herbicide | Trade name | Dose rate (g/ha) |
|---|---|---|
| Diuron (1960) | Diuron | 2000–3000 |
| Isoproturon (1975) | Arelon | 1500–2500 |
| Chlorotoluron (1971) | Dicurane | 1400–3500 |
| Linuron (1963) | Linuron | 1100–4500 |

Figure 2.6 a,b Inhibitors of photosynthesis at PSII.

broad-leaved and some grass weeds in cereal crops. This selectivity at crop level is due to rapid metabolism of the ureas in species such as wheat and barley, enabling these crops to tolerate the herbicide at the dose needed to kill weeds. Bentazone, widely used for weed control in legumes and cotton, is detoxified in these crop species but not by susceptible weeds. Metamitron and metribuzin are metabolised to non-toxic forms in sugar beet, thus allowing their use in this crop for weed control. Atrazine is detoxified in maize, and has been employed for weed control in this crop since the 1960s.

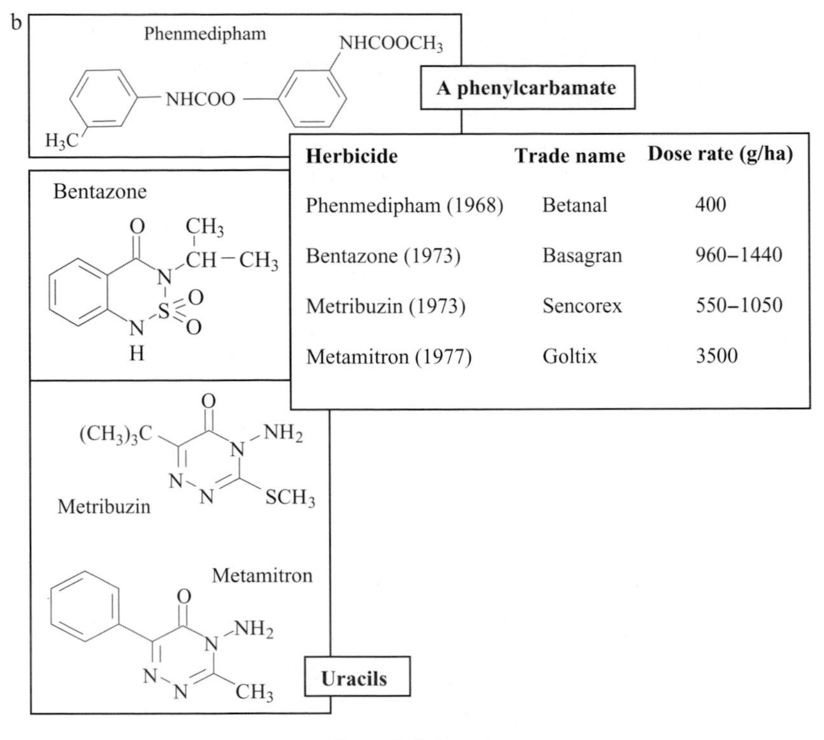


| Herbicide | Trade name | Dose rate (g/ha) |
|---|---|---|
| Phenmedipham (1968) | Betanal | 400 |
| Bentazone (1973) | Basagran | 960–1440 |
| Metribuzin (1973) | Sencorex | 550–1050 |
| Metamitron (1977) | Goltix | 3500 |

Figure 2.6 *(cont.)*.

## 2.3.2 Inhibition by herbicides that divert electron flow at photosystem 1

The bipyridylium compounds diquat and paraquat (Figure 2.7) have a different site of action to those herbicides which inhibit electron flow in photosynthesis. Bipyridylium compounds act by capturing electrons from photosystem I (PSI) (Figure 2.8). Diquat and paraquat are cations (e.g. $PQ^{2+}$) and can accept an electron from PSI to become reduced (e.g. to $PQ^{+}$).

The paraquat ($PQ^{+}$) and diquat ($DQ^{+}$) free radicals do not themselves cause tissue damage. The unstable free radicals are rapidly re-oxidised by molecular oxygen (abundant of course in chloroplasts) to produce an activated oxygen species and $PQ^{2+}$, which can then accept in turn another electron from PSI. The principal activated oxygen species formed are the superoxide free radical $O_2^{-}$, peroxide ($H_2O_2$) and hydroxyl radicals ($OH^{-}$). These are all highly potent biological oxidants and quickly initiate lipid peroxidation of unsaturated fatty acids resulting, as with PSII inhibitors, in membrane degradation.

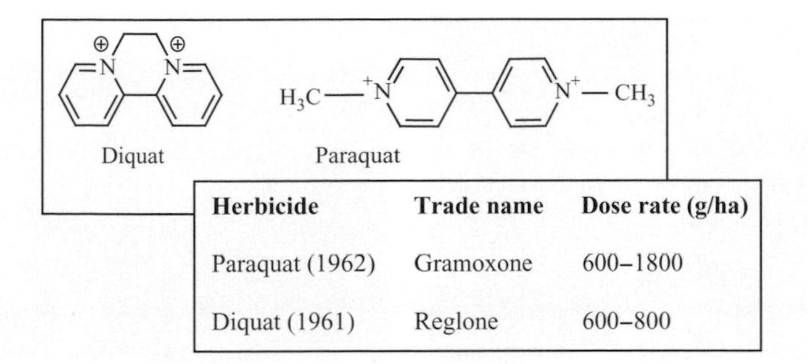


| Herbicide | Trade name | Dose rate (g/ha) |
|---|---|---|
| Paraquat (1962) | Gramoxone | 600–1800 |
| Diquat (1961) | Reglone | 600–800 |

Figure 2.7 Inhibitors of photosynthesis at PSI.

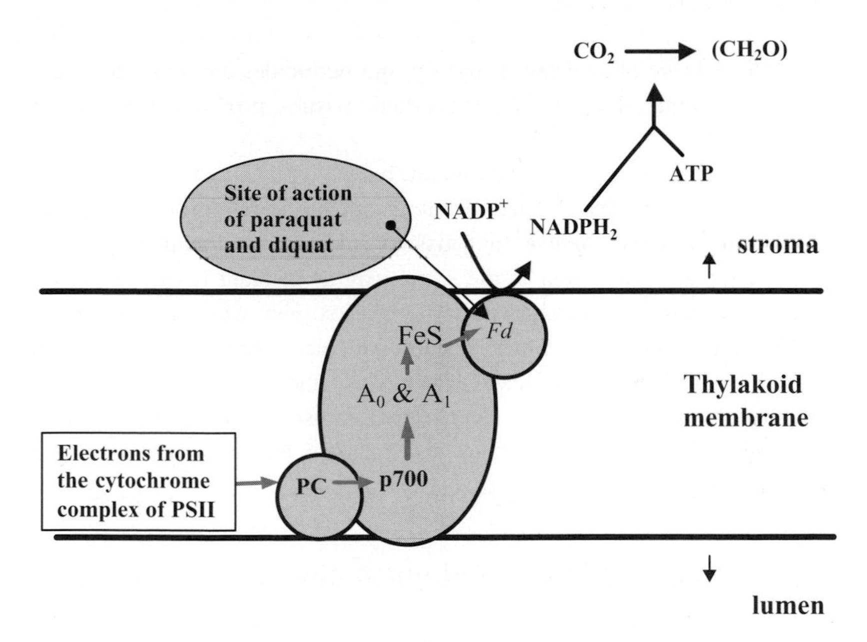


Figure 2.8 Electron flow in photosystem I and the site of inhibition by bipyridylium herbicide. Electrons from PSII pass via PC, plastocyanin, to p700, the chlorophyll reaction centre of photosystem I; $A_0$, $A_1$ and FeS are electron carriers and Fd, ferredoxin, accepts electrons from FeS.

Treated tissues viewed under the electron microscope show initially intergranal spaces in the chloroplast, and this is quickly followed by membrane disintegration in the chloroplast and elsewhere. The continual production of activated oxygen species in paraquat-treated plants, especially under strong summer light, can lead to very quick tissue death – less than a day in some cases (Figure 2.9). Paraquat though is not translocated and thus good foliage cover is

Figure 2.9 The action of paraquat on plants one day (left) and four days after treatment (right).

essential to achieve plant death. Bipyridylium herbicides are non-selective in their action, killing all actively photosynthetic tissues, providing they can get to these.

The PSI inhibitors, especially paraquat, have a high acute toxicity to many non-target species, and have been associated with human fatalities in many parts of the world. In human beings, lung tissues selectively accumulate paraquat where it undergoes reduction in the same way as in plant tissues to form a free radical capable of reacting with molecular oxygen, which is of course in plentiful supply. As in the chloroplast, the paraquat cation is re-formed with the simultaneous formation of reactive oxygen species such as superoxide and hydrogen peroxide. These anions can then peroxidise lipids in cell membranes of the lung, which severely influences gas exchange, leading in turn to respiratory failure and in many cases death.

## 2.4 Inhibitors of pigment biosynthesis

In addition to direct interference with electron transfer, inhibition of the synthesis and function of the chlorophyll and carotenoid pigments, which are essential components for photosynthesis, may lead to plant death. Compounds developed in the 1990s that inhibit pigment biosynthesis are more potent than photosynthetic electron transfer inhibitor herbicides, with dose rates as low as 15 grams of active ingredient per hectare. These herbicides act by inhibiting the function of key enzymes in pathways of pigment biosynthesis.

### 2.4.1 Protox inhibitors

Several herbicides block the synthesis of porphyrins and tetrapyrroles, which are precursors of chlorophyll. The diphenyl ether herbicides, introduced in

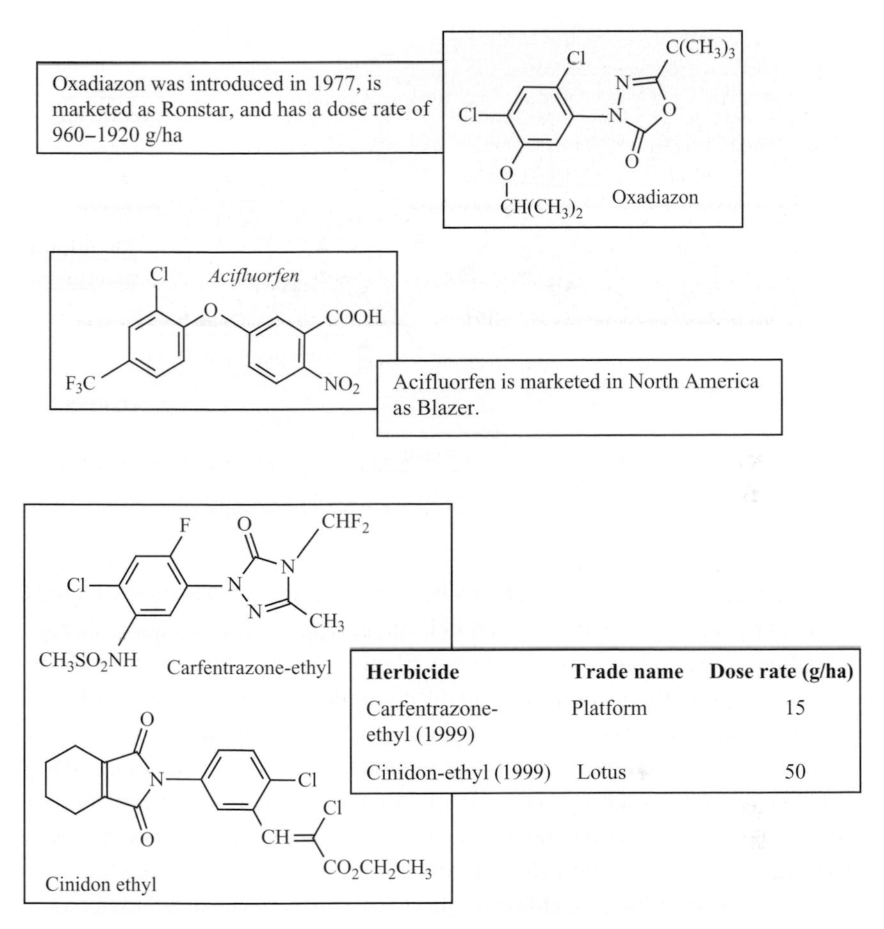


| Herbicide | Trade name | Dose rate (g/ha) |
|---|---|---|
| Carfentrazone-ethyl (1999) | Platform | 15 |
| Cinidon-ethyl (1999) | Lotus | 50 |

Figure 2.10 Herbicides that inhibit protoporphyrinogen oxidase (protox).

the 1960s, and others such as oxadiazon are now known to interfere with chlorophyll synthesis. Modern developments include acifluorfen used to control weeds in soybean, peanut and cotton, and cinidon-ethyl and carfentrazone-ethyl (Figure 2.10), both used to control broad-leaved weeds in cereals and soybeans. The latter compound is effective at astonishingly low dose rates: as low as 5 g of active ingredient per hectare.

These herbicides inhibit the enzyme protoporphyrinogen oxidase (usually abbreviated to protox – Figure 2.11) in plant chloroplasts (Dayan and Duke, 1996, 1997). The inhibition results in an accumulation of the compound protoporphyrinogen IX, which leaks out of chloroplasts, and is converted in plant cytoplasm to protoporphyrin IX. In light, protoporphyrin IX becomes photoactivated and generates highly reactive oxygen radicals such as singlet

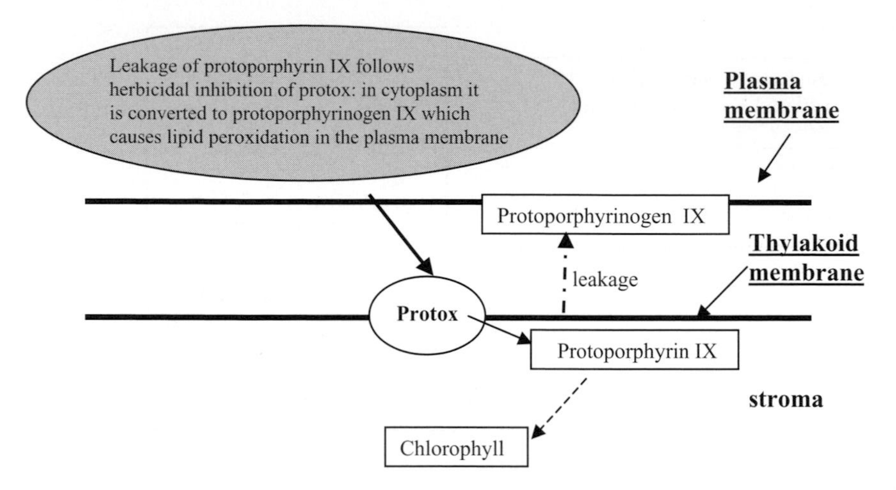


Figure 2.11 Site of action of protox-inhibiting herbicides.

oxygen and, as with direct photosynthetic inhibitors, peroxidation of lipids occurs, here particularly in the plant cell membrane, with consequent disruption of membrane structure and function. Plants wilt, become necrotic and die. The selectivity of these compounds to different plant species is due to different rates of metabolism. For example, cinidon-ethyl is broken down to non-toxic metabolites more rapidly in wheat than in target weeds, which are thus killed.

Other organisms also synthesise porphyrins, which are precursors of cytochromes as well as haemoglobin. Indeed the activity of protox enzymes extracted from the mitochondria of fungal and animal cells is inhibited by protox-inhibitor herbicides. However, these herbicides have passed toxicological scrutiny prior to their approval: they have low mammalian toxicity and thus selectivity clearly exists. The selectivity of protox-inhibitor herbicides is almost certainly due to rapid metabolism to non-toxic molecules, as well as lack of uptake followed by excretion in animal species.

### 2.4.2 Inhibitors of carotenoid biosynthesis

Some of the enzyme steps leading to carotenoid synthesis occur only in plants, some fungi and bacteria, and thus compounds that specifically inhibit these enzymes have few, if any, acute toxic effects on other organisms. Inhibition of carotenoid biosynthesis in plants may result in an excess of oxygen free radicals such as singlet oxygen, again leading to oxidative degradation of chlorophyll as well as peroxidation and destruction of photosynthetic membranes. Initial symptoms often include whitening of young tissues and the compounds are

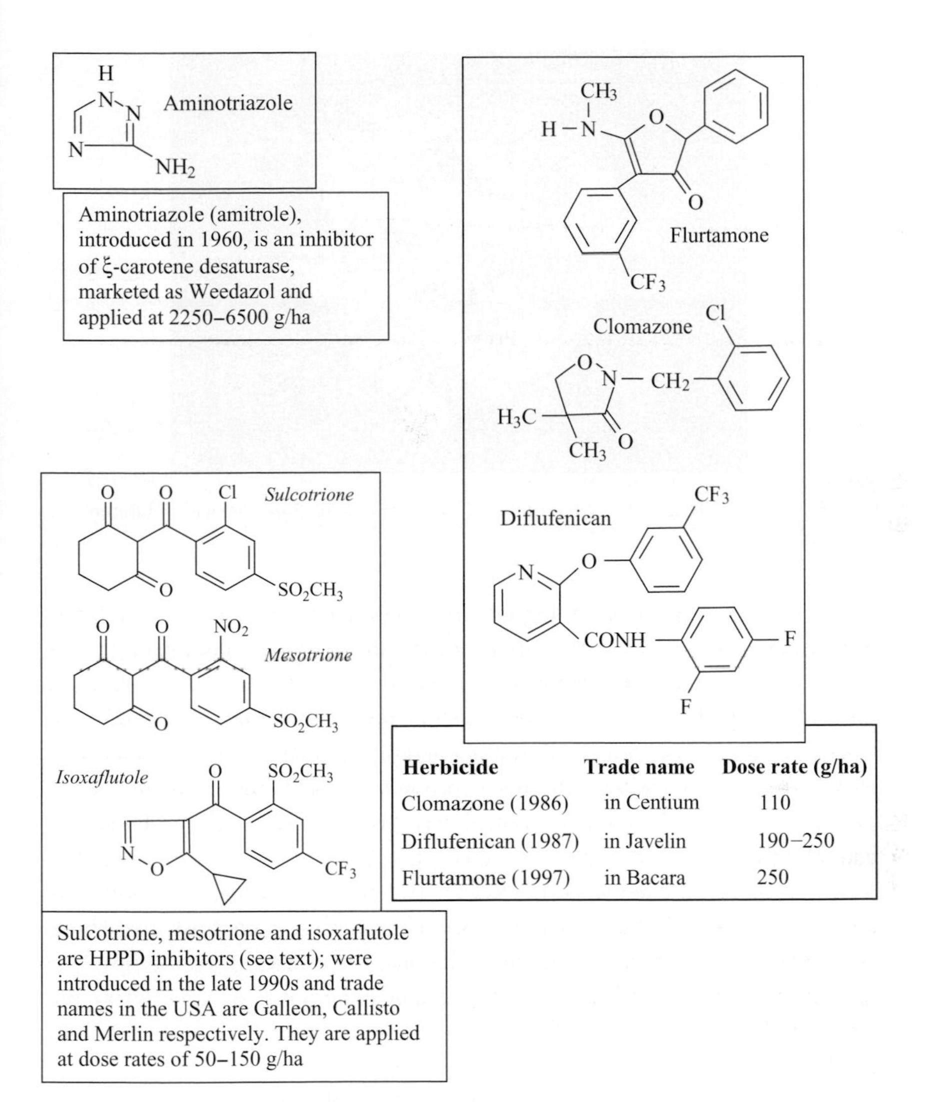


| Herbicide | Trade name | Dose rate (g/ha) |
|---|---|---|
| Clomazone (1986) | in Centium | 110 |
| Diflufenican (1987) | in Javelin | 190–250 |
| Flurtamone (1997) | in Bacara | 250 |

Figure 2.12 Inhibitors of carotenoid biosynthesis.

often collectively referred to as the bleaching herbicides (Boger and Sandmann, 1998).

Aminotriazole (amitrole, Figure 2.12) was introduced in the 1960s and is used for total vegetation control, principally in orchards and non-crop situations. It acts by inhibiting the activity of zeta-carotene desaturase. Some herbicides act by blocking the activity of the enzyme phytoene desaturase (Bramley and Pallett, 1993) that is involved in the transformation of phytoene to zeta-carotene

Figure 2.13 Bleaching symptoms of the carotenoid biosynthesis inhibitor diflufenican.

in the complex pathway of carotenoid biosynthesis (Figure 2.14). Accumulation of phytoene, the colourless precursor of the carotenoids, occurs leading to the typically pronounced bleaching or whitening of young, developing tissues in plants (Figure 2.13). Bleached tissues contain no pigmentation as the absence of carotenoids prevents normal chloroplast and pigment development. Old leaves remain green when treated with these compounds, and hence most are applied pre-emergence to soil or used to control germinating weed seedlings. Diflufenican, often in mixture with flurtamone (Figure 2.12), is widely used in the UK as a pre- and post-emergence herbicide, and at the recommended dose rate selectively controls broad-leaved weeds in cereal crops. In contrast, clomazone is used to control weeds in legumes and some other broad-leaved crops. The mechanism of selectivity of both diflufenican and clomazone is not clear, but metabolism in cereals and broad-leaved crops respectively appears the most likely explanation.

Herbicides have been discovered that act by blocking the synthesis of a quinone cofactor in carotenoid biosynthesis (Prisbylla *et al.*, 1993). These compounds, developed in the late 1990s, include sulcotrione, mesotrione and isoxaflutole (Figure 2.12), which are used for control of weeds in maize at considerably lower doses than, for example, triazine herbicides such as atrazine. Isoxaflutole is converted to diketonitrile on uptake by plants and this compound inhibits the enzyme hydroxyphenylpyruvate dioxygenase (HPPD) in plants. HPPD is an essential step in the synthesis of homogentisic acid, a precursor of plastoquinone-9, which is itself a cofactor necessary for the action of phytoene desaturase. As noted above, inhibition of the latter enzyme in the biosynthetic

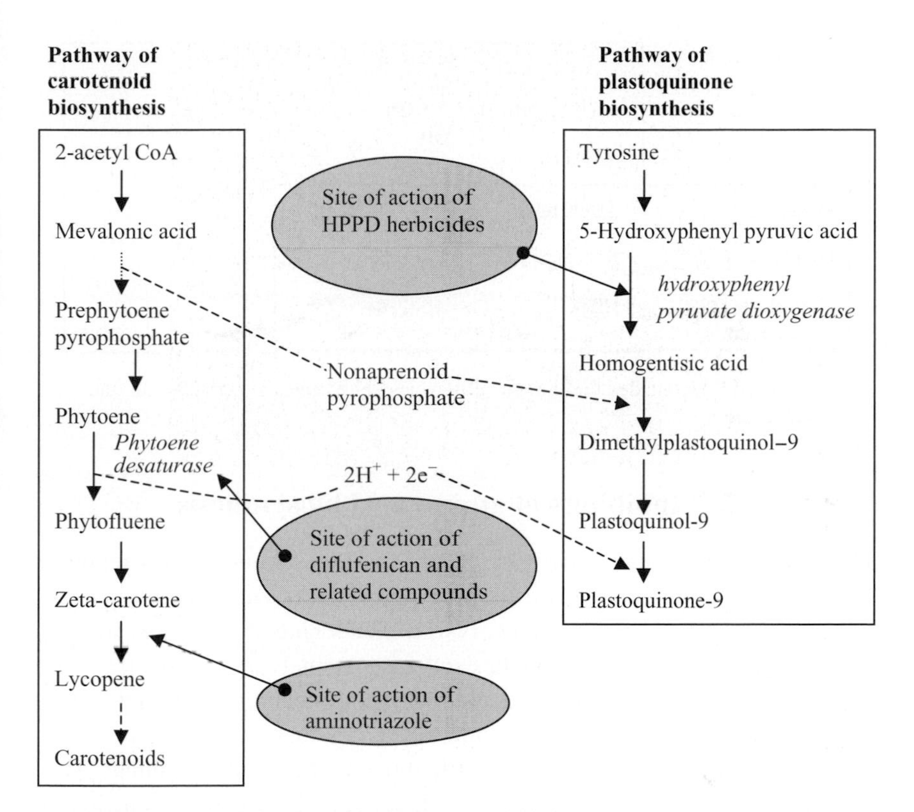


Figure 2.14 Site of action of herbicides that inhibit carotenoid biosynthesis. The diagram illustrates the relationship between the pathways of carotenoid and plastoquinone biosynthesis.

pathway to carotenoids causes pronounced bleaching of susceptible plants, and, as with phytoene desaturase inhibitors such as diflufenican, this is a prominent symptom of the herbicidal action of HPPD inhibitors. Selectivity of these herbicides is due to their rapid breakdown in maize, and much slower degradation in susceptible weed species.

HPPD also occurs in other organisms and is involved in the breakdown of the aromatic amino acid tyrosine. In fact HPPD inhibitors are used in the treatment of tyrosinaemia – a rare human disorder characterised by accumulation of hepatotoxic tyrosine-derived metabolites. Herbicides which inhibit HPPD in plants will also block HPPD isolated from liver and other tissues of mammals, and selectivity, like that of protox inhibitors, appears to be due to rapid metabolism and excretion of these compounds in non-target species. The toxicity of these herbicides to non-target species, including human beings, is low.

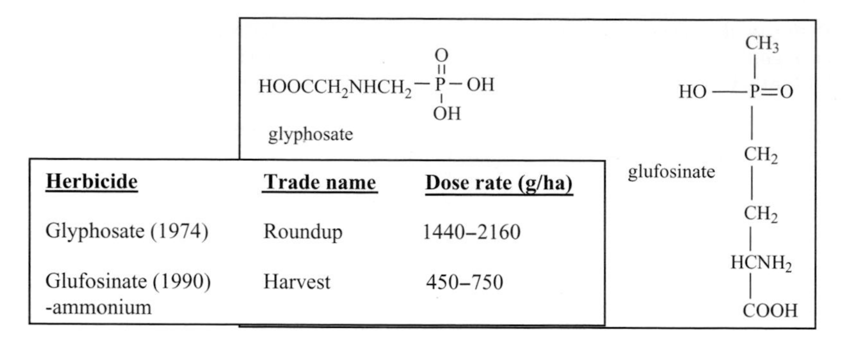

| Herbicide | Trade name | Dose rate (g/ha) |
|---|---|---|
| Glyphosate (1974) | Roundup | 1440–2160 |
| Glufosinate (1990) -ammonium | Harvest | 450–750 |



Figure 2.15 Herbicides that inhibit amino acid biosynthesis – glyphosate and glufosinate.

## 2.5 Inhibitors of amino acid biosynthesis

Plants are the principal dietary source of 9 of the 20 amino acids required by animals for protein synthesis. These so-called essential amino acids – leucine, isoleucine, histidine, valine, lysine, methionine, theonine, tryptophan and phenylalanine – are not synthesised by mammals and most other animals. The fact that their biosynthetic pathways only occur in plants and micro-organisms means that chemicals that inhibit steps in these pathways, as well as being effective herbicides, may pose little threat to the health of animals. This has proved to be true and, paradoxically, herbicides that inhibit production of amino acids essential for the health of human beings are, to humans and other animals, some of the most toxicologically benign chemicals used in agriculture. Furthermore, some of these compounds are very potent herbicides, being effective at doses of a few grams per hectare.

### 2.5.1 Glyphosate and glufosinate

The remarkable compound glyphosate (Figure 2.15) developed in the early 1970s is the most extensively used pesticide in the world, and its use has increased with the advent of glyphosate-resistant crop cultivars. It is foliar applied, and primarily used as a herbicide for total vegetation control of annual as well as perennial weeds, giving particularly good control of many deep-rooted species. It has proved invaluable in systems where minimal soil tillage is desirable, allowing removal of surface vegetation prior to sowing without the risk of soil erosion. In terms of environmental effects and risks to human health, glyphosate is one of the most benign of all pesticides (Baylis, 2000).

Glyphosate, uniquely among herbicides, inhibits a key step in the biosynthesis of the aromatic amino acids phenylalanine, tyrosine and tryptophan

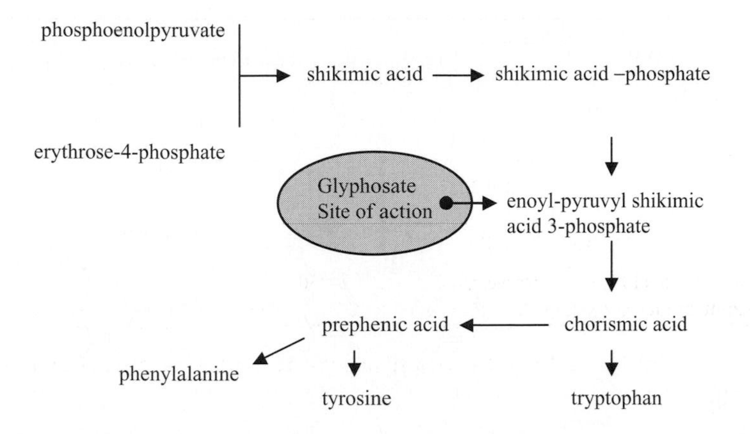


Figure 2.16 Part of the aromatic amino acid biosynthetic pathway in plants and the site of inhibition by glyphosate.

(Figure 2.16). The actual and highly specific site of inhibition is the enzyme 5-enoyl-pyruvyl shikimic acid 3-phosphate synthase, more commonly known as EPSP synthase. Glyphosate binds to this enzyme, causing a slight configurational change that prevents the substrate gaining access to the active site. The pathway to production of aromatic amino acids is blocked, with subsequent effects on protein synthesis. Although highly systemic, glyphosate exerts its effects rather slowly compared to those of, for example, the hormone herbicides and paraquat, and this is due to the time taken for the plants to use up existing reserves of aromatic amino acids. Once these are utilised, the inability of the plant to synthesise further supplies impairs protein synthesis and in most cases death ensues.

The ammonium salt of glufosinate is widely used for total weed control and, as with glyphosate, its use has increased with the development of genetically modified crop cultivars. Glufosinate prevents the incorporation of ammonium into glutamate by glutamine synthase. The formation of the amino acid glutamine is thus inhibited. Death probably results from the toxic levels of ammonium that accumulate, but membrane function may also be impaired in plants. Glutamine synthase occurs in plants, animals and bacteria but glufosinate has a relatively low acute toxicity to mammals, and it appears that this may be due to rapid metabolism and excretion in these non-target organisms.

### 2.5.2 Inhibitors of acetolactate synthase

In the early 1980s a group of herbicides was discovered and developed that allowed weed control at remarkably low concentrations of around 10–30 g of

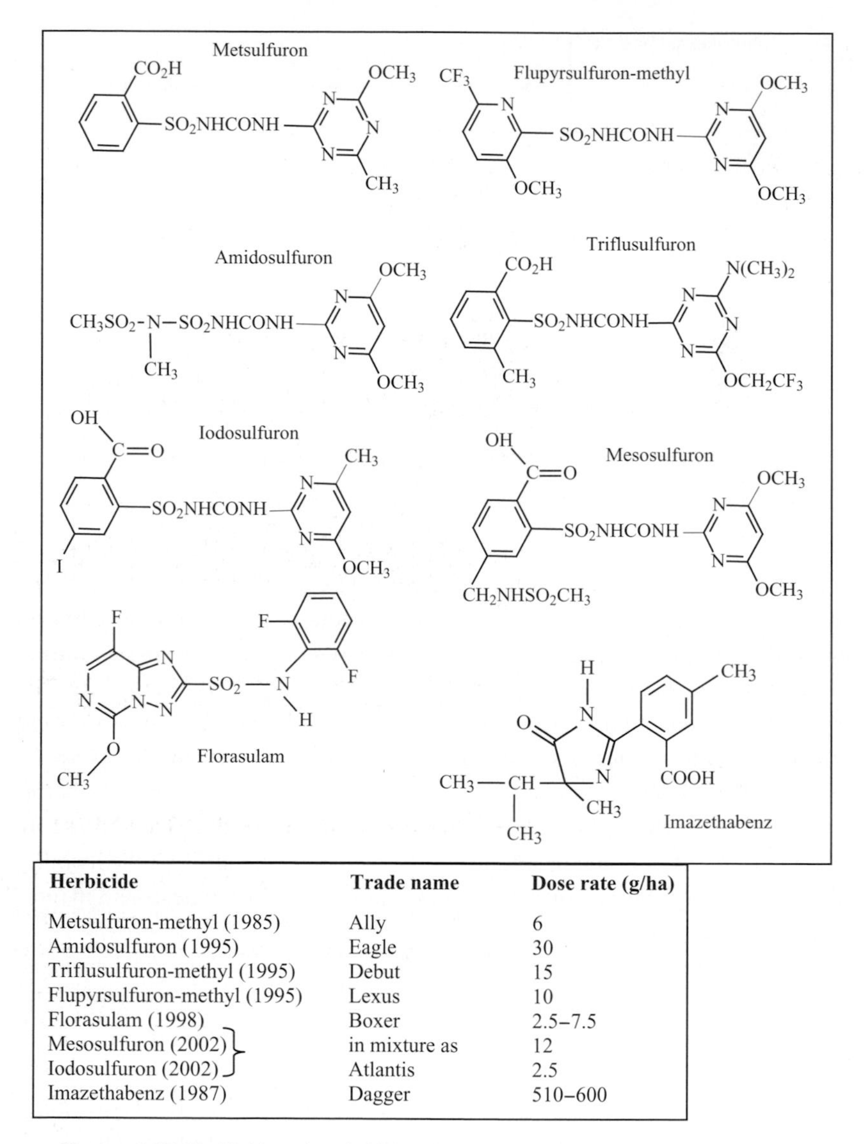


| Herbicide | Trade name | Dose rate (g/ha) |
|---|---|---|
| Metsulfuron-methyl (1985) | Ally | 6 |
| Amidosulfuron (1995) | Eagle | 30 |
| Triflusulfuron-methyl (1995) | Debut | 15 |
| Flupyrsulfuron-methyl (1995) | Lexus | 10 |
| Florasulam (1998) | Boxer | 2.5–7.5 |
| Mesosulfuron (2002) } | in mixture as | 12 |
| Iodosulfuron (2002) } | Atlantis | 2.5 |
| Imazethabenz (1987) | Dagger | 510–600 |

Figure 2.17 Herbicides that inhibit amino acid biosynthesis – acetolactate synthase inhibitors.

active ingredient per hectare. These sulphonyl urea compounds (Figure 2.17) were found to be very potent inhibitors of a key enzyme in the biosynthetic pathway to the branched-chain amino acids valine, leucine and isoleucine (Figure 2.18). The enzyme acetolactate synthase only occurs in plants and some bacteria and so the acute risk to other organisms from these compounds is very

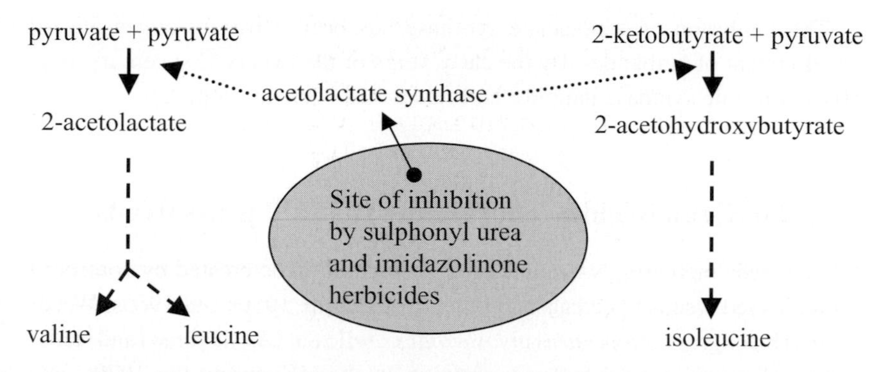


Figure 2.18 Site of inhibition of sulphonylurea and imidazolinone herbicides.

low (Hawkes, 1993). This, allied to their very low dose, makes them among the least hazardous of all agricultural chemicals. Many are soil-applied and inhibit plant growth soon after germination. Growth initially ceases once the plant's reserves of branched-chain amino acids run out, and chlorosis, necrosis and death soon follow. Their very potent nature led to some initial problems of persistence and damage to broad-leaved crops following cereals, but most of the modern sulphonyl-urea herbicides are of moderate to low persistence.

As an added bonus, selectivity at the crop level has been achieved with sulphonylurea herbicides. This selectivity is due to rapid breakdown in tolerant crops. For example, metsulfuron-methyl and thifensulfuron-methyl are rapidly degraded in wheat, as is the triazolopyrimidine compound florasulam, thus allowing selective weed control. Triasulfuron is similarly metabolised in sugar beet, enabling selective control of weeds in this crop. Furthermore, the introduction of a mixture of mesosulfuron-methyl and iodosulfuron-methyl-sodium has enabled control of problem monocotyledonous weeds such as blackgrass in cereals. To achieve selectivity, a herbicide safener (2.9) mefenpyr diethyl, which enhances the production of detoxifying enzymes in the crop, is added to the formulated product of mesosulfuron-methyl and iodosulfuron-methyl.

The imidazolinone herbicides (Figure 2.17) are completely unrelated in terms of chemical structure to the sulphonylureas, but also inhibit acetolactate synthase. These compounds are not as potent as sulphonylureas and consequently higher doses are needed to achieve weed control. Crop selectivity is also evident with this group of herbicides and compounds such as imazamethabenz are inactivated in wheat and maize, allowing control of weeds in these crops. Like sulphonylureas, these compounds present few acute risks to the health of mammals and other animals.

The inhibition of acetolactate synthase has been a fruitful target site for development of herbicides. By the early years of the twenty-first century, over 50 acetolactate synthase inhibitor herbicides had been developed.

## 2.6 Graminicides – herbicides that kill grass weeds

Grass weeds increasingly occupied the ecological niche created by control of broad-leaved weeds in cereals and other crops in the 1950s and 1960s. Weeds such as blackgrass (*Alopecurus myosuroides*), wild oat (*Avenu fatua*) and sterile brome (*Bromus sterilis*) became common in the UK during the 1970s, and their development was favoured by systems of cereal monoculture and reduced cultivation. Self-sown wheat and barley also became problems in following crops. The use of herbicides capable of killing grass weeds in broad leaf and, more remarkably, in cereal crops increased considerably in the late twentieth century.

The thiocarbamate herbicides such as triallate, and the chloracetamide compounds such as alachlor, propachlor and metazachlor (Figure 2.19), which also control broad-leaved weeds, were the first compounds to be extensively used for selective control of grass weeds. These herbicides are soil-applied and are used at relatively high doses of 2–3 kg/ha. Much of the selectivity of the thiocarbamate herbicides is related to application technique. For example, triallate is incorporated into the top 2.5 cm of soil where it kills germinating grass seeds. Wheat and barley are somewhat tolerant of triallate, and are also protected to a large extent during emergence by the leaf sheaths surrounding their coleoptiles.

Chloracetamide and some other herbicides such as the anilide compound flufenacet appear to block the synthesis of very-long-chain fatty acids (VLCFAs) necessary for, among other processes, formation of the plant cuticle (Boger, Matthes and Schmalfuss, 2000). The reasons for selectivity of the chloracetamide herbicides have been the subject of considerable debate, and it seems likely that metabolism and/or inactivation of the herbicide may occur in tolerant species.

The principal compounds used for control of grass weeds in crops are the aryloxyphenoxypropionates (commonly known as 'fops') and the cyclohexanediones (or 'dims') – Figure 2.19. The phenylpyrazolin herbicide pinoxaden is also recommended for grass weed control in cereals, and particularly of wild oat and ryegrass. The 'fops', 'dims' and pinoxaden are used after crop emergence and are effective at dose rates of 100–300 g/ha. These herbicides inhibit an early step in the biosynthesis of fatty acids in plants (Harwood, 1999). The primary target site is the plastid stromal enzyme acetyl Co-A carboxylase (ACCase), and inhibition leads to disruption of membrane synthesis (Figure 2.20). Growth of susceptible plants quickly ceases, usually within 48 h of application, and death follows in 1–2 weeks.

The target acetyl Co-A carboxylase catalyses a key reaction in all organisms, and thus compounds which inhibit the activity of this enzyme in plants might be expected to have acute effects on other organisms. However, these herbicides have a very low acute toxicity to non-target species. Two forms of the ACCase target enzyme appear to exist in plants, one in the plastids being responsible for fatty acid synthesis. Furthermore, the plastid ACCase appears to occur in two distinct forms, one in grasses and another in broad-leaved species.

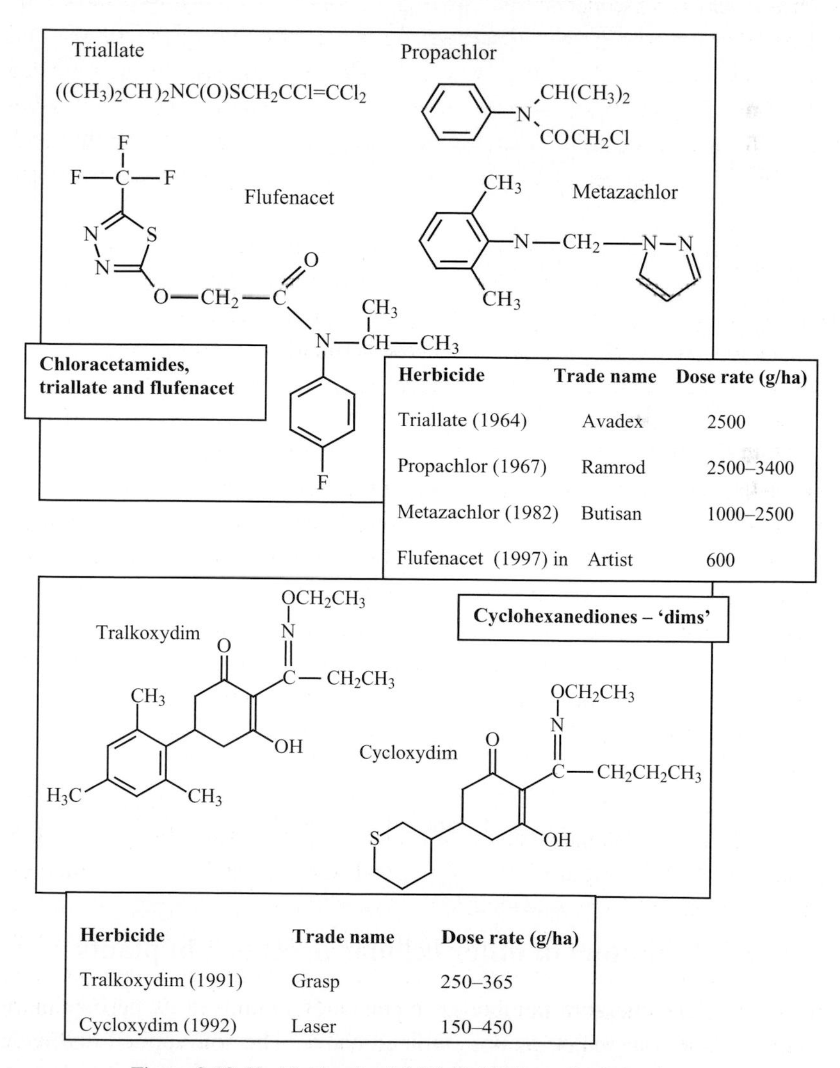


| Herbicide | Trade name | Dose rate (g/ha) |
|---|---|---|
| Triallate (1964) | Avadex | 2500 |
| Propachlor (1967) | Ramrod | 2500–3400 |
| Metazachlor (1982) | Butisan | 1000–2500 |
| Flufenacet (1997) in | Artist | 600 |

| Herbicide | Trade name | Dose rate (g/ha) |
|---|---|---|
| Tralkoxydim (1991) | Grasp | 250–365 |
| Cycloxydim (1992) | Laser | 150–450 |

Figure 2.19 Herbicides that inhibit lipid biosynthesis in plants.

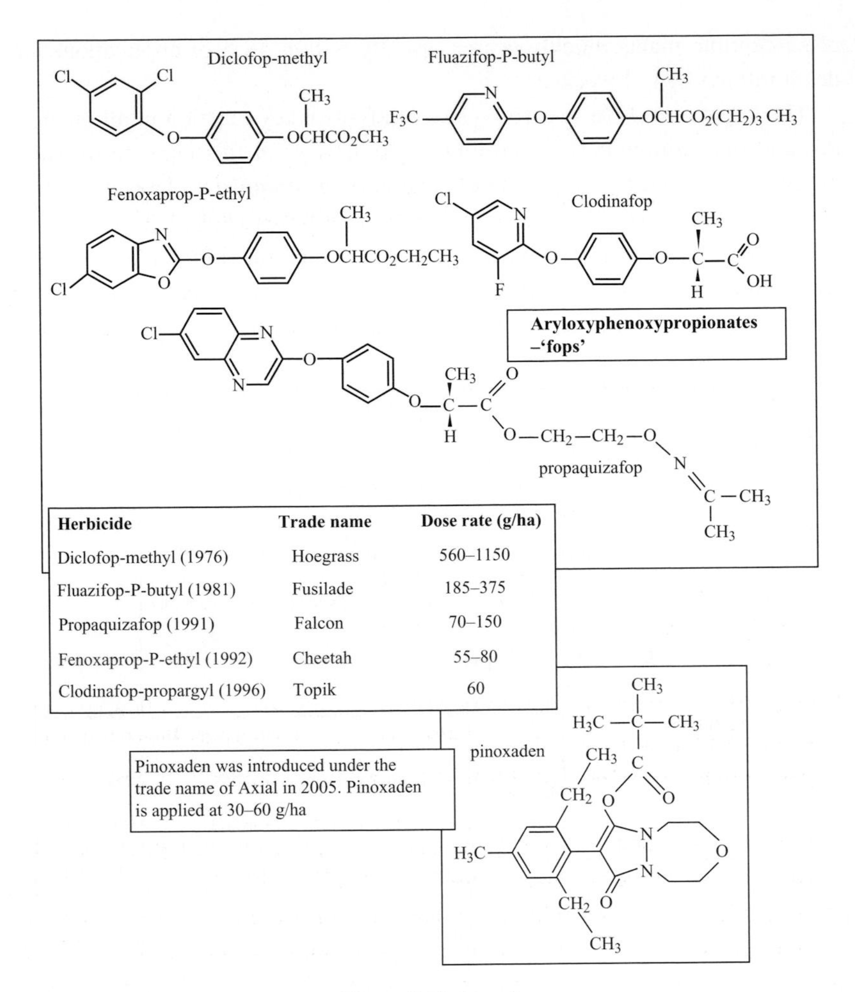


| Herbicide | Trade name | Dose rate (g/ha) |
|---|---|---|
| Diclofop-methyl (1976) | Hoegrass | 560–1150 |
| Fluazifop-P-butyl (1981) | Fusilade | 185–375 |
| Propaquizafop (1991) | Falcon | 70–150 |
| Fenoxaprop-P-ethyl (1992) | Cheetah | 55–80 |
| Clodinafop-propargyl (1996) | Topik | 60 |

Figure 2.19 (*cont.*).

Herbicides that block ACCase specifically inhibit the enzyme that occurs in grasses with little effect on ACCase in other organisms. At their recommended rate of application, fops and dims have no effect on ACCase from plant genera other than Graminae, and do not appear to inhibit ACCase from mammalian tissue (Shaner, 2003).

## 2.7 Inhibitors of other cellular processes in plants

There are a considerable number of compounds employed as herbicides in addition to the four major groups outlined above. The soil-applied herbicide

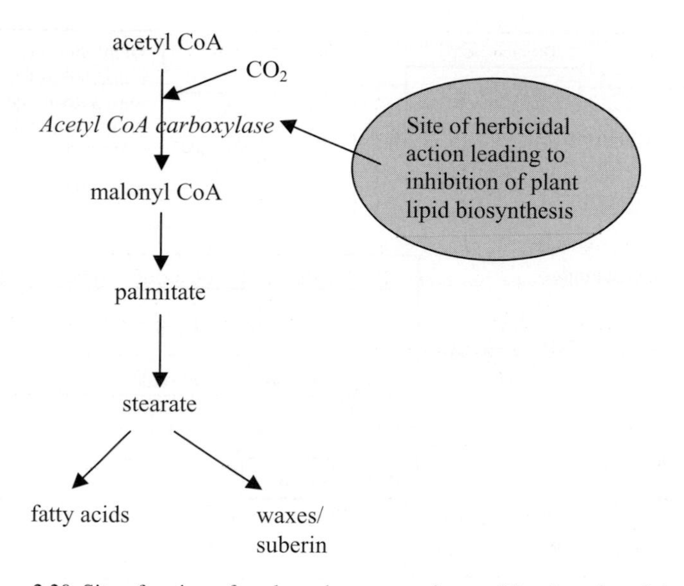


Figure 2.20 Site of action of aryloxyphenoxypropionate ('fops') and cyclohexanedione ('dims') herbicides.

dichlobenil is used for long-term weed control in shrub and tree plantings, and appears to block the synthesis of cellulose, a process that only occurs in plants and some fungi. It has the highest recommended dose rate of any herbicide used in the UK.

The dinitroaniline herbicides trifluralin and pendimethalin (Figure 2.21) are also mainly soil-applied and widely used for control of weeds in a range of crops. Dinitroaniline herbicides inhibit the synthesis of tubulin proteins in plants. Tubulin proteins are major structural constituents of microtubules that occur in the cytoplasm of cells. The spindle, which forms during cell division, consists of microtubules and inhibition of tubulin synthesis leads to interference with mitosis and cell division. Plant growth is retarded and ultimately death occurs. Whereas the molecular basis and consequent mechanisms of selectivity are well documented for the benzimidazole fungicides that inhibit tubulin polymerisation in sensitive species (3.7.1), the precise mechanism of action and basis of selectivity of herbicides which prevent tubulin formation are not clear, even though some of these molecules have been used for over 40 years.

The carbamate herbicides such as chlorpropham and the amides propyzamide and isoxaben (Figure 2.21) also disrupt cell division. Their principal use is for long-term control of weeds in tree and bush crops. These are also mainly soil-applied, although chlorpropham is also used to suppress sprout formation in stored potatoes.

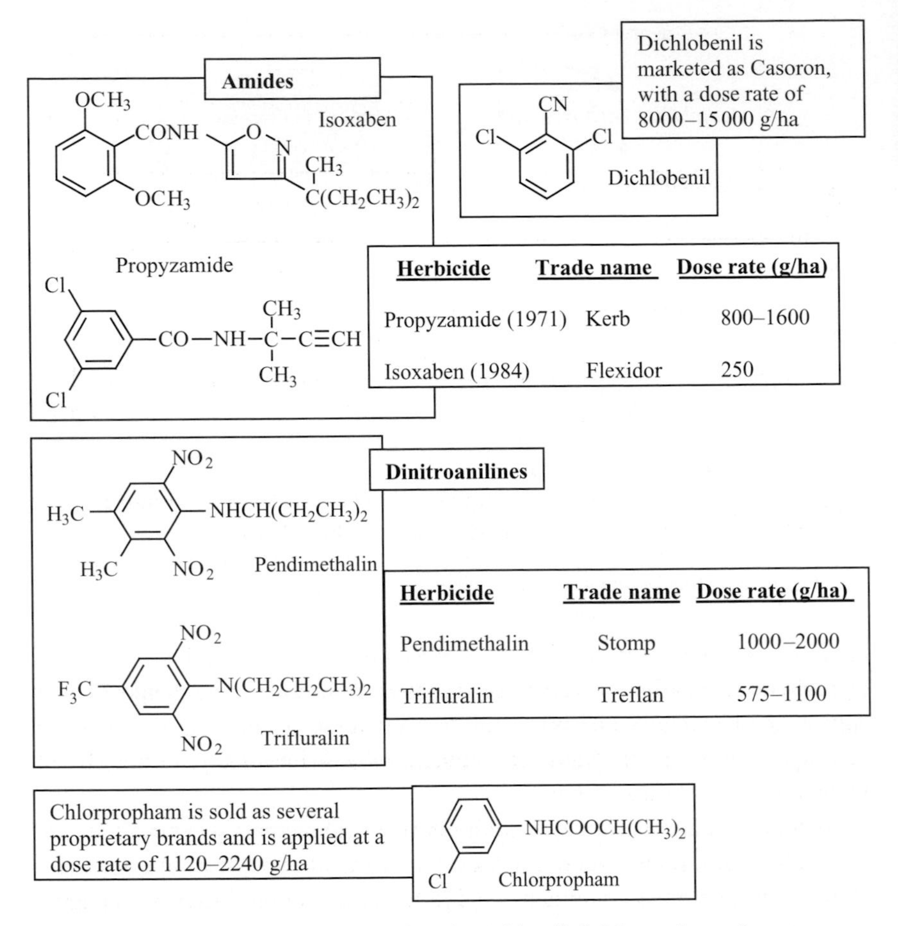


| Herbicide | Trade name | Dose rate (g/ha) |
|---|---|---|
| Propyzamide (1971) | Kerb | 800–1600 |
| Isoxaben (1984) | Flexidor | 250 |

| Herbicide | Trade name | Dose rate (g/ha) |
|---|---|---|
| Pendimethalin | Stomp | 1000–2000 |
| Trifluralin | Treflan | 575–1100 |

Figure 2.21 Herbicides that interfere with cell division and growth.

Finally, some inorganic compounds are used for long-term control of vegetation on non-agricultural land. These include ammonium sulphamate and sodium chlorate, which are applied in high volume. Despite their use for over 50 years the precise mode of action of these inorganic salts is not clear, nor is the basis of their selectivity to plants, although they have a low acute mammalian toxicity.

## 2.8 Selectivity, mode of action and herbicide resistance

Resistance occurs when the dose of herbicide initially recommended for weed control fails to achieve this effect. The means by which herbicides exert their effects is frequently by inhibiting single enzymes in target species, which in turn

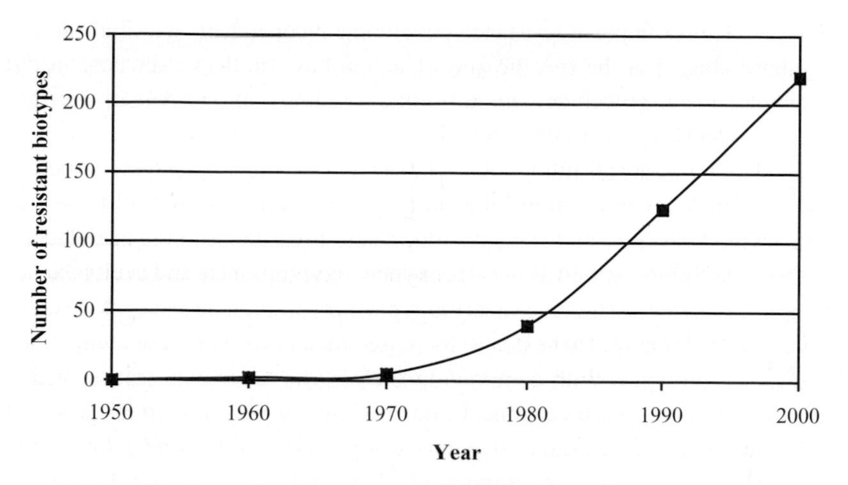


Figure 2.22 The increase in number of herbicide-resistant weeds worldwide (adapted with permission from Heap, I. M. (1999). International Survey of herbicide-resistant weeds: lessons and limitations. *Proceedings of the Brighton Crop Protection Conference, Weeds*, 769–776).

are coded by single genes. Such inhibitors include the PSII herbicides, inhibitors of acetolactate synthase and the graminicides. Resistance to such specifically acting compounds may be linked to either mutations leading to alterations at the target site of the herbicide, or increased metabolism and detoxification of the compound within the plant.

Within weed populations, sensitivity to herbicides may vary, even at the time of introduction of these compounds. Less sensitive biotypes may be effectively removed by herbicides and the ecological niche thus created may be occupied by more resistant biotypes of the same species. Establishment of herbicide-resistant weed populations is often linked to the continuous use of the same herbicide or herbicides with identical modes of action, a practice that maintains selection pressure in favour of resistant biotypes. Resistance is now a worldwide phenomenon with over 200 species of plants having developed resistance to herbicides that initially controlled them (Figure 2.22).

One of the best-documented and researched examples of herbicide resistance is that of triazine-resistant weeds. Over 60 triazine-resistant biotypes now occur worldwide, including both broad-leaved and grass weeds. In many cases the D1 protein of triazine-resistant weeds differs from that in triazine-susceptible biotypes by a single amino acid, but this is sufficient to prevent binding of the herbicidal molecule to the plastoquinone electron acceptor $Q_B$ binding site on the D1 protein. The amino acid substitution is in turn dictated by a single base change in the chloroplast gene coding for the D1 protein in resistant biotypes. This example of resistance development dependent on a single amino acid/base change at the specific site of action has parallels elsewhere in the pesticide sphere, as well as with antibiotic resistance in microorganisms that infect human beings and other animals.

Nucleotide base substitution and consequent changes in enzyme structure and/or configuration, resulting in a failure of the herbicide to bind to its target enzyme, may account for the development of resistance to the acetolactate synthase inhibitors, as well as the aryloxyphenoxypropionate and cyclohexanedione graminicides. Resistance to aryloxyphenoxypropionate and cyclohexanedione herbicides may also be due to increased metabolism of these compounds in resistant biotypes. With compounds such as the latter two groups, which act at the same target biochemical site in plants, a very important practical consideration is that resistance which develops to one compound acting at the site usually extends to other compounds that act at the same site. For example resistance has developed to aryloxyphenoxypropionate compounds such as fenoxaprop-ethyl in populations of blackgrass (*Alopecurus myosuroides*) in the UK and this resistance also extends to cyclohexanedione herbicides. This is the phenomenon of cross-resistance and where it happens, selection of a herbicide with a different biochemical site of action may be required to control resistant biotypes. Cross and particularly multiple resistance may also be due to the enhancement of metabolic detoxification mechanisms that may be effective against a range of herbicidal compounds.

## 2.9 Herbicide safeners

These are compounds that are added to herbicide formulations or used separately as seed dressings to improve selectivity by reducing damage to crop plants whilst allowing weed control (Hatzios, 2000), and have been primarily developed for use in monocotyledonous crops including maize, wheat and rice. Initially safeners were developed to minimise the effects of thiocarbamate herbicides in maize, thus allowing these compounds to be used in a selective manner. Compounds have now been developed that are capable of minimising the effects of aryloxyphenoxypropionates ('fops'), cyclohexanediones ('dims'), sulphonylureas and imidazolinone herbicides. Here safeners have extended the use of molecules exhibiting poor selectivity between crops and weeds, but which have favourable toxicological and environmental profiles. Safeners include a variety of chemical compounds, and some are shown in Figure 2.23.

Safeners appear to promote the detoxification of herbicides in crop plants, in most cases by enhancement of detoxifying enzymes such as cytochrome P450 monooxygenases and glutathione-*S*-transferases. The precise mechanism by

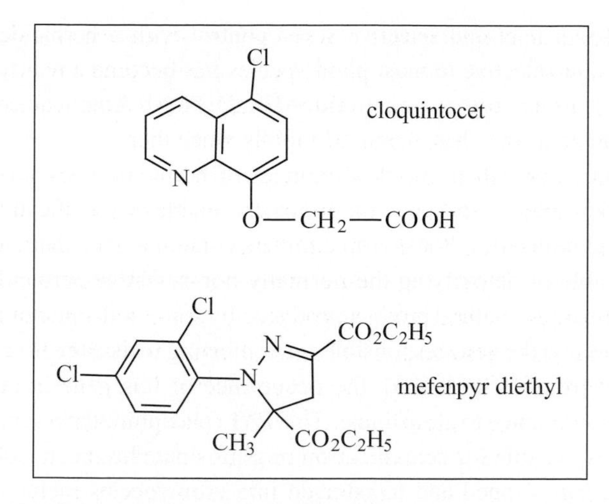


Figure 2.23 Two examples of herbicide safeners. Cloquintocet-mexyl is used in a mixture with clodinafop-propargyl, and mefenpyr-diethyl in mixture with fenoxaprop-ethyl and mesosulfuron-methyl/iodosulfuron-methyl-sodium. The herbicides are all used to control grass weeds in cereals.

which the action of these enzymes is enhanced is not known, but safeners may indirectly influence the regulation of genes responsible for cytochrome P450 enzymes (Davies, 2001).

## 2.10 Selectivity, mode of action and the development of herbicide-resistant crops

The identification and characterisation of individual enzymes as specific herbicide targets in plant-specific pathways of metabolism have been followed by the location and cloning in some cases of the genes that code for these target enzymes. For example, the gene responsible for production of EPSP synthase, the target site of glyphosate, has been isolated and cloned from several bacterial and plant species. Gene sequences associated with acetolactate synthase, the target site of sulphonylurea and imidazolinone herbicides, have been similarly characterised.

Strains of bacteria that synthesise aromatic amino acids have been discovered that are resistant to glyphosate, and this resistance is due to minor differences in the base sequence (compared to sensitive strains) of the gene coding for EPSP synthase. Glyphosate-resistant plants have been produced by insertion of genes from glyphosate-resistant bacteria. Genes coding for glyphosate resistance have been transferred into a number of crop species including oilseed rape, soya bean, maize and sugar beet and selective weed control with a herbicide formerly regarded as non-selective to most plant species has become a reality. The first glyphosate-resistant crops were introduced in the North American continent in 1996, and the area sown has increased rapidly since then.

A separate approach in the development of herbicide-resistant crops has been to introduce genes that code for enzymes capable of specifically detoxifying herbicidal molecules. Some crop cultivars contain a gene that codes for an enzyme capable of detoxifying the normally non-selective herbicide glufosinate. Glufosinate is a natural product produced by some soil-inhabiting *Streptomyces* species and the gene responsible for conferring resistance in crop species also isolated from these. Indeed, the occurrence of this gene in the bacteria explains their tolerance to glufosinate. This PAT (phosphinothrin acetyl transferase) gene responsible for detoxification of glufosinate has been isolated from *S. hygroscopicus*, cloned and transferred into crop species including oilseed rape, maize, soya beans and sugar beet. These crops can therefore detoxify glufosinate. Selective weed control can be achieved in these genetically modified crops by one or two sprays of glufosinate each growing season. Glufosinate-tolerant cultivars of oilseed rape, soya bean and maize were released on the North American market in 1997, and since then areas sown with these crops have dramatically expanded.

## 2.11 New herbicides and novel mechanisms of selectivity

Development of resistance in weed species to several of the major classes of herbicide has increased the need to develop compounds with novel modes of action that may control weeds resistant to currently available products. By the end of the twentieth century about 20 target sites had been exploited to achieve selective inhibition of plant growth by herbicides. Empirical screening of large numbers of compounds, as outlined in the first chapter of this book, is still the principal means of new compound discovery and in most cases the nature of the target site has been elucidated at, or in some cases after, the discovery and commercial development of the herbicide.

This situation may well change with advances in the structure and function of plant genomes. The complete gene sequences of some plant species are now known, and this knowledge is likely to be applied to the design of herbicides. At least 30 000 genes appear to be present in most plant species: the products of a considerable number of these are likely to be essential for healthy growth, and thus may be useful targets for herbicide development (Berg *et al.*, 1999; Hay, 1999; Saari, 1999; Cole, Pallett, and Rodgers, 2000).

Genes can be selectively removed from plants and the consequent effect on growth observed. If the so-called gene knock-out proves lethal, then inhibition of the function of the protein coded for may also be lethal. If the protein or its function is unique to plants, or can be inhibited without effects on non-target species, it may prove an ideal target for development of a selective herbicide. Indeed in recent years, several enzyme steps that occur in plant species have been the subjects of commercial patents as herbicide targets. Furthermore, patents have also been filed of compounds that have been developed through plant genomic studies. There seems little doubt that further mechanisms of selectivity discovered as a consequence of studies on plant genomics will be described, and likely targets for investigation include further enzymes in the complex pathways of carotenoid and chlorophyll biosynthesis, as well as associated pathways leading to the production of secondary metabolites in plants (Berg, Tietjen, Wollweber and Hain, 1999).

## 2.12 Plant growth regulators

Chemicals that are used to change the rate of growth and development of plants are commonly referred to as plant growth regulators, often abbreviated to PGRs. Such substances may influence germination, vegetative growth and plant maturity, reproduction, senescence as well as post-harvest preservation (Nickell, 1994). Their effects are primarily due to reductions in cell elongation and division. More specifically, PGRs are used in the following:

- Initiation, maintenance or termination of dormancy
- Promotion of adventitious root formation in stem cuttings
- Control of plant size
- Control of fruit set and development
- Prevention of deterioration in store after harvest.

The concepts of dose and response are well illustrated by compounds used to regulate plant growth, with some acting as regulatory agents at low dose but which may cause plant death at higher doses. The early dinitrophenol herbicide dinitroorthocresol (DNOC) was formerly used at lower doses as a plant growth regulator, and as seen earlier (2.2) many of the auxin herbicides were developed from research into compounds that modified plant growth. Indeed many of the commercially available PGRs are either preparations or derivatives of naturally occurring plant hormones, or synthetic chemicals that influence the activity of plant hormones.

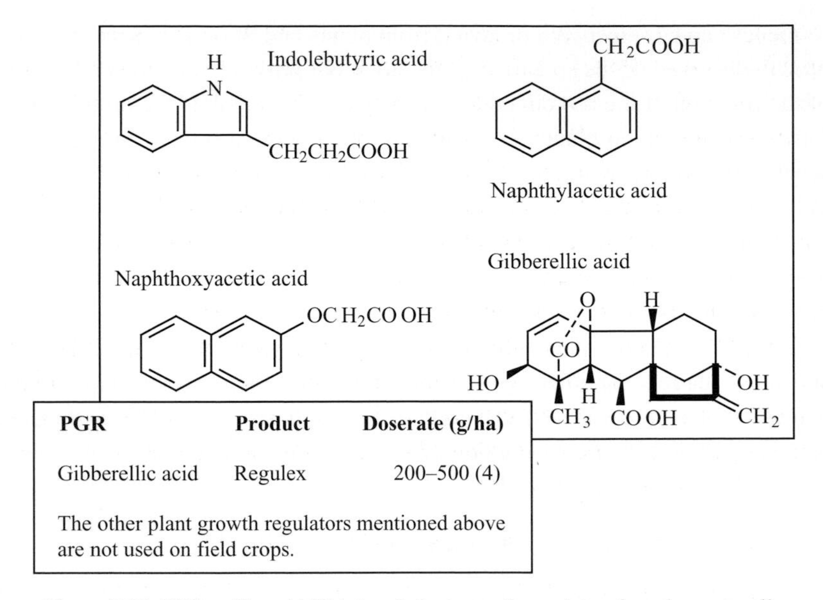


| PGR | Product | Doserate (g/ha) |
|---|---|---|
| Gibberellic acid | Regulex | 200–500 (4) |

The other plant growth regulators mentioned above are not used on field crops.

Figure 2.24 Gibberellic acid ($GA_3$) and plant growth regulators based on naturally occurring hormones.

## 2.12.1 PGRs based on plant hormones

One of the first commercial uses of PGRs was the use of auxin derivatives (Figure 2.24) such as naphthylacetic acid (NAA) and indolebutyric acid (IBA) to initiate cell division and root formation in stem and leaf cuttings of ornamental horticultural species. These compounds were introduced in the 1930s and, despite much research since, no better root promoters have been discovered.

In ideal conditions for fruit formation, apple trees may produce three to four times the amount of fruit that can develop to adequate size and quality. To avoid the production of small unmarketable apples, fruit thinning may be undertaken. Manual thinning is clearly a long and laborious procedure and synthetic auxins such as NAA have been used successfully as post-bloom thinners. Similar synthetic auxins such as 2-naphthoxyacetic acid have been used to stimulate development of fruits in plants.

However, gibberellins are the preferred hormonal compounds employed for fruit set. One of the principal uses of gibberellins in the USA, Japan and elsewhere is to promote the development of fruit in seedless grapes. Such parthenocarpic fruit will develop without the need for pollination, but application of gibberellins to such plants is needed to promote the translocation of nutrients to the developing fruit. Gibberellins are also used to improve fruit quality by stimulating epidermal and cuticular development in apples, which enhances the appearance of the skin and reduces fruit cracking. With citrus, gibberellins applied to immature fruit can prolong storage life after harvest, and they are routinely used in the USA for this purpose.

Gibberellins are widely used to stimulate the germination of malting barley in the brewing industry. They induce the production of $\alpha$-amylase and thus enhance the conversion of starch to soluble sugars. Gibberellins have been used to break dormancy in potato tubers and more widely in seeds such as those of some trees, and again the mobilisation of storage reserves results from application of these plant hormones.

The selectivity of the naturally occurring plant hormones may be equated with that of the auxin herbicides. It seems likely that receptor sites which respond to the presence of these compounds and which initiate physiological responses resulting from their application only occur in plant tissues.

### 2.12.2 Synthetic plant growth regulators

Most synthetic PGRs have been discovered by chance in screening studies for new herbicides and fungicides. Their precise mode of action and basis of selectivity are closely linked to the mechanisms of growth regulation by endogenous plant hormones. Synthetic PGRs that stimulate the release of ethylene or inhibit gibberellin biosynthesis are particularly widely used.

Ethylene is a naturally occurring growth regulator that affects both ripening and abscission. The synchronous development and abscission of, for example, fruit crops has obvious advantages to growers. The ethylene-releasing agent ethephon (Figure 2.25) is used in situations as diverse as thinning of apple, plum and cherry fruit, ripening of tomato fruit, and promotion of latex yield in rubber trees. Ethylene receptors appear to exist in plant cells and, on binding to these, a series of cellular reactions is initiated that leads to physiological changes such as tissue softening, promotion of colouring (for example in tomato) and abscission.

Several synthetic PGRs block the biosynthesis of gibberellins causing a shortening of internode length in plants, and this inhibition in height may be reversed by exogenous application of gibberellins (Rademacher, 2000). Synthetic PGRs have found widespread use as straw-shortening agents in cereal crops, where they prevent stem breakage caused by heavy wind and/or rainfall which can leading to collapse of the plant, the condition known as lodging. They are also used extensively in commercial horticulture to restrict the growth of fruit trees, reducing pruning costs and making harvesting easier, and also to produce compact ornamental pot and bedding plants. PGRs are also applied to turf and hedging to reduce the frequency of mowing and trimming.

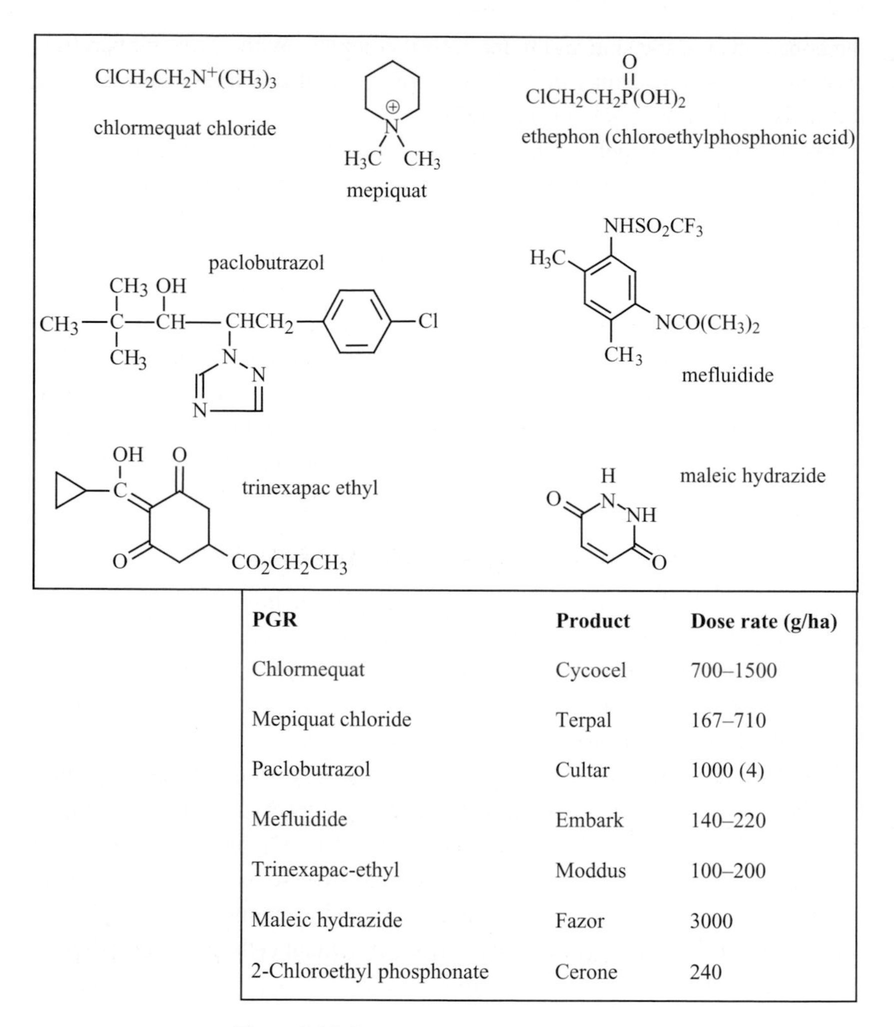


| PGR | Product | Dose rate (g/ha) |
|---|---|---|
| Chlormequat | Cycocel | 700–1500 |
| Mepiquat chloride | Terpal | 167–710 |
| Paclobutrazol | Cultar | 1000 (4) |
| Mefluidide | Embark | 140–220 |
| Trinexapac-ethyl | Moddus | 100–200 |
| Maleic hydrazide | Fazor | 3000 |
| 2-Chloroethyl phosphonate | Cerone | 240 |

Figure 2.25 Synthetic plant growth regulators.

The synthetic gibberellin inhibitors are structurally diverse (Figure 2.25). The principal compounds used to reduce lodging in cereals include chlormequat, mepiquat chloride and trinexapac-ethyl. The triazole paclobutrazol is used to produce dwarf ornamental horticultural species and also to reduce the height of fruit trees. The triazole compounds are closely related to the fungicidal triazoles, and as such give some protection against attack by plant pathogens. Indeed some triazole fungicides, such as tebuconazole (3.7.2), as well as offering disease control also assist with canopy management through shortening of stems in crops such as oilseed rape. Ancymidol and uniconazole, another triazole, are

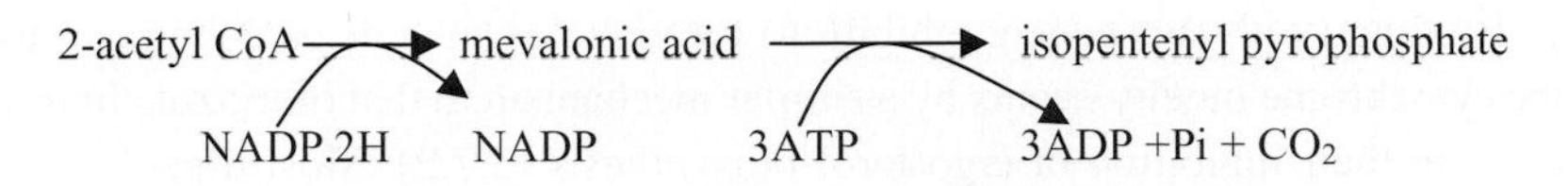


Condensation of 4 isopentenyl phosphate molecules by the enzyme prenyl transferase gives the compound geranylgeranyl pyrophosphate (GGPP)

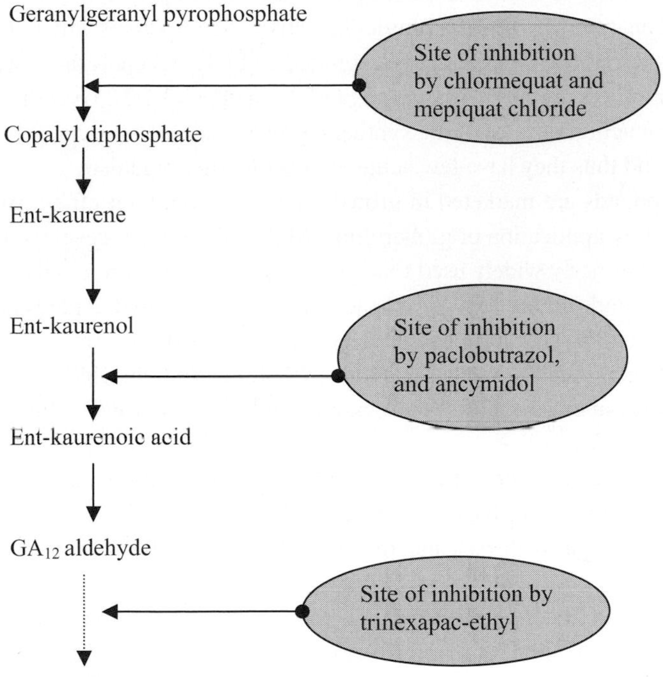


Figure 2.26 Sites of action of plant growth regulators that inhibit gibberellin biosynthesis.

used in the USA to dwarf pot plants. Ancymidol and trinexapac-ethyl are used to reduce the growth of turf on golf greens, as well as plant growth on roadside verges.

These compounds block gibberellin biosynthesis at several sites, with some compounds such as chlormequat and mepiquat chloride inhibiting the enzymes copalyl diphosphate synthase and ent-kaurene synthase, and others inhibiting later steps in the biosynthetic pathway (Figure 2.26). Ancymidol and paclobutrazol act as inhibitors of monooxygenases that catalyse the oxidation of ent-kaurene to ent-kaureonic acid, both intermediates in the pathway of gibberellin biosynthesis.

Binding (with subsequent inhibition) by triazoles such as paclobutrazol to the cytochrome moeity occurs by a similar mechanism to that of triazole fungicides in their inhibition of ergosterol biosynthesis (3.7.2). Most triazoles are mixtures of enantiomers, molecules of the same chemical structure but slightly different shape. Enantiomers of triazole compounds display an extraordinary degree of selectivity. The *S* enantiomers principally act as growth regulators, whereas the *R* enantiomers possess fungicidal activity. Compounds with potent *S* enantiomers have been commercially exploited as PGRs whereas those with potent *R* enantiomers are marketed as fungicidal triazoles (3.7.2). Overall, the primary sites of action of most of the synthetic growth regulators occur only in plant tissues, and thus they have few acute effects on other organisms.

Other compounds are marketed in growth retardants, but their effects may not be reversed by application of gibberellins. Maleic hydrazide, developed in the 1940s, was formerly widely used to suppress sprout formation in potatoes, which is clearly undesirable for the pre-pack market. Maleic hydrazide is also used to suppress grass growth in parks, cemeteries and roadside verges and in these situations is frequently used with another PGR, mefluidide, which retards seed development in grasses. These compounds inhibit apical growth, allowing lateral buds to develop.

The precise molecular sites of action of some of these compounds are not known, and thus neither is the basis of their selectivity. Maleic hydrazide appears to inhibit nucleic acid synthesis, but the basis of its selectivity to plants is obscure.

Very few new PGRs have been brought to commercial fruition since the early 1990s. However, some of the existing compounds are very widely used, with chlormequat being, in terms of amount applied, the most extensively used crop protection chemical in the UK (Table 1.2).

# 3
# Fungicides

## 3.1 Introduction

Fungicides are primarily used to control plant pathogens, but are also employed for protection of timber from attack by wood-rotting fungi, and have other minor uses, for example in wallpaper pastes and paints to prevent mould growth on wallcoverings. Fungicides are also used to prevent or cure fungal infections of human beings and other animals.

Plant pathogens affect both the yield and quality of crops (Figure 3.1). In western Europe, yields of cereals may be drastically reduced by pathogens such as the leaf spot *Mycosphaerella graminicola*, more commonly known as *Septoria tritici*, on wheat, and leaf blotch (*Rhynchosporium secalis*) on barley and powdery mildews of the genus *Blumeria*, formerly known as *Erysiphe*, on both. *Phytophthora infestans*, the causal agent of late blight, may cause devastating losses in potato crops. Fruit crops may be attacked by scab (*Venturia* spp.) as well as powdery mildew, and fruit may be rotted in storage by a range of fungi. Protected crops frequently suffer from the grey mould pathogen *Botrytis cinerea*, resulting in losses of both yield and quality. In the developed world, demands by multiple retailers of fruit and vegetables for high-quality, blemish-free produce has required a high degree of control of diseases such as scabs and grey mould.

Plant diseases may also cause serious losses in subtropical and tropical countries. Yields of cereals may be reduced by rust fungi such as *Puccinia graminis f. sp. tritici* (black stem rust). *Magnaporthe grisea*, rice blast, is a major disease in South-East Asia, Africa and Latin America. Pathogens such as *Mycosphaerella fijiensis* and *M. musicola*, the causal agents of Sigitoka diseases, *Hemieia vastatrix* (coffee rust), *Colletotrichum kahawae* (coffee berry disease), and *Fusarium* bark disease can cause major yield reductions in banana and coffee respectively.

Late blight of Potato (*Phytophthora infestans*)

*Mycosphaerella graminicola* (*Septoria tritici*) on wheat

Neck rot of lettuce caused by *Botrytis cinerea* (Crown copyright)

Apple scab (*Venturia inaequalis*)

Figure 3.1 Effects of plant pathogens.

It is important to note that plant diseases may be controlled by non-chemical methods of crop hygiene and, particularly in the case of biotrophs, pathogens that subsist only on their host plants, by cultivars into which genes have been incorporated that confer resistance to these organisms. Indeed, some viral diseases of plants and many bacterial pathogens are controlled solely by non-chemical means. In some cases, deployment of resistant cultivars may reduce the need for fungicide application or even eliminate the need for chemical control of fungal pathogens. However, chemical control is almost essential to prevent severe loss in yield from devastating diseases such as potato blight. Fungicides greatly assist in preventing colonisation of stored foods such as grain, fruit and vegetables by organisms that may rot such foodstuffs and which may also produce toxic metabolites whilst doing so.

Currently extensive use of fungicides occurs in the intensive farming systems of western Europe. Fungicides are routinely applied to field crops and cereals such as wheat and barley may receive two or more applications during each growing season. Potatoes may receive ten or more applications of fungicides in a single growing season to deter attacks by late blight. Where fungicides are used extensively on fruit and vegetable crops to prevent rots, scabs, grey mould and other diseases that may affect the appearance of crop produce, several applications may be made during the growing season, often close to harvest. Consequently, the potential for contamination of products such as fresh fruit and vegetables by fungicide residues is higher than for most other pesticides.

Fungicides may act in a protectant (protective) or systemic manner. The selective action of protectant fungicides is principally based on uptake of the active ingredient by fungal spores and hyphae on the surface of plants. Protectant fungicides do not usually enter plants, where the cuticle is an effective barrier to penetration. Also, they do not readily penetrate the skin and external surfaces of most animals. However, good coverage by protectants is essential to prevent attack by pathogens and many of these compounds are applied at high dose rates. Furthermore, these fungicides are subject to weathering, and may be washed off leaf surfaces in the field. Reapplication may then be required: consequently these compounds may present higher risks of environmental contamination than systemic fungicides.

Curative and systemic fungicides enter plant tissues and thus may be unaffected by weathering. Some may penetrate only the cuticle and epidermal layers but will cure existing infections there and in this respect are often termed eradicant. Others may have translaminar activity, moving throughout leaves to which they are applied. Some, particularly those used as seed dressings, may be translocated to all parts of the plant. Virtually all systemic fungicides move in the xylem and hence pass upwards in the translocation stream. There are as yet few phloem-mobile systemic compounds that move downwards in plants to offer protection to major crops from attack by root diseases.

The selectivity of systemic fungicides is based primarily on the interference with specific biochemical processes in fungi (Lyr *et al.*, 1998). These processes may be absent in other organisms or not inhibited within these organisms by the fungicide. Some fungicides are selective due to rapid detoxification in non-target organisms. Given the similarities of metabolism in plants and fungi, the development of compounds that can selectively eliminate fungal pathogens within plants represents a remarkable achievement by pesticide scientists.

The persistence of systemic compounds after application to within plant tissues allows extended control of pathogens, and a lower risk to the environment compared with multiple applications of protectant compounds. As noted, systemicity may allow the control of already existing infections: however, persistence in treated plants should not extend to the situation where concern may be caused to consumers from residues remaining on the crop. Again, this is particularly important with fresh produce such as fruits and vegetables.

It is important to note that there is no universal fungicide, although the strobilurins (3.7.3) have a very wide spectrum of activity. Different fungicides are needed to control fungi from different taxonomic groups. However, compounds are frequently mixed to extend the antifungal spectrum of commercial products. The variation in antifungal activity of fungicides can largely be explained by their mechanism of action at the cellular level. This knowledge also serves to explain the general lack of acute toxicity of most fungicides to plants, human beings and other animals.

## 3.2 Early protectant fungicides

Elemental sulphur and compounds of copper were the first agricultural fungicides and are still used today. Sulphur is particularly effective against the powdery mildew fungi as well as exerting control of some pest species, notably mites. Its selectivity in fungi is based on uptake and bioaccumulation by powdery mildews to the extent that it kills these pathogens. Why this accumulation occurs in powdery mildews and mites, and not in other organisms, is not clear. The actual toxic action may occur through inhibition of respiration, where sulphur may compete with oxygen as a terminal receptor in the respiratory chain. Sulphur, although applied at high doses of around 5 kg/ha, is highly prone to weathering and is generally used in drier areas and on glasshouse crops for the control of powdery mildews. For field crops it is generally applied in fungicide mixtures where, in addition to its preventative action against powdery mildews, it helps in providing sulphur for plant nutrition, particularly where crops can no longer rely on sulphur present in the atmosphere from coal-fired power station discharges!

Copper compounds (Figure 3.2) were first investigated for their fungicidal potential in France during the mid eighteenth century as seed dressings to alleviate the destructive effects of smut fungi. Almost a century later the first commercial fungicide, Bordeaux mixture, was marketed in France for the control of downy mildew on grapevines. Copper sulphate is the fungitoxic compound in Bordeaux mixture, and has proved consistently effective in controlling a range of pathogens, most notably potato blight and related fungi. These fungi bioaccumulate copper during periods of leaf wetness. The copper

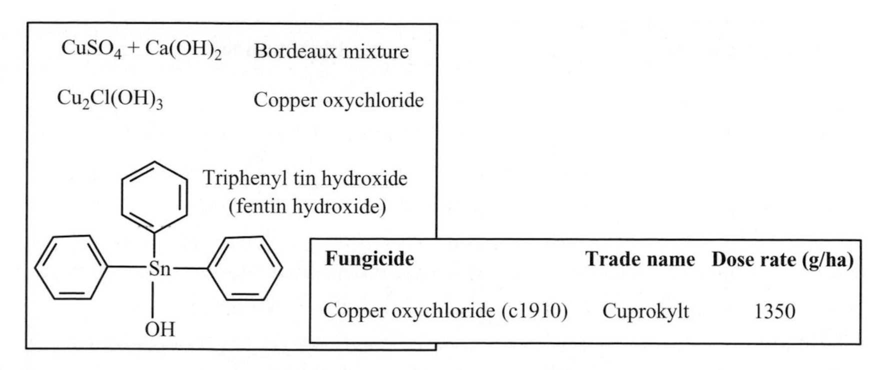


Figure 3.2 Copper and tin-based fungicides.

cation after accumulation is present in relatively high concentrations and may attach to negatively charged sites on molecules in the fungal cytoplasm. These include hydroxyl ($OH^-$), carboxyl ($COO^-$), sulphydryl or thiol ($SH^-$) and amino ($NH^-$) groups and the attachment (chelation) by copper ions causes disruption of the structure and function of proteins, and thus has pronounced effects on enzyme activity. Fungal metabolism is severely disrupted and the pathogen dies. Copper compounds are potent inhibitors of spore germination (Corbett, Wright and Baillie, 1984).

Bordeaux mixture in some circumstances, for example where plant growth is rapid and cuticles correspondingly thin, has proved phytotoxic. Hence, compounds of lower toxicity to plants such as copper oxychloride (Figure 3.2) and copper-dithiocarbamate complexes are the principal copper-based fungicides used as foliar sprays.

Since both copper and sulphur are considered 'natural' compounds they are approved for use in organic systems. Sulphur has virtually no long- or short-term effects on non-target organisms. Indeed it has no effect on many groups of fungi. However, copper is toxic to many forms of life in high concentrations.

Mercury-based fungicides were banned in the EU from 1990, and any problems of toxicity to non-target species through bioaccumulation of these compounds have largely disappeared. Tin-based fungicides (Figure 3.2) were prohibited in the EU from 2003, but are still used in some countries as end-of-season sprays for potato blight, as well as in timber treatments. These compounds are effective at much lower doses than copper fungicides, and are persistent in timber. However, tin-based fungicides are general respiratory inhibitors and, although used at much lower doses than copper-based fungicides, care is needed during application by personnel handling and applying them.

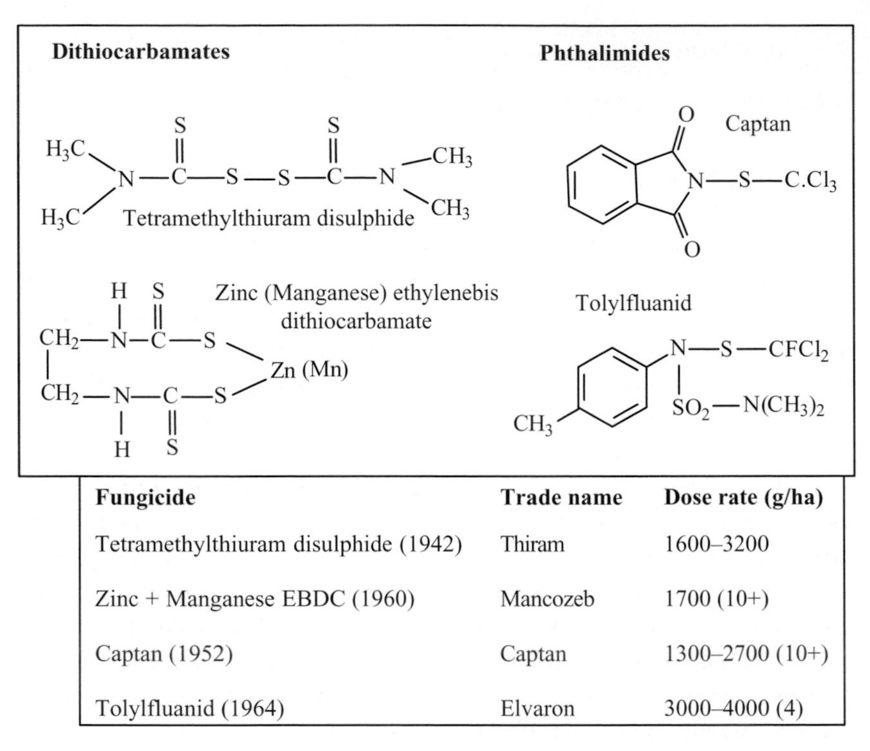


| Fungicide | Trade name | Dose rate (g/ha) |
|---|---|---|
| Tetramethylthiuram disulphide (1942) | Thiram | 1600–3200 |
| Zinc + Manganese EBDC (1960) | Mancozeb | 1700 (10+) |
| Captan (1952) | Captan | 1300–2700 (10+) |
| Tolylfluanid (1964) | Elvaron | 3000–4000 (4) |

Figure 3.3 Organosulphur protectant fungicides.

## 3.3 Organosulphur protectant fungicides

In terms of tonnages applied, these are the most extensively used fungicides in the world. They are cheap, stable and have a wide antifungal spectrum that includes potato blight, many leaf spot fungi and rusts; but, paradoxically, in view of the effects of elemental sulphur, they have little effect on powdery mildews. Organosulphur compounds are applied to many crops, especially in developing countries, and also to crop produce in storage. This has led to concerns being expressed about their potential to cause adverse long-term effects to health. Few environmental problems have been associated with the use of these compounds, although crops such as potatoes may receive between 10 and 20 sprays for control of blight (*Phytophthora infestans*) in a single season.

Most of the organosulphur compounds are derivatives of dithiocarbamic acid (Figure 3.3). The first organosulphur to be exploited as a fungicide was tetramethylthiuram disulphide or thiram, which has been widely used as a seed dressing and on crops such as lettuce for control of the grey mould pathogen *Botrytis cinerea*. The metallic dithiocarbamates were introduced in the late

1940s, but were largely superseded by the ethylenebisdithiocarbamates in the 1950s. The compounds Maneb and Mancozeb, a mixture of Maneb and Zineb, are used extensively for the control of potato blight, leaf spot pathogens and some rust fungi.

As with copper-based fungicides, the acute selective activity of these compounds seems to be mainly due to uptake into fungal spores and hyphae, but not into non-target organisms. When ethylenebisdithiocarbamates are taken up by target pathogens, they may break down in fungal cytoplasm to toxic compounds, with the most prominent being ethylene thiourea and ethylene diisothiocyanate. These toxicants readily attach to thiol ($SH^-$) groups, and enzymes that are rich in thiols, such as dehydrogenases, may be particularly disrupted. Inhibition of respiration consequently occurs in susceptible fungi, which then die (Corbett, Wright and Baillie, 1984).

The selectivity of the organosulphur phthalimide fungicide Captan, widely used for control of apple scab, is similarly based on attachment to thiol groups after uptake by fungal spores and hyphae on plant surfaces. As with the dithiocarbamates, conversion to a more active toxicant, in this case thiophosgene, may occur after uptake into the cytoplasm of fungi. As with the dithiocarbamates on potatoes, Captan may be sprayed on fruit several times in a season. The related fungicide tolylfluanid, widely used for control of the grey mould pathogen *Botrytis cinerea*, may have a similar mode of action.

## 3.4 Other organic protectant fungicides

The chlorinated hydrocarbon compound chlorothalonil (Figure 3.4) is extensively used in fungicide mixtures for the control of cereal diseases in Europe. This is yet another compound that after uptake probably binds to thiol groups in the cytoplasm of susceptible fungi, and in turn disrupts the structure and function of proteins leading to enzyme malfunction and hyphal death.

Few protectant fungicides have been released since the 1960s. The pyrrolnitriles were introduced in the late 1980s and were developed from a natural antimicrobial substance produced by the soil-inhabiting bacterium *Pseudomonas pyrrocinia*. Of these, fludioxinil (Figure 3.4) is used as a foliar fungicide for the grey mould pathogen *Botrytis cinerea*. The compounds appear to interfere with the initial stages of glucose metabolism in sensitive fungi, but the basis of selectivity is not known. These compounds, like older protectants, do not appear to readily penetrate the external surfaces of plants and animals, and this may partially explain their lack of toxicity to non-target species. This may also partially explain the lack of toxicity of the protectant fungicide fluazinam

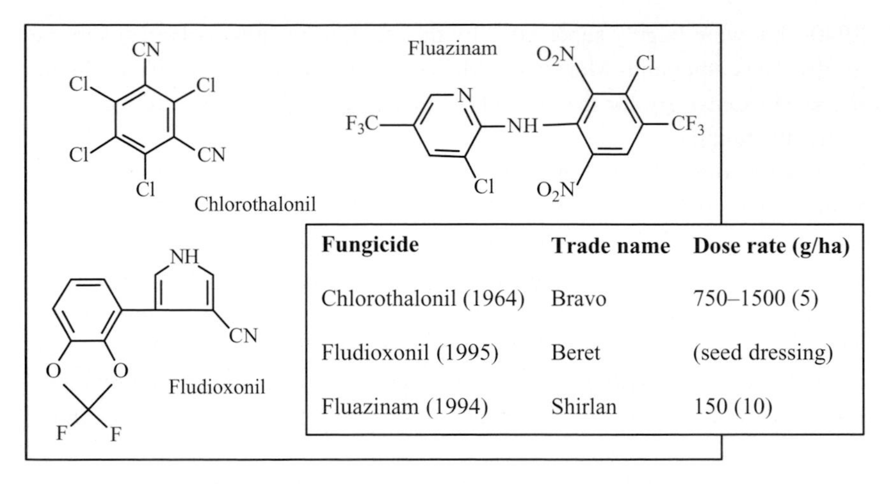


| Fungicide | Trade name | Dose rate (g/ha) |
|---|---|---|
| Chlorothalonil (1964) | Bravo | 750–1500 (5) |
| Fludioxonil (1995) | Beret | (seed dressing) |
| Fluazinam (1994) | Shirlan | 150 (10) |

Figure 3.4 Chlorothalonil, fluazinam and fludioxonil.

(Figure 3.4) to non-target organisms. Fluazinam, released to the market in the early 1990s, is used widely in Europe to control potato blight, and is applied at much lower doses and less frequently than some of the older protectants used for this purpose. The compound is a respiratory inhibitor, but the extent to which it inhibits this process in non-target species is not clear. The basis of selectivity may be due to rapid breakdown in non-target organisms.

The dicarboximide fungicides (Figure 3.5) introduced in the early 1970s have been used for control of the grey mould pathogen *Botrytis cinerea* and related fungi. Their mode of action has been the subject of considerable debate, with initial research pointing to a blockage of respiration in sensitive fungi. Subsequent studies in the 1990s indicate that the mode of action may involve inhibition of protein kinases associated with protein phosphorylation, but the basis of selectivity of the dicarboximide fungicides remains obscure. The use of vinclozolin has been restricted, and this is linked to its potential as an endocrine-disrupting substance (10.4).

The basis of selectivity of some protectant fungicides that have been on the market for many years is not known. Guazatine and hymexazol are used for control of seedling diseases in cereals and sugar beet respectively. Dithianon and dodine are employed for control of foliar diseases of fruit crops such as apple and pear: both fungicides are bioaccumulated by fungal spores and hyphae. Dithianon is yet another thiol inactivator, and dodine appears to interfere with membrane structure in fungi. Selectivity of these compounds is probably linked to lack of uptake by non-target organisms. Dinocap (Figure 4.15) is a respiratory inhibitor widely marketed to amateur (hobby) horticulturalists in the UK for

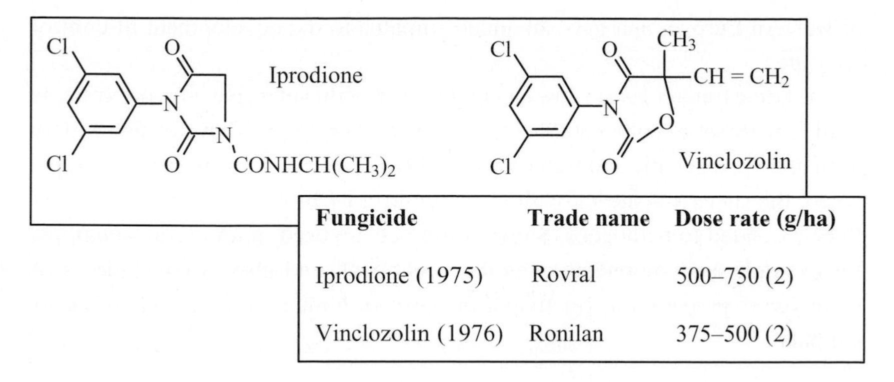


| Fungicide | Trade name | Dose rate (g/ha) |
|---|---|---|
| Iprodione (1975) | Rovral | 500–750 (2) |
| Vinclozolin (1976) | Ronilan | 375–500 (2) |

Figure 3.5 Dicarboximide fungicides.

control of powdery mildew fungi as well as mites (4.3) on fruit crops and ornamentals.

## 3.5 Systemic fungicides

Whereas systemic herbicides and insecticides were developed in the 1940s, systemic fungicides did not appear until the mid 1960s, and this may have been at least in part due to the difficulties in producing molecules of sufficient selectivity to kill target fungi without harming crop species. However, with a few notable exceptions such as potato blight and the cereal smut and rust fungi, the actual damage caused by pathogens to crops was not generally appreciated when compared to the readily observed depradations of insects and the competitive effects of weeds. The appreciation of yield and quality reductions that could result from, for example, attack by powdery mildews and other fungi in cereal crops led to an increased interest in control. This interest in fungal disease control was further developed in the EU with a move from the early 1970s onwards to extensive autumn sowing of cereals and a consequent rise in the importance of pathogens carried on crop debris, and which could survive the winter on young autumn-sown plants. Diseases formerly of minor importance, such as the leaf spot and glume blotch fungi *Mycosphaerella graminicola* (*Septoria tritici*) and *Stagonospora nodorum* (*Septoria nodorum*) on wheat, *Rhynchosporium* leaf spot and net blotch, *Pyrenophora teres*, on barley, as well as stem base and root diseases such as eyespot (*Tapesia yallundae*, formerly *Pseudocercosporella herpotrichoides*) and take-all (*Gauemannomyces graminis*) increased in importance in the intensive cereal production systems of western Europe, and gave an added impetus to the development of control measures.

Systemic fungicides are now an integral part of the intensive growing systems used for cereals and other arable crops in western Europe. They are also widely used in field vegetable and fruit crops worldwide as well as in glasshouse crops, where the success achieved with insect control by biological methods has yet to be extended to pathogens. Some systemics are used in non-crop situations, for example as components of timber treatment, and also as constituents of antimycotic preparations for fungal diseases of human beings, pets and other animals.

## 3.6 Systemic fungicides with a narrow spectrum of activity

For centuries the smuts, with their systemic development from seed infection leading to their appearance at ear emergence (Figure 3.1), were a major problem, particularly for cereal growers. Although hot water treatments developed in the 1920s served to partially alleviate seed infection, the development of the carboxamide (oxathiin) fungicides oxycarboxin and carboxin (Figure 3.6) in the mid 1960s reduced the cereal smuts to the status of minor pathogens. These remarkable compounds kill only basidiomycete fungi at low doses with little effect at such concentrations on plants, animals or indeed non-basidiomycete fungi. This extraordinary degree of specificity is due to the affinity of the compounds for part of the succinate dehydrogenase (SDH) complex of sensitive fungi. This enzyme is part of the tricarboxylic acid cycle in all organisms and its activity, and thus respiration, is blocked by carboxin and oxycarboxin at low dose, but only in rusts, smuts and related fungi. A slight difference in the amino acid structure of the SDH enzyme in sensitive fungi compared to SDH in other organisms allows binding of the fungicides at low dose, malfunction of this critically important enzyme and death of the pathogen.

The control of smut fungi by carboxin represents a major success story of the agrochemical industry. Carboxin is highly selective, applied at very low dose as a seed dressing (thus with virtually no risk of contamination of grain from the following harvest) and does not persist in the environment. Paradoxically, its low dose through use as a seed dressing and high efficacy against smut fungi has led to little profit to the manufacturing company when compared to returns from fungicides applied to foliage and stored products.

In the early 2000s, further developments in carboxamide fungicides have taken place (Figure 3.6). Silthiofam also has a narrow spectrum of activity with the principal fungal target being the take-all root disease of cereals

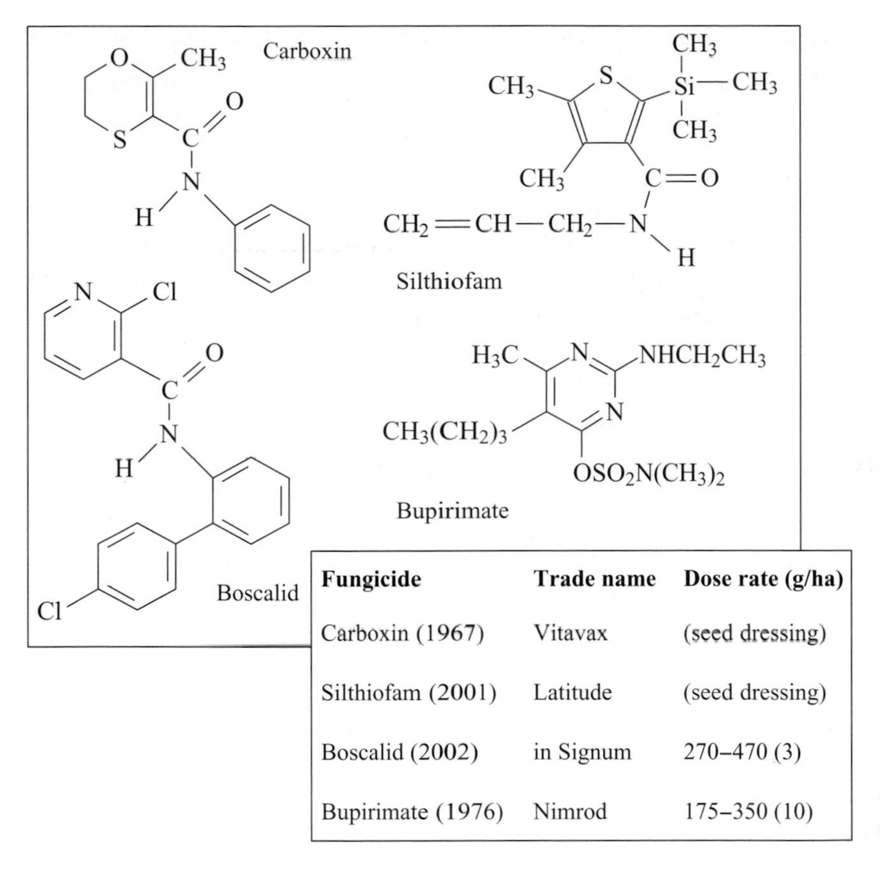


| Fungicide | Trade name | Dose rate (g/ha) |
|---|---|---|
| Carboxin (1967) | Vitavax | (seed dressing) |
| Silthiofam (2001) | Latitude | (seed dressing) |
| Boscalid (2002) | in Signum | 270–470 (3) |
| Bupirimate (1976) | Nimrod | 175–350 (10) |

Figure 3.6 Carboxamides and bupirimate – systemic fungicides with a narrow spectrum of activity.

that has proved very difficult to control by chemical means or the use of resistant cultivars. Silthiofam, like the carboxamides used for the control of smut fungi, appears to affect energy production in hyphae of the take-all fungus, but the mechanisms of selectivity are not clear. Boscalid, unlike other carboxamide compounds, has a wide spectrum of activity and is used for the control of the grey mould *Botrytis cinerea*, powdery mildews and leaf spots in vegetable crops. The basis of selectivity is not clear, but boscalid, like carboxin and oxycarboxin, inhibits SDH in sensitive fungal species.

The hydroxypyrimidine fungicides have a spectrum of activity as narrow as the carboximides, but their target is the powdery mildews. The only hydroxypyrimidine now used extensively is bupirimate (Figure 3.6). The selectivity of these fungicides appears to be due to inhibition of the enzyme adenosine

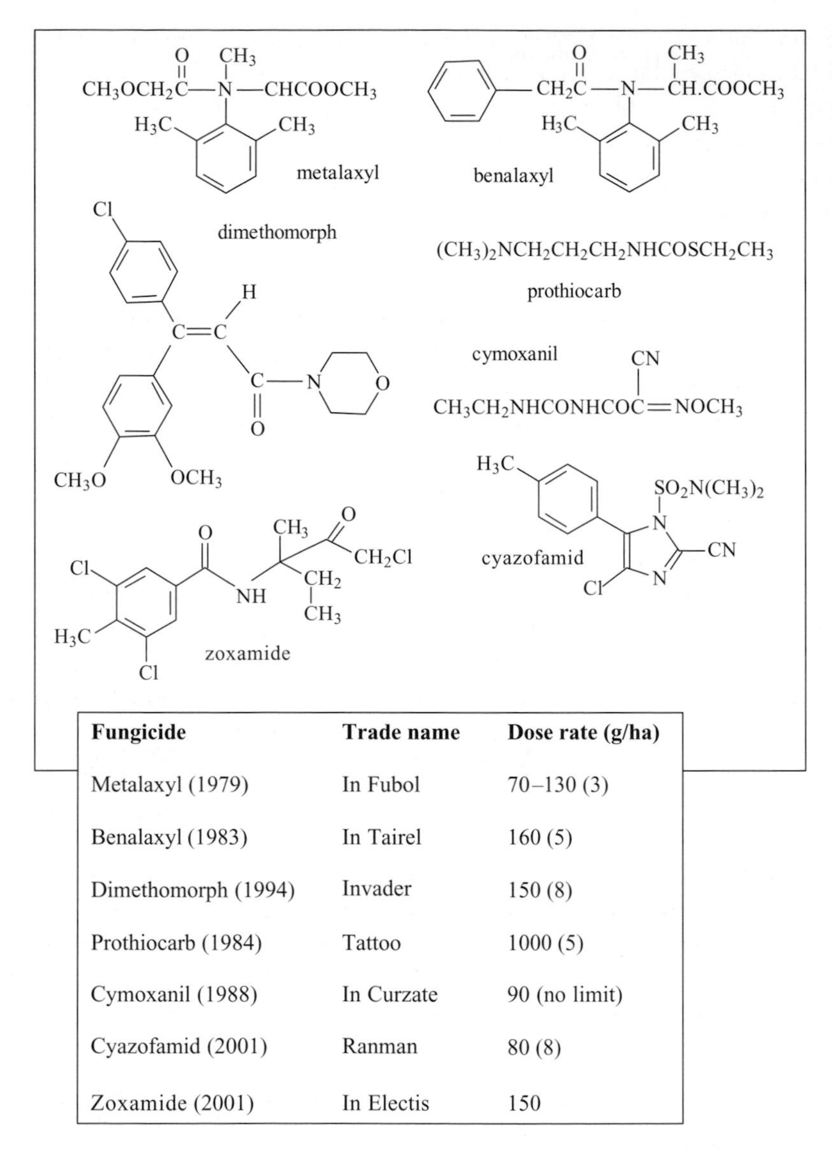


| Fungicide | Trade name | Dose rate (g/ha) |
|---|---|---|
| Metalaxyl (1979) | In Fubol | 70–130 (3) |
| Benalaxyl (1983) | In Tairel | 160 (5) |
| Dimethomorph (1994) | Invader | 150 (8) |
| Prothiocarb (1984) | Tattoo | 1000 (5) |
| Cymoxanil (1988) | In Curzate | 90 (no limit) |
| Cyazofamid (2001) | Ranman | 80 (8) |
| Zoxamide (2001) | In Electis | 150 |

Figure 3.7 Phenylamide and other fungicides for oomycete pathogens.

deaminase in sensitive fungi. The precise mechanism of selectivity is not clear, but adenosine deaminase may not be present in plants and be of a slightly different structure in other fungi and non-target species.

The phenylamide fungicides metalaxyl and benalaxyl (Figure 3.7) were introduced in the late 1970s and offered the prospect of control of root-attacking species of the genera *Pythium* and *Phytophthora*, potato blight and the related downy mildew fungi without recourse to soil drenching or repeated foliar application of protectant fungicides. Phenylamides are readily translocated in the xylem, and inhibit RNA polymerase activity in the ribosomes of sensitive fungi, probably by binding to a receptor in these organelles. Selectivity may be explained by differences in receptor morphology, such that the phenylamides only bind to receptors in the ribosomes of sensitive fungi.

Pesticide manufacturers see fungicides for potato blight and downy mildew control as a profitable area and several systemic compounds have been developed in addition to the phenylamide group. Cymoxanil and dimethomorph (Figure 3.7) must have a different mode of action to phenylamides since these both inhibit phenylamide-resistant strains of oomycete fungi. Dimethomorph may interfere with cinnamic acid metabolism in sensitive fungi, but the mode of action of cymoxanil and the basis of selectivity of both these compounds is not known. Prothiocarb is slowly translocated in plants; its precise mode of action and mechanism of selectivity remains unclear but it may interfere with membrane function in sensitive fungal species.

Further developments in the early twenty-first century have included the introduction of zoxamide and cyazofamid, primarily for control of potato blight and downy mildew fungi. In sensitive fungi, zoxamide interferes with microtubule biosynthesis in a similar manner to the benzimidazoles (3.7.1). Interestingly, cyazofamid inhibits electron transport in sensitive fungi in a similar manner to the strobilurin fungicides (3.7.3), but at a different site in the respiratory chain. The basis of selectivity of cyazofamid may be due to binding of the fungicide to its active site only in oomycete fungi.

Strains of potato blight resistant to systemic fungicides developed for control of oomycete pathogens have arisen widely, most notably to the phenylamide compounds such as metalaxyl. To reduce the risk of resistance, most curative and systemic fungicides for potato blight are marketed in mixtures with dithiocarbamate protectant compounds (3.3)

## 3.7 Systemic fungicides with a broad spectrum of activity

The principal broad-spectrum systemic fungicides include the benzimidazoles, sterol biosynthesis inhibitors and strobilurins. A consistent trend with these compounds has been the development of molecules effective at low dose and which have a benign environmental and toxicological profile.

### 3.7.1 Benzimidazole fungicides

Thiabendazole, the first commercially developed benzimidazole fungicide, is used to control storage rots of fruits and vegetables (Figure 3.8). Subsequently

| Fungicide | Trade name | Dose rate (g/ha) |
|---|---|---|
| Thiabendazole (1968) | Storite | (applied to fruit) |
| Benomyl (1969) | Benlate | 250–550 |
| Thiophanate-methyl (1971) | Cercobin | 500–1000 |
| Fuberidazole (1970) | in Baytan | (seed dressing) |
| Carbendazim (1974) | Bavistin | 125–500 (2) |
| Mebendazole | Vermox | (antihelminthic) |

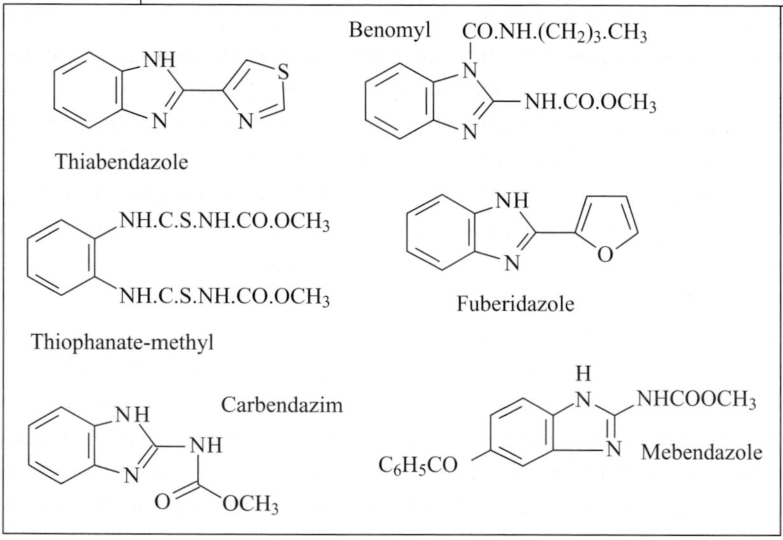

Figure 3.8 Benzimidazole fungicides.

developed molecules include benomyl, now withdrawn in the EU, thiophanate-methyl and fuberidazole. The latter compounds are converted to the active toxicant methyl-2yl benzimidazole carbamate (MBC) in water and/or on uptake by plants, and the group is sometimes referred to as MBC-generating fungicides. MBC itself is marketed as Carbendazim.

When introduced, these fungicides proved active against many ascomycotine and deuteromycotine fungi at very low concentrations. Their lack of acute toxicity to plants and animals, and non-target fungi can be explained by the biochemical mode of action of these compounds, through interference with nuclear division.

Mitosis and meiosis are disrupted by benzimidazoles due to inhibition of formation of microtubules. These are highly flexible cylindrical cellular structures of which the most prominent and well known are the spindle fibres where chromosome division and separation occur during nuclear division. Within a few minutes of exposure to MBC, almost all cytoplasmic microtubules disappear in sensitive fungi. MBC binds to β tubulin, one of the two principal proteins that make up microtubules (the other being α tubulin). The binding prevents incorporation of the β tubulin into microtubules and thus stops their assembly. Without microtubular assembly, spindles cannot form and nuclear division is halted. These fungicides are not inhibitors of spore germination, but hyphal growth cannot proceed without nuclear division. Growth is thus prevented and ultimately sensitive fungi may die. The activity of benzimidazoles is almost identical to that of colchicine, which inhibits microtubule assembly in vertebrates, thus making the latter a potent toxin to some mammalian species.

All eukaryotic organisms including plants, animals and fungi have microtubules composed of tubulins. The structure of tubulins in eukaryotes is remarkably similar. All are polypeptides of around 450 amino acids, with a structural similarity of 80–90% in plants, animals and fungi. Nevertheless it is the slight differences in amino acid structure at different sites in the tubulin molecules from different species that appear to determine whether benzimidazole fungicides (as well as colchicine and other antimitotic agents) are effective.

The genes for tubulin synthesis have been isolated and cloned for several species, including fungi. A single base change may result in an amino acid substitution with a consequent alteration in the binding site of benzimidazole fungicides: MBC may not bind, and then hyphal growth is not inhibited. Such subtle differences are also responsible for the lack of binding of these fungicides to the tubulins of plants, most animals and indeed non-target fungi, and the consequent lack of acute toxic effects on many of these organisms.

However, benzimidazoles will bind at low dose to the tubulins of not only fungi but also some invertebrate species, particularly trematodes and cestodes. Thus, some of these compounds (such as mebendazole and fenbendazole) have been developed as therapeutic agents for treatment of trematode and cestode parasites in mammals. Flukes, tapeworms and parasites of a similar nature can be effectively treated with these drugs. Indeed mebendazole is used for the treatment of tapeworms in human beings. This use is only possible because of the discriminatory activity of benzimidazoles in binding to the tubulins of the parasites and not those of the host – a classic example of toxicological selectivity.

The spectrum of activity of benzimidazoles also extends to some non-target species of a distinctly beneficial nature, most notably earthworms. Localised soil drenches with benzimidazole fungicides may reduce earthworm populations and these compounds are widely used for control of earthworms on grassed areas of valued recreational or amenity use such as golf and bowling greens, where earthworm excreta or casts may severely affect performance on these sports surfaces.

Table 3.1 *The nucleotide sequence of the β tubulin gene, and associated sensitivity/resistance to benzimidazole fungicides, of two strains of the barley leaf spot pathogen* Rhynchosporium secalis

| Strain of *R. secalis* | Nucleotide sequence of β tubulin gene (*amino-acid codon*) 197 | 198 | 199 | 200 | 201 | Benzimidazole response |
|---|---|---|---|---|---|---|
| 810.39.02 | GAT | GAG (*glu*) | ACC | TTC | TGT | Sensitive |
| 765.03.01 | GAT | GGG (*gly*) | ACC | TTC | TGT | Resistant |

A single base substitution – glycine for glutamate, prevents binding of the fungicide and confers resistance. Adapted from Butters, J.A., Kendall, S.K., Wheeler, I.E. and Holloman, D.W. (1995). Tubulins: lessons from existing products that can be applied to target new antifungals. In: Antifungal Agents. Discovery and Mode of Action. Eds Dixon, G.K.; Copping, L.C. and Holloman, D.W. Bios Scientific. Pp. 131–141.

Resistant strains of fungi formerly controlled by benzimidazole fungicides are now widespread. Development of resistance is closely associated with the highly specific mode of action of these compounds. Single base changes in the genes coding for tubulin structure in fungi result in a change in amino acid structure of the tubulin molecule, to which benzimidazole fungicides will not bind (Table 3.1). Continued use of these fungicides has enabled fungicide-resistant mutants to become common, in some cases to the extent that the benzimidazoles are no longer effective, and so the use of these compounds is declining.

### 3.7.2 Inhibitors of sterol biosynthesis

These extensively used antifungal compounds have applications in human and veterinary medicine as well as agriculture. In the early twenty-first century, the sterol biosynthesis inhibitors (SBIs) accounted for about 25% of global fungicide sales in agriculture. SBI fungicides were introduced in the early 1970s, and their development has continued since, with the emphasis being towards increased potency: effective at low dose rates, with few side-effects on non-target organisms and the environment in general (Baldwin and Corran, 1995).

These compounds act by blocking the synthesis of fungal sterols, notably ergosterol. Sterols are present in the cell membranes of all higher plants, animals and fungi, where they form part of the phospholipid bilayer of membranes, in which they have a stabilising role. The principal sterol in fungi is ergosterol; in the cell membrane of many plants it is stigmasterol; and in mammals and

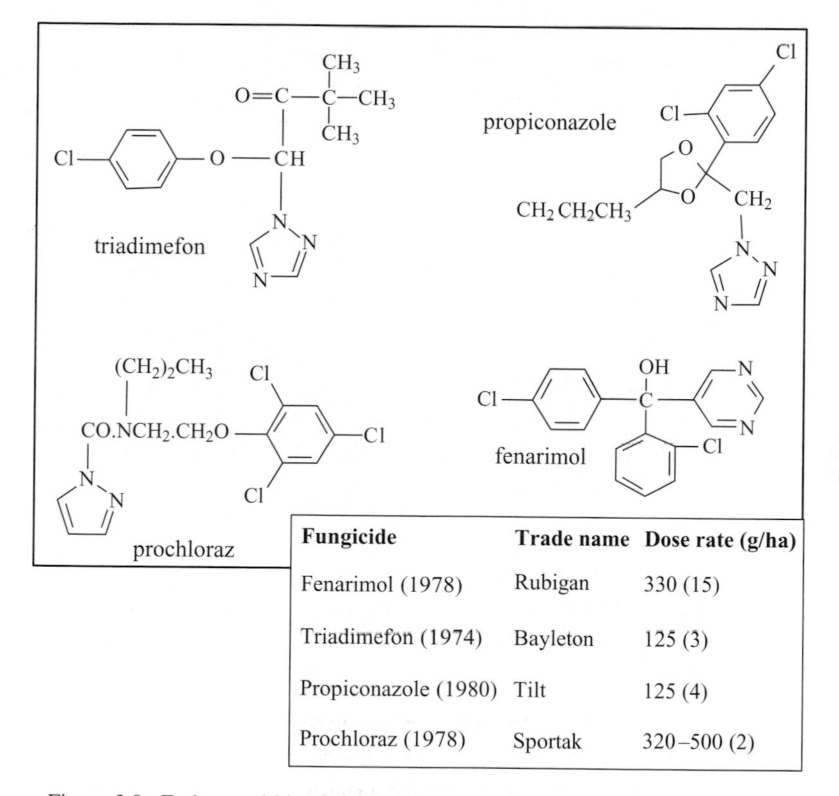


| Fungicide | Trade name | Dose rate (g/ha) |
|---|---|---|
| Fenarimol (1978) | Rubigan | 330 (15) |
| Triadimefon (1974) | Bayleton | 125 (3) |
| Propiconazole (1980) | Tilt | 125 (4) |
| Prochloraz (1978) | Sportak | 320–500 (2) |

Figure 3.9 Early sterol biosynthesis inhibitors – the pyrimidine fenarimol, two early triazoles and the imidazole prochloraz.

some other animals, cholesterol. The biosynthetic pathways to these sterols in different organisms share some common steps, but also vary at certain stages.

Sterol-biosynthesis-inhibiting fungicides act by blocking specific enzyme steps in the pathway of ergosterol biosynthesis. Fungi that belong to the Ascomycotina (e.g. powdery mildews), Basidiomycotina (rusts and smuts) and Deuteromycotina (e.g. eyespot) synthesise ergosterol, and fungicides that inhibit the formation of this compound control many diseases in these taxonomic groups. Oomycete pathogens, such as the downy mildews, *Pythium*, a major cause of damping-off disease and *Phytophthora infestans*, potato blight, do not synthesis ergosterol and so fungicides that inhibit sterol biosynthesis are ineffective against them.

Most of the SBI fungicides act by preventing the oxidative demethylation of an intermediate in the pathway of ergosterol biosynthesis. The pyrimidines (fenarimol, Figure 3.9), imidazole (imazilil, prochloraz), and all triazoles (Figures 3.9 and 3.10) inhibit demethylation at the $C_{14}$ atom of the biosynthetic

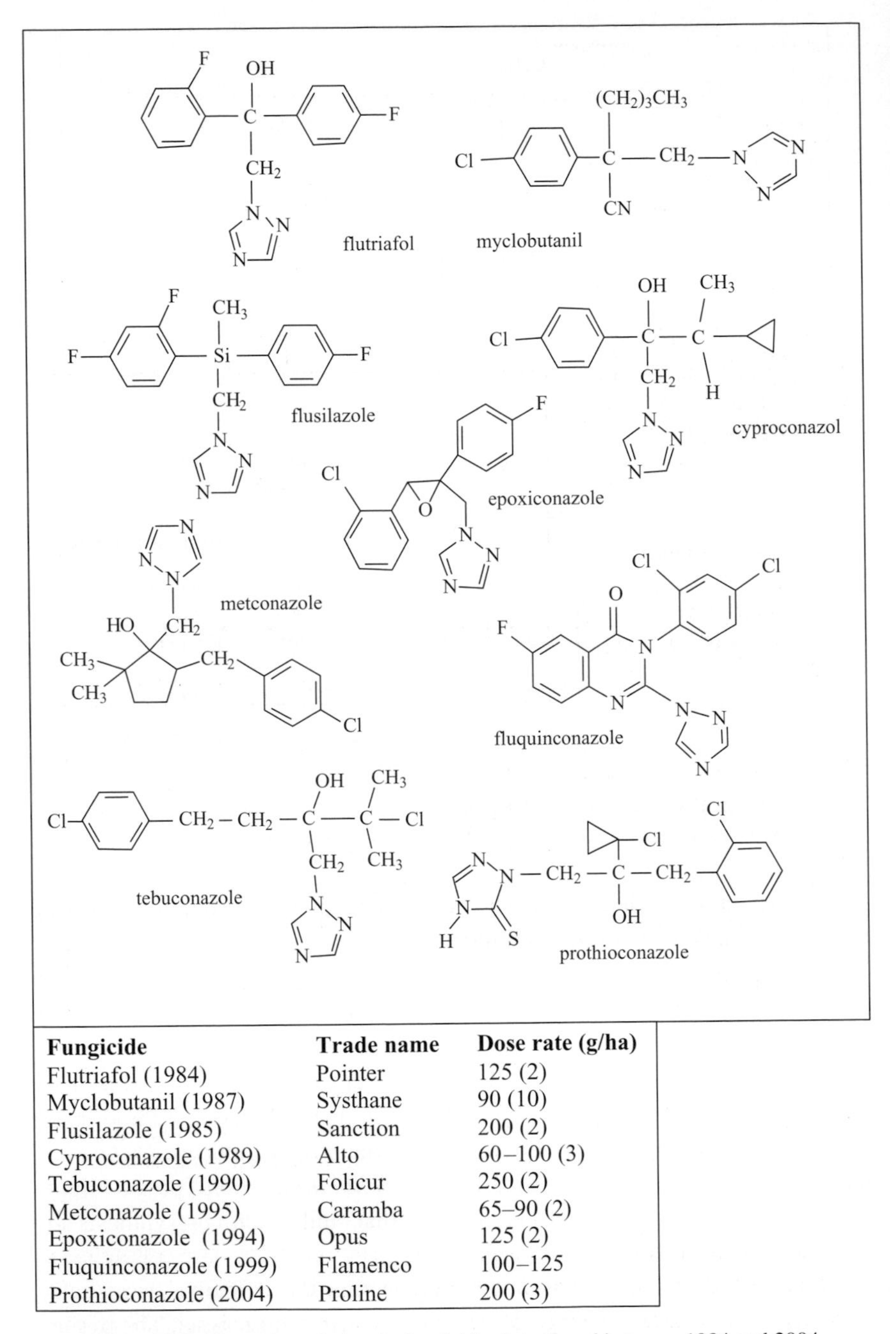


| Fungicide | Trade name | Dose rate (g/ha) |
|---|---|---|
| Flutriafol (1984) | Pointer | 125 (2) |
| Myclobutanil (1987) | Systhane | 90 (10) |
| Flusilazole (1985) | Sanction | 200 (2) |
| Cyproconazole (1989) | Alto | 60–100 (3) |
| Tebuconazole (1990) | Folicur | 250 (2) |
| Metconazole (1995) | Caramba | 65–90 (2) |
| Epoxiconazole (1994) | Opus | 125 (2) |
| Fluquinconazole (1999) | Flamenco | 100–125 |
| Prothioconazole (2004) | Proline | 200 (3) |

Figure 3.10 Examples of triazole fungicides introduced between 1984 and 2004.

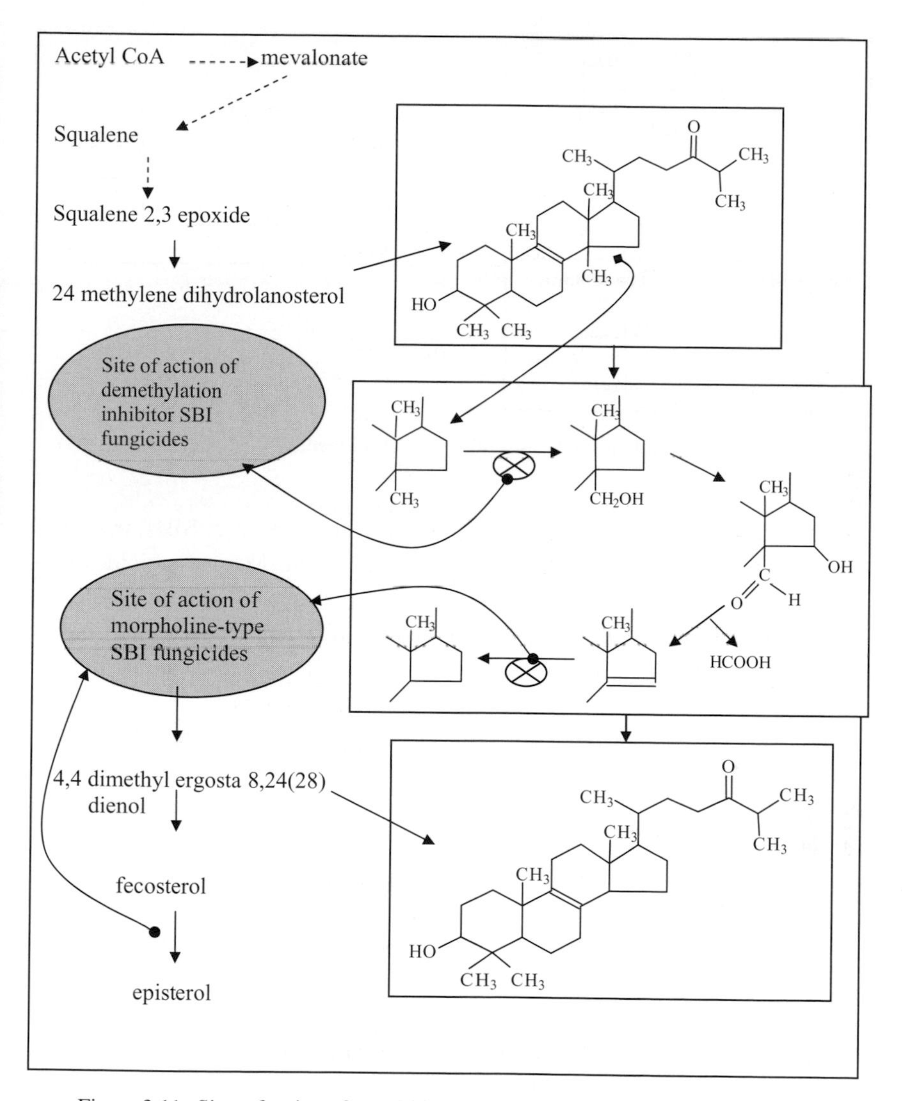


Figure 3.11 Sites of action of sterol-biosynthesis-inhibiting fungicides. The pathway to ergosterol biosynthesis is shown only in part, and focuses on the inhibition of demethylation of the ergosterol precursor 24-methylene-dihydrolanosterol.

intermediate 24-methylene-dihydrolanosterol (Figure 3.11). These fungicides are consequently referred to as demethylation inhibitors (DMIs). When sterol synthesis is blocked, assembly of fungal cell membranes is impaired, with consequent leakage of metabolites from fungal cytoplasm. Growth ceases and hyphal death follows.

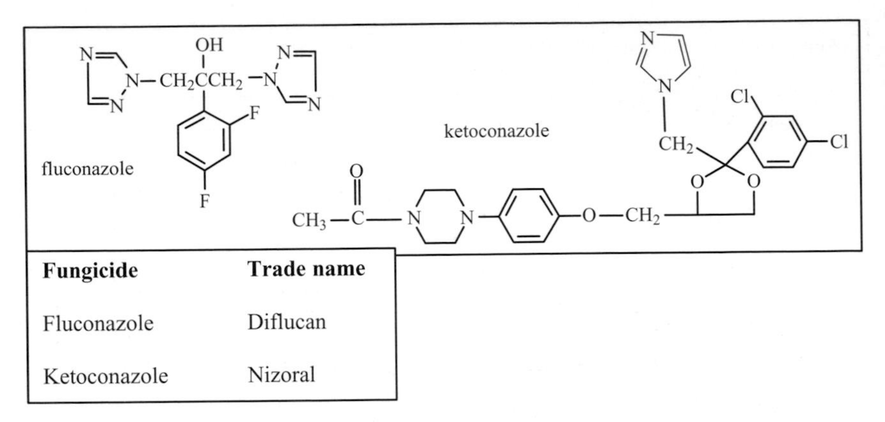


| Fungicide | Trade name |
|---|---|
| Fluconazole | Diflucan |
| Ketoconazole | Nizoral |

Figure 3.12 Sterol biosynthesis inhibitors used in medical treatment of fungi.

The selective action of the most extensively used group of SBIs, triazoles, is associated with their stereochemical configuration. Many fungicides that inhibit sterol biosynthesis are manufactured as mixtures of isomers, where the molecule may be present in two or four slightly different configurations. This property is shared with many other molecules, and its significance is that the different isomers may react differently in biological systems.

Of course, SBIs marketed for control of fungi contain potent fungicidal isomers. However, some of the isomers of triazole SBIs can have effects on the synthesis of plant sterols as well as interfering with gibberellin biosynthesis in plants, and thus may cause stunting. Triadimenol, the first triazole to be used widely as a seed dressing in agriculture, may temporarily stunt young plants, and this is due to one of its isomers having plant growth regulatory activity. This property has been exploited commercially. The triazole compound paclobutrazol was developed from studies with the short-lived triazole fungicide diclobutrazol, and is used as a dwarfing agent for pot plants and shrubs (2.12.2).

The SBIs, including triazole isomers, used in agriculture have few adverse effects on mammals. Indeed, like the benzimidazole fungicides, there are SBIs that are used to treat fungal infections of human beings as well as pets and other domesticated animals (Figure 3.12). The triazole fluconazole is a standard treatment for throat and vaginal fungal infections and the imidazole ketoconazole is present in some shampoo preparations to combat the fungi that contribute to dandruff.

There are a few fungicides that inhibit sterol biosynthesis in fungi by blocking steps in the pathway other than $C_{14}$ demethylation. These include the morpholine fungicide fenpropimorph, the related compound fenpropidin and the spiroketal amine spiroxamine (Figure 3.13). They are applied at much higher

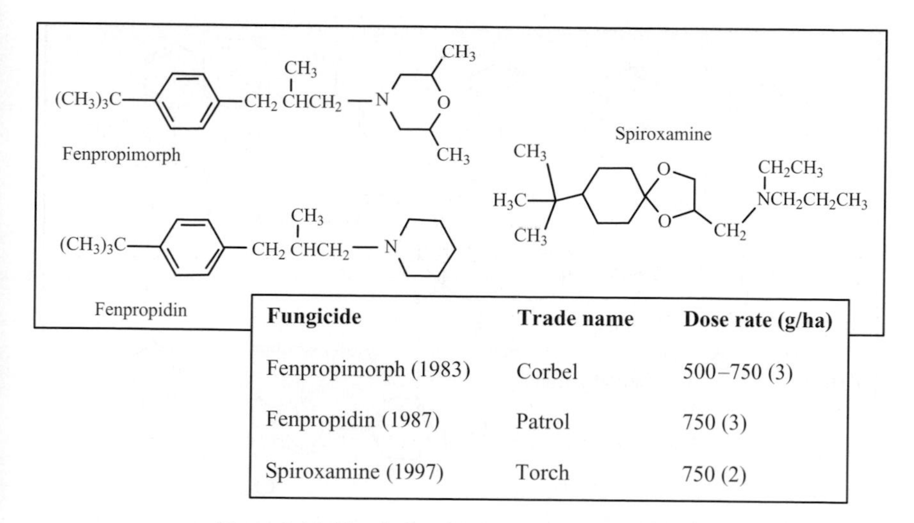


| **Fungicide** | **Trade name** | **Dose rate (g/ha)** |
|---|---|---|
| Fenpropimorph (1983) | Corbel | 500–750 (3) |
| Fenpropidin (1987) | Patrol | 750 (3) |
| Spiroxamine (1997) | Torch | 750 (2) |

Figure 3.13 Morpholine fungicides and spiroxamine.

rates than DMIs. These compounds inhibit susceptible fungi at two points in the pathway to sterol biosynthesis (Figure 3.11). Inhibition of $\Delta^8$-$\Delta^7$ isomerisation may occur and an accumulation of $\Delta^8$ sterols occurs. Inhibition of reduction of the double bond at position 14 introduced subsequent to the $C_{14}$ demethylation reaction may also occur. Fenpropidin is a very potent inhibitor of this latter step, and fenpropimorph is a highly potent inhibitor of both steps. The fact that morpholines inhibit biosynthesis of ergosterol at different points in the pathway to that inhibited by DMIs has considerable practical significance. Morpholines are effective against DMI-resistant strains of fungi and have proved useful alternatives to the DMI fungicides in spray programmes.

These fungicides, like DMIs, are generally non-toxic in an acute sense to most non-target species. However, like some triazole isomers, morpholines can also interfere with sterol biosynthesis and/or gibberellin metabolism in some plant species. Indeed these fungicides have severe effects on seed germination in cereals, and this has prevented their use as seed dressings for these crops.

### 3.7.3 Strobilurins and other systemics that inhibit electron transport

The first strobilurins (Figure 3.14), introduced to the market in the late 1990s, were derivatives of naturally occurring antifungal compounds produced, paradoxically, by fungi – notably the basidiomycotine species *Strobilurus tenacellus*. There are considerable similarities between the development of

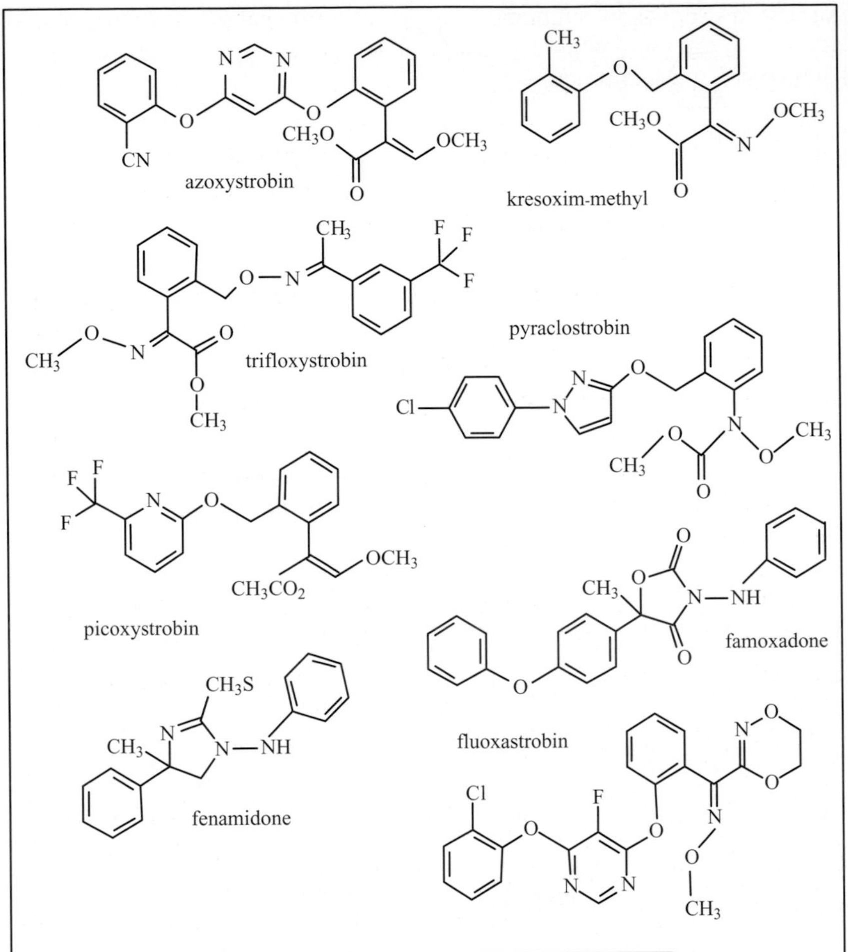

| Fungicide | Trade name | Dose rate (g/ha) |
|---|---|---|
| Azoxystrobin (1996) | Amistar | 250 (1) |
| Kresoxim-methyl (1996) | In Ensign | 150 (2) |
| Trifloxystrobin (1999) | Twist | 250 (2) |
| Pyraclostrobin (2002) | In Comet | 250 (2) |
| Picoxystrobin (2002) | Acanto | 250 (2) |
| Famoxadone (1998) | In Tanos | 175 (12) |
| Fenamidone (2001) | In Sonata | 150 (6) |
| Fluoxastrobin (2004) | In Fandango | 150 (2) |

Figure 3.14 Strobilurins and related fungicides.

strobilurins and that of antibiotics. The actual antifungal compounds produced by *S. tenacellus* are not very stable, particularly in light. However, some synthetic derivatives have proved stable and several have been developed as commercial products. The parallel studies undertaken independently by ICI (now Syngenta) and BASF research scientists during the development and optimisation of strobilurin activity from the initial natural product lead molecules are remarkable in their similarity. Each only became fully aware of the other's studies at the time of patenting. The development and adoption of these fungicides has been rapid, and in the early years of the twenty-first century, azoxystrobin was the world's best-selling systemic fungicide (Bartlett *et al.*, 2002).

The strobilurins have the widest spectrum of activity of any group of agricultural fungicides. They are the only systemics to control fungi from all taxonomic groups, although other fungicides may give better control of individual species. In Europe they are used to control fungi as taxonomically diverse as powdery mildews, *Septoria* leaf spot on cereals, downy mildews and potato blight. Strobilurins inhibit respiration in sensitive fungi. They interfere with the function of the cytochrome $bc_1$ complex on the inner membrane of mitochondria by binding to the so-called Qo complex of cytochrome b (Figure 3.15). The passage of electrons along the respiratory electron transport chain is disrupted and energy production in the form of ATP is drastically reduced in the mitochondria of sensitive fungi with consequent cessation of hyphal growth.

Strobilurins can block respiration in isolated mitochondria of plants and animals as well as fungi, and thus clearly appear to have the potential to be toxic to non-target organisms and indeed the acute toxicity to mammals of some of the strobilurin natural product leads such as myxothiazol is high. However, the selectivity of the commercial strobilurin fungicides is based on rapid breakdown in non-target organisms before they can block respiration. In fact, the strobilurin fungicides have a very low acute toxicity to non-target organisms, and present few risks to the environment, degrading rapidly and completely in soils. Paradoxically, one of the problems that has arisen with strobilurins is connected to their breakdown in plant tissues, since attempts to enhance their penetration with adjuvants have in some cases resulted in greater detoxification in the cytoplasm of treated plants, and a lower degree of pathogen control.

The synthetic oxazolidinedione famoxadone, used in mixture with cymoxanil (3.6) for control of potato blight, and the imidazolinone fenamidone (Figure 3.14) have the same mode of action as the strobilurins, and like these have a broad spectrum of fungicidal activity. They inhibit respiration in sensitive fungi, and the basis of selectivity, as with natural-product derivative strobilurins, appears to be due to detoxification of the molecule by non-target organisms.

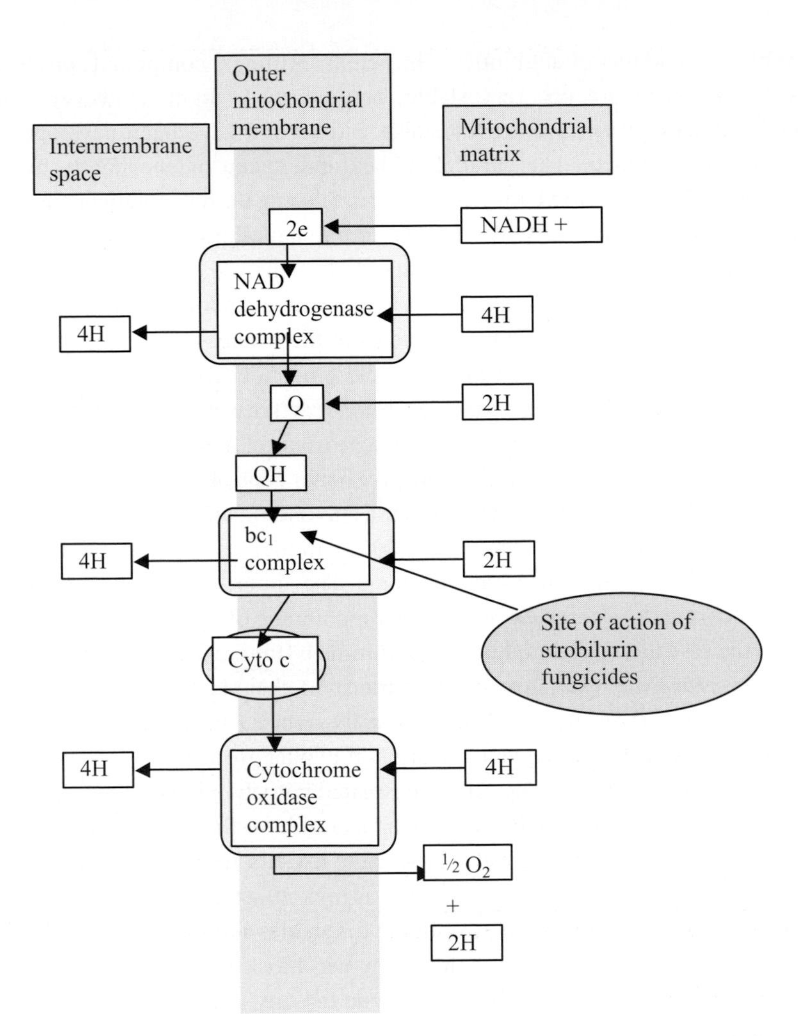


Figure 3.15 Site of action of strobilurin fungicides in the electron transport chain associated with the outer mitochondrial membrane.

### 3.7.4 Other systemic and curative fungicides

The 1990s was a notable decade for development of systemic fungicides and in addition to the strobilurins other compounds introduced include the anilinopyrimidines, quinoxyfen and fenhexamid (Figure 3.16). One of the anilinopyrimidine fungicides, pyrimethanil, is widely used in mainland Europe for control of apple scab, *Venturia inaequalis*, and grey mould, *Botrytis cinerea*, on grapevine. Mepanipyrim is also used for control of grey mould on fruit. Cyprodinil is

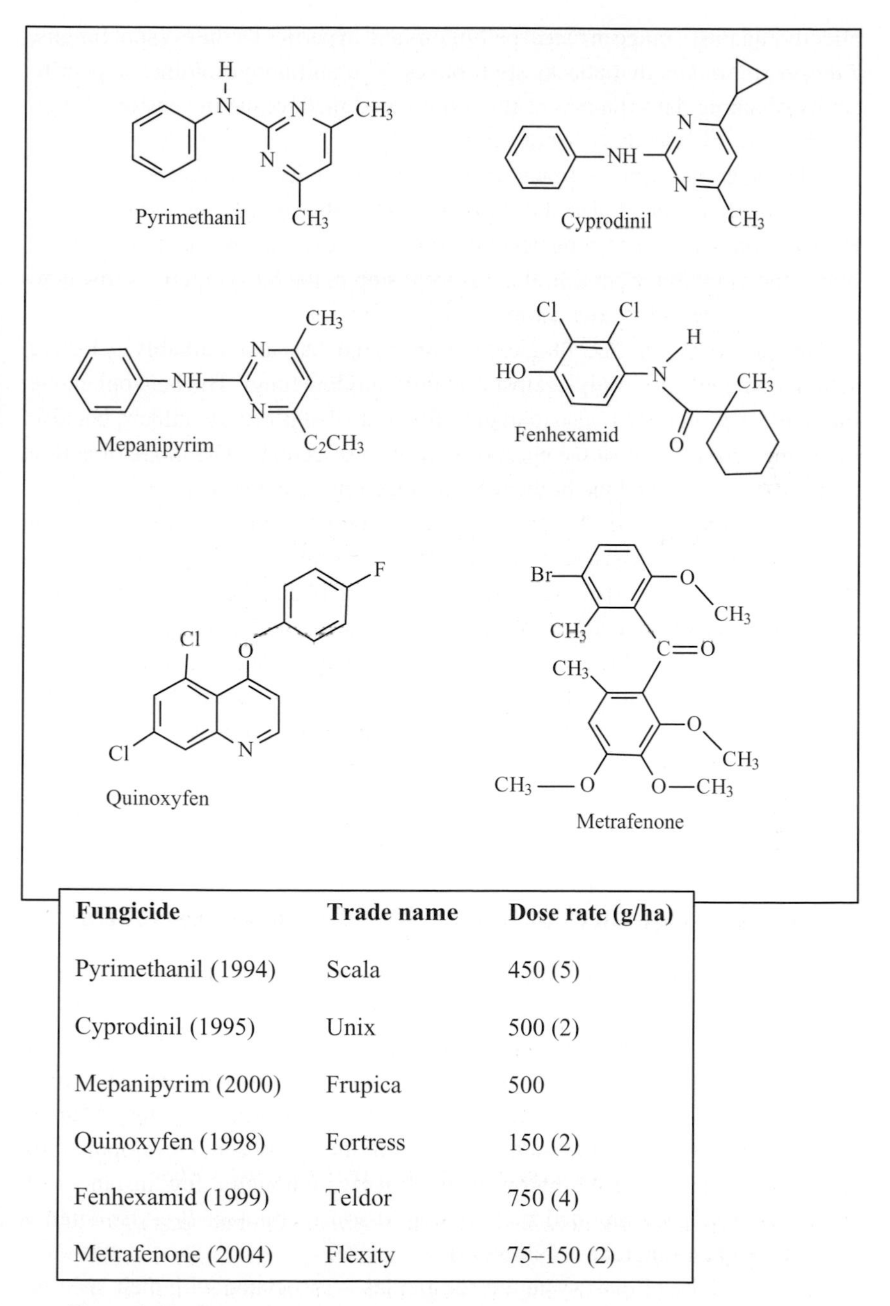


| Fungicide | Trade name | Dose rate (g/ha) |
|---|---|---|
| Pyrimethanil (1994) | Scala | 450 (5) |
| Cyprodinil (1995) | Unix | 500 (2) |
| Mepanipyrim (2000) | Frupica | 500 |
| Quinoxyfen (1998) | Fortress | 150 (2) |
| Fenhexamid (1999) | Teldor | 750 (4) |
| Metrafenone (2004) | Flexity | 75–150 (2) |

Figure 3.16 Systemic and curative fungicides of the 1990s and early 2000s – anilinopyrimidines, quinoxyfen, fenhexamid and metrafenone.

effective against a range of cereal pathogens and in particular the eyespot fungus, *Tapesia yallundae*, that attacks stem bases. The anilinopyrimidines appear to act by blocking the synthesis of the amino acid methionine in sensitive fungi, but the basis of selectivity is not yet clear.

The mode of action and mechanisms of selectivity of fenhexamid, which like pyrimethanil mentioned above gives good control of the grey mould fungus *Bortytis cinerea*, are as yet unclear, although fenhexamid appears to be a sterol biosynthesis inhibitor but acts at a different step in the biosynthetic pathway to triazoles (3.7.2) and related compounds.

The phenoxyquinoline fungicide quinoxyfen has a remarkably selective action, being effective only against powdery mildew fungi. The benzophenone fungicide metrafenone is also marketed for control of powdery mildew, but also has some activity against the eyespot pathogen of cereals. The mode of action of these molecules and the basis of their selectivity are not clear.

Finally, some chemical compounds can modify plant responses to fungal attack. Much research has been carried out on derivatives of salicylic acid, which can induce resistance to pathogens when applied to plants. The benzothiadiazole compound acibenzolar-*S*-methyl, a synthetic analogue of salicylic acid, has been developed and introduced as a plant therapeutic agent in some countries. This compound has no fungicidal activity in vitro, but appears to bind to the same site in plants as salicylic acid and in doing so stimulates a resistance response that is systemic in nature, and is effective against a range of pathogens in hosts such as cereals.

## 3.8 Consequences of selectivity – fungicide resistance

Instances of failure of fungicides to control diseases for which they were originally marketed have become common. The development of resistance has occurred in target fungi of all major systemic fungicide groups – carboxamides, phenylamides, benzimidazoles, SBIs and most recently strobilurins. Most protectant fungicides act as non-specific poisons following uptake by target fungi, by inhibiting the function of many enzymes in the cytoplasm. Development of resistance to this multisite inhibition is unlikely, although a few instances of resistance have been reported such as that in some strains of *Botrytis cinerea* to the protectant fungicide tolylfluanid.

The selectivity of most systemic fungicides is associated with their specific site of action, but this is also the principal contributing factor to the development of resistant strains in target fungal populations. The chances of genetic variation arising in many fungi is high due to the production of huge numbers of spores, and such variation is likely to include the development of resistance to specifically acting fungicides. As noted earlier with the benzimidazole compounds (Table 3.1), a single base change resulting in an alteration to the binding site of the fungicide within fungal cytoplasm may be sufficient to make the pathogen resistant. In the early twenty-first century, a similar situation has arisen with the strobilurin fungicides where resistance has developed rapidly in populations of *Blumeria* (*Erysiphe*) *graminis* and *Septoria tritici* on wheat, and powdery mildew of cucurbits, *Sphaerotheca fuliginea*. The mechanism of resistance involves a point mutation ($G_{143}A$) in the gene coding for cytochrome b within mitochondria. A single base change from guanine to cytosine resulted in the substitution of alanine for glycine at amino acid position 143 in the cytochrome b protein. Strobilurins and related fungicides cannot bind to the target cytochrome b protein in the presence of an amino acid other than glycine at position 143, and so their fungicidal activity is lost (Heaney, Hall, Davies and Olaya, 2000).

In field situations, selection pressure exerted by the presence of a large supply of host tissue treated with a single specifically acting compound may favour the rapid establishment of fungicide-resistant strains of pathogens. The use of single molecules with a high degree of selectivity and specificity such as benzimidazoles and strobilurins in sprays has promoted the development of resistant strains of pathogens. The high sporulating capacity of most fungi, noted above, will further ensure the rapid spread of these resistant strains.

There is considerable concern about the erosion of effectiveness of fungicides due to the development of resistance. In view of the costs of developing new antifungal products, the importance of conserving the benefits offered by site-specific fungicides is well recognised. Antiresistance measures have been proposed by the industry-coordinated Fungicide Resistance Action Committee (FRAC) for all of the major fungicide groups. Furthermore, identification of resistance risks is a requirement for registration of new active ingredients in Europe and the USA.

# 4

# Insecticides and other compounds that control invertebrate pests

## 4.1 Introduction

By virtue of their mobility, fecundity and adaptable nature, insects are the most ubiquitous of invertebrate pests, causing problems to crops, clothing, property and the health of human beings and other animals in most parts of the world (Gullan and Cranston, 2000). The effects of insects are wide-ranging, from the direct destructive attack on crops and crop produce as well as timber and clothing, to the indirect effects of transmitting diseases of crops, livestock and human beings (Figure 4.1). The damage caused by insects is often immediately apparent, and it is not surprising that attempts to control infestations have been documented throughout recorded history.

Damage caused by insects is primarily associated with their feeding habits. Defoliation of crops, their consumption in storage, destruction of natural fabrics such as cotton and linen, and weakening of timber in wooden materials by wood-borers are obvious symptoms of insect attack. In many cases it is the larval stage of the organisms concerned that cause the damage, although mature adults such as locusts can inflict catastrophic crop losses. In addition to direct consumption, excreta and the cast skins from insect moults may also contaminate crops, produce and other materials.

Insects transmit a wide range of diseases of both plants and animals. A few fungal, several bacterial and many viral diseases of plants are carried by insect vectors: many protozoan, as well as some bacterial and viral diseases of human beings and other animals are similarly transmitted (Table 4.1 and Lounibos, 2002). The efficiency of transmission is again linked to feeding mechanisms, with the evolution in several insect orders of mouthparts capable of sucking plant and body fluids, and in doing so transmitting disease.

In affluent societies of the late twentieth and early twenty-first centuries, some of the more subtle and minor effects of insect infestation have come

Aphids on leaf midrib. Photograph by kind permission of Chris Terrell-Nield

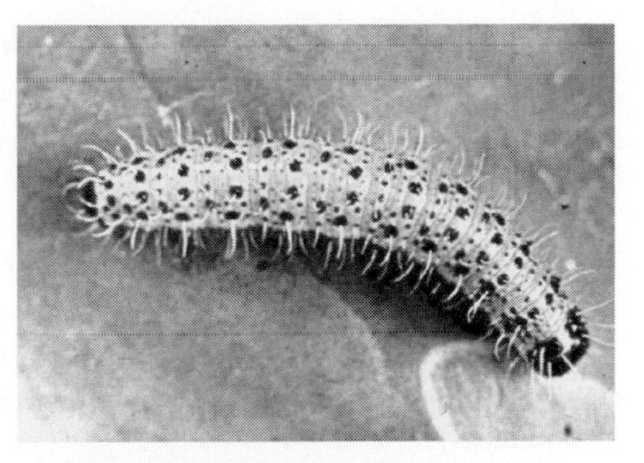

Caterpillar larva of the cabbage white butterfly. Photograph by kind permission of Chris Terrell-Nield

Damage caused by vine weevil. Photograph by kind permission of Chris Terrell-Nield

The effects of potato cyst nematodes. Crown copyright

Figure 4.1 Invertebrates of economic importance.

to be regarded as unacceptable. The minor blemishes that for example thrips may cause on fruit, vegetables and flower crops are deemed unacceptable by supermarkets and multiple retailers. Exclusion of insects such as fleas from homes and pets is a desirable aim, and local incursions by wasps and other stinging insects are often viewed with apprehension.

Other invertebrate pests include representatives of the crustaceans, arachnids, nematodes and gastropod molluscs. Barnacles, from the crustacean group of Arthropods, colonise the hulls of boats, and in this respect have proved a problem for centuries. In the arachnid phylum, acarine pests include house dust mites, spider mites that infest fruit crops and disfigure ornamental plants, and blood-sucking ticks that attack human beings and other warm-blooded animals. Mites may cause allergenic reactions by contact, for example in the case of house dust mites, and by ingestion, such as mites present in stored

Table 4.1 *Some examples of human diseases carried by insects*

| Disease | Causal agent | Vector |
|---|---|---|
| ***Helminth parasites*** | | |
| Dog tapeworm | *Dipylidium caninum* | Dog flea |
| Filariasis | *Microfilaria* spp. | Mosquites *Aedes, Culex* and *Anopheles* spp. |
| Onchocerciasis (river blindness) | *Onchocerca volvulus* | African black flies, e.g. *Simulium damnosum* |
| ***Protozoal parasites*** | | |
| Malaria | *Plasmodium vivax* | Mosquitos (*Anopheles* spp.) |
| Sleeping sickness | *Trypanosoma gambiense* | Tsetse flies (*Glossina* spp.) |
| Amoebic dysentery | *Entamoeba histolytica* | House fly (*Musca domestica*) |
| ***Bacterial and rickettsial diseases*** | | |
| Bubonic plague | *Yersinia pestis* | Rat flea (*Xenopsylla cheopsis*) |
| Endemic typhus | *Rickettsia prowazekii* | Body louse (*Pediculus humanus*) |
| ***Viral diseases*** | | |
| Yellow fever | | Mosquito – *Aedes aegypti* |
| Dengue | | Mosquito – *Aedes aegypti* |
| West Nile virus | | Mosquito – *Culex pipiens* |

grain and whose remains and excreta may end up in flour and bread (Arlian, 2002).

Soil-inhabiting nematodes include the cyst eelworms, which may cause yield reductions in potatoes (Figure 4.1) and sugar beet, and other genera that may transmit viral pathogens in vines and other fruit crops. Gastropod molluscs such as slugs and snails may cause severe defoliation of young crop plants, reduce the market value of crops such as potatoes through tuber damage, and may also disfigure ornamental horticultural plants. Pesticides used to kill mites, nematodes and gastropods are closely related to those used for insect control, and are considered along with insecticides.

Physical removal of insects from humans, animals and plants was the principal form of control until the early nineteenth century, when a number of botanical and inorganic insecticides, such as pyrethrum and arsenical preparations respectively, became widely used especially on crops. A few inorganic insecticides are still employed in the UK and elsewhere and the use of some is expanding, paradoxically in organic systems of agriculture and horticulture. Sulphur, as mentioned in the first chapter, is probably the oldest known

Table 4.2 *Principal insecticide groups: dose rates for agricultural use*

| Insecticide group | Average application rate (g/ha) |
|---|---|
| Organochlorines | 3000 |
| Organophosphates | 1800 |
| Methylcarbamates | 1500 |
| Pyrethroids | 50 |
| Benzoylureas | 110 |
| Neonicotinoids | 150 |

effective pesticide, and is toxic to mites, although the basis of this selectivity is not fully understood. Boric acid, introduced in the 1930s for control of household pests such as cockroaches and ants, is still used for this purpose. It is a stomach poison and also affects the cuticular waxes of insects. Boric acid is of low toxicity to human beings at the dose used, as is disodium octaborate, which is still widely employed as a timber treatment against wood-boring insects.

Arsenical compounds such as Paris green, copper and lead arsenates, which act by uncoupling oxidative phosphorylation from the electron transport chain, formerly provided good control of some insect and other invertebrate pest species. Their unspecific inhibitory effect on respiration has led to their prohibition on safety grounds, both to humans and the environment. The inorganic fluorides used widely in the 1920s and 1930s have been withdrawn for similar reasons.

Most currently used insecticides and other chemicals for control of invertebrate pests in both agriculture and public health act through interference with the nervous system. As such they present few problems of phytotoxicity, but of all pesticides they present the greatest acute risk to the health of human beings and fauna in the environment. Four major groups of insecticides dominate the world market: the pyrethroids, organophosphates, methylcarbamates and neonicotinoids (Table 4.2). The organochlorines, whose use is now prohibited for agricultural purposes in many parts of the world, are also neurotoxins. As with other pesticides, major trends in insecticide development have been towards production of compounds with lower application rates (Table 4.2) offering greater safety to human beings and other non-target organisms, and which are overall environmentally benign (Casida and Quistad, 1998).

Unlike herbicides and fungicides, the mechanism of action at the target site of insecticides is not the principal determinant of selectivity, although the insect growth regulator compounds (4.4) and to some extent the neonicotinoids (4.2.3) exhibit selectivity at the biochemical site of action. Insecticides may inhibit the functions of isolated nerve cells from mammals and other animals as well as insects. The selective action of most insecticides is linked to either the dose, aided in some cases by rapid breakdown in mammals and other non-target organisms, or low penetration and uptake through skin and other body surfaces. In some cases, with compounds of high mammalian toxicity, safety (as above with the arsenicals) may depend upon avoidance of contact with the pesticide in its concentrated form, either through formulation as, for example, granules or use of protective clothing whilst spraying.

## 4.2 Insecticides that act on the nervous system

These compounds either interfere with the passage of impulses along the principal cells of the nervous system, the neurones, or prevent the transmission of the impulse across the synapse between nerve cells. Neurones generally consist of a cell body, several short branches or dendrites, and a long axon, down which nerve impulses pass. The passage of an impulse down an axon is facilitated by a moving or propagated depolarisation of its membrane, often referred to as an action potential. Transmission is dependent upon a change in membrane potential and is consequently referred to as voltage dependent.

At rest, the concentration of sodium ions in the fluid surrounding a nerve cell is much higher than inside the cell itself. Microelectrodes show that a transmembrane potential of about –60 millivolts (mV) is usually evident when impulses are not being transmitted along nerve cells: this is the resting potential (Figure 4.2). This electrical gradient is maintained across the membrane of the neurone by ion pumps. The development of the action potential involves a transient depolarisation with a local influx of sodium ions, which involves a change in the electrical gradient.

This influx of sodium occurs due to the opening of specific ion channels in the membrane of the neurone (Figure 4.3). Ion channels are protein complexes that span membranes, forming water-filled pores through which inorganic ions pass. Their opening involves a change in voltage across the neuronal membrane and they are thus said to be voltage-gated (Zlotkin, 1999). Specific channels exist for individual ions. Influx of sodium through its ion channel constitutes the rising phase of the action potential. At its peak sodium flow into the neurone stops, since at this time the sodium ion channel undergoes a slight change in

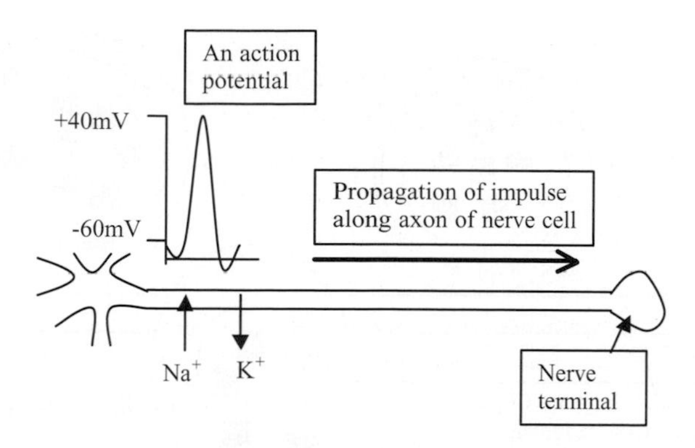


Figure 4.2 The movement of an impulse along an axon of a nerve cell.

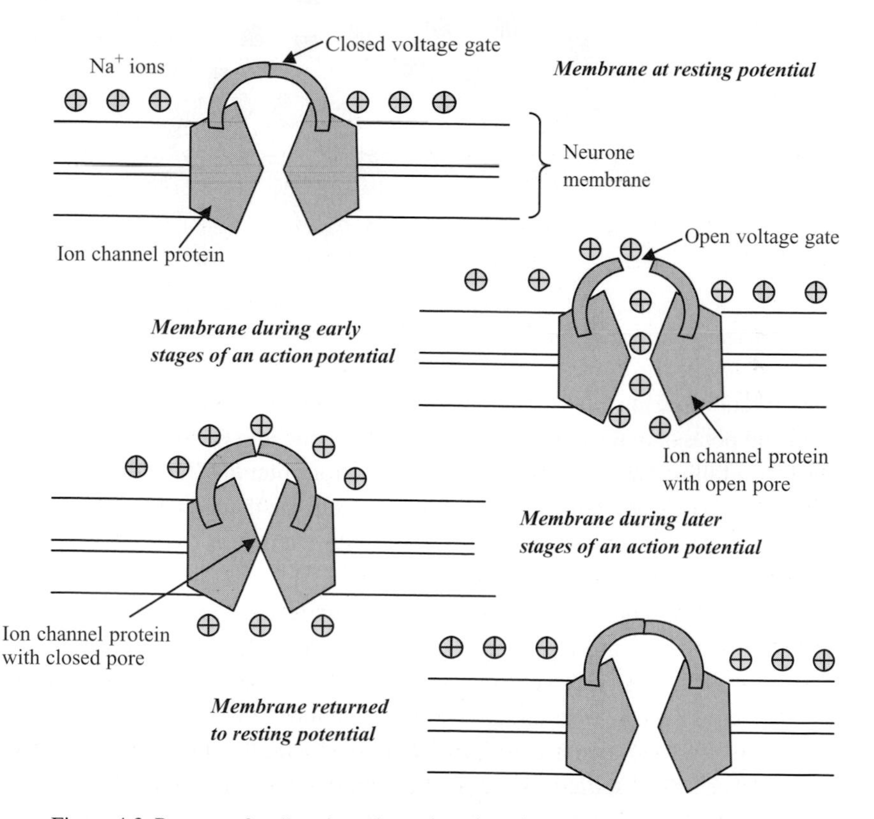


Figure 4.3 Passage of sodium ions through an ion channel into a nerve cell during an action potential.

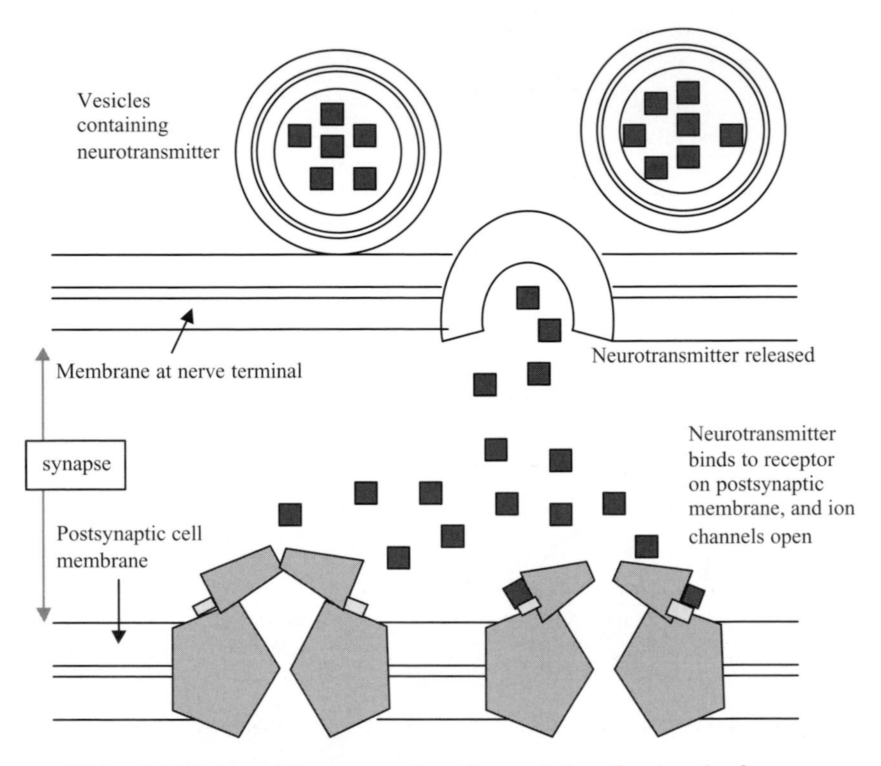


Figure 4.4 Diagrammatic representation of transmission of an impulse from an axon terminus to an adjacent nerve cell across a synapse.

configuration and ion movement through the channel is blocked. Channels then open and potassium ions move out from the neurone into the surrounding fluid during the falling phase or downstroke of the action potential. The concentration of potassium ions is usually higher within the neurone than outside. The action potential is propagated down the axon to the nerve terminal where the impulse may be transmitted across a synapse.

After transmission, a $Na^+/K^+$ exchange pump restores the original ion balance in the cytoplasm of the nerve cell and surrounding fluid. The sodium channel itself is a large protein that spans the nerve membrane, and in mammals contains at least nine target sites where natural toxins, insecticides and drugs may bind and interfere with transmission of the impulse. The composition and function of sodium channels are remarkably similar in invertebrate as well as vertebrate species.

Transmission of nerve impulses across the synaptic gap or cleft between nerve cells is dependent on chemical messengers released from vesicles within the terminal parts of axons (Figure 4.4). When nerve impulses reach the

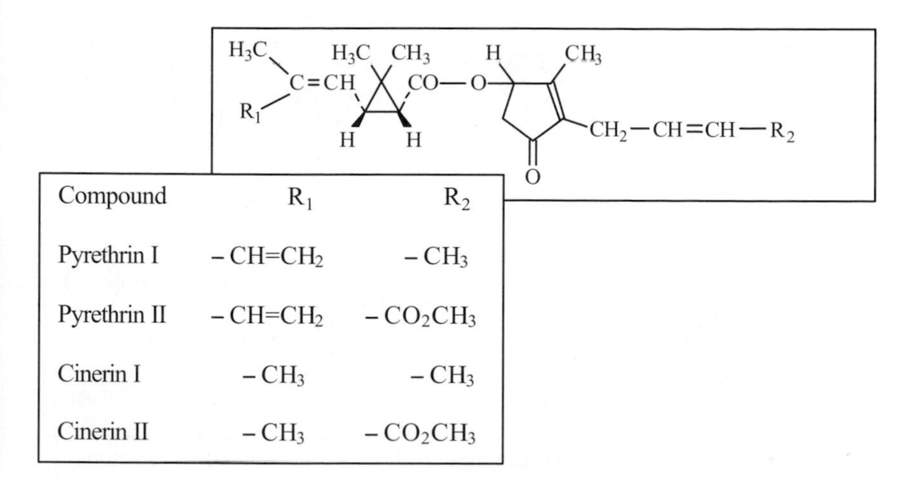


| Compound | $R_1$ | $R_2$ |
|---|---|---|
| Pyrethrin I | $-CH=CH_2$ | $-CH_3$ |
| Pyrethrin II | $-CH=CH_2$ | $-CO_2CH_3$ |
| Cinerin I | $-CH_3$ | $-CH_3$ |
| Cinerin II | $-CH_3$ | $-CO_2CH_3$ |

Figure 4.5 Natural pyrethrins.

terminals, these vesicles fuse with the presynaptic membrane. Small quantities of neurotransmitter substances are thus released into the synaptic cleft and these diffuse to receptors on the membrane of the adjacent nerve cell where an electrical change is provoked following binding of the neurotransmitter to transmitter-gated ion channels. The chemical signal gives rise to an electrical response, the membrane of the recipient nerve cell is depolarised, and the impulse propagated. For the system to function efficiently, the neurotransmitter must then be removed or destroyed.

The most well-known neurotransmitter in insects as well as other animals including human beings is acetylcholine, which is hydrolysed after transmission of the impulse and released from its postsynaptic receptor by the enzyme acetylcholinesterase. Acetylcholine is the principal neurotransmitter at neuromuscular junctions, where an axon terminates on a muscle fibre. Other neurotransmitters that are primarily involved in transmission between nerve cells include gamma-aminobutyric acid (GABA), glutamic acid, octopamine, dopamine and serotonin.

### 4.2.1 Pyrethrins and pyrethroid insecticides

Pyrethrins are naturally occurring compounds present in the flower heads of certain species of *Chrysanthemum*, and have been extracted and used commercially since the middle of the nineteenth century, although the use of pyrethrum is recorded in China well before this time. They are a mixture of four insecticidal compounds (Figure 4.5). Pyrethrins control a wide range of insect species

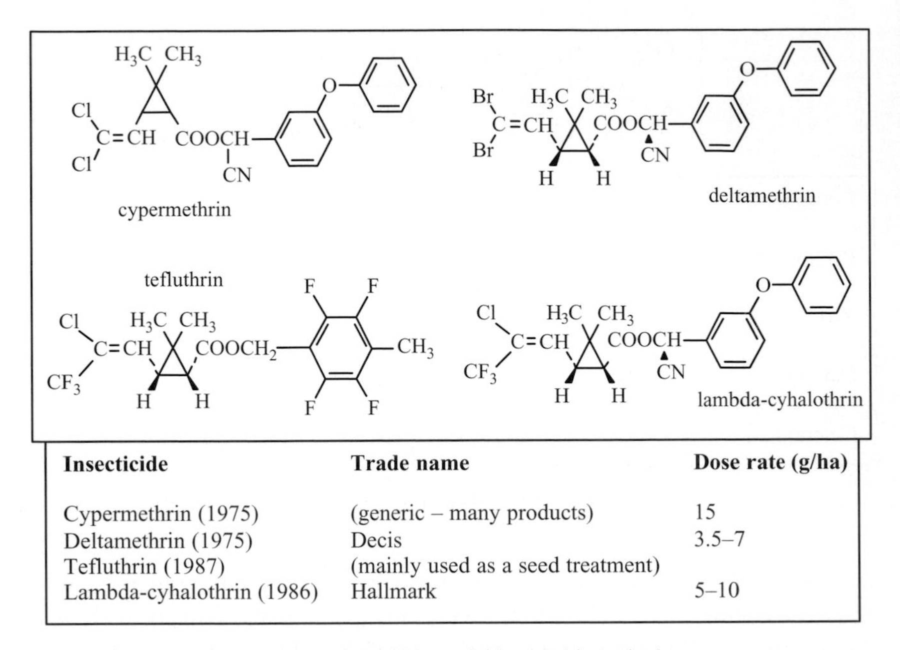


| Insecticide | Trade name | Dose rate (g/ha) |
|---|---|---|
| Cypermethrin (1975) | (generic – many products) | 15 |
| Deltamethrin (1975) | Decis | 3.5–7 |
| Tefluthrin (1987) | (mainly used as a seed treatment) | |
| Lambda-cyhalothrin (1986) | Hallmark | 5–10 |

Figure 4.6 Pyrethroid insecticides used in agriculture.

and act swiftly at low doses when compared to organophosphate and methyl-carbamate insecticides (4.2.2). Rapid knockdown of insects is a very attractive feature of pyrethrins, as is their low mammalian toxicity and lack of persistence. However, the problem with natural pyrethrins is that they break down readily in light.

Research in Japan by the Sumitomo Corporation, and by Elliott and his co-workers at the Rothamsted Experimental Station in England over a period of 30 years from the mid 1940s led to the development of the light-stable pyrethroids. Their fundamental research identified the sites in natural and synthetic compounds that were associated with instability to light, and the synthetic molecules emerging from this work formed the largest sector, in terms of sales, of the insecticide market during the 1990s (Figure 4.6). Pyrethroids are extensively used in public and animal health, amenity and household situations as well as grain stores (Figure 4.7), and a large number of commercial products are available for these purposes (Perrin, 1995).

The first synthetic pyrethroid to be marketed was allethrin in the late 1940s. The second generation of pyrethroids appeared in the mid 1960s and included tetramethrin, resmethrin and bioresmethrin. Since these compounds retain a degree of instability in light, they are mainly used in public health. Permethrin

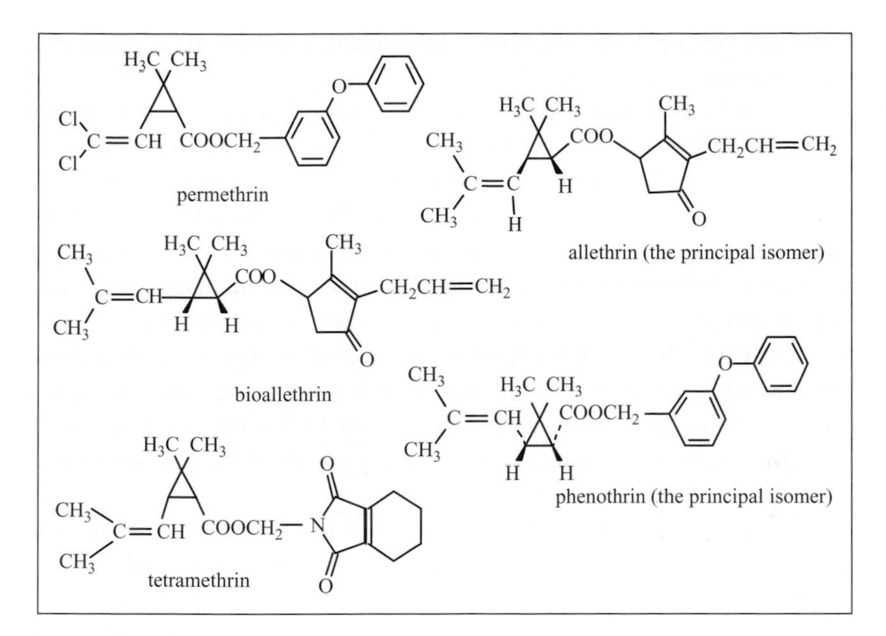


Figure 4.7 Pyrethroid insecticides principally used in public health, fabric and timber treatment.

has been withdrawn in the UK for agricultural purposes, but is still used in public health. In the early twenty-first century, cypermethrin was the most extensively used insecticide in the UK (Table 1.2), being deployed in a range of crops for control of aphids and other foliar pests. The so-called fourth generation pyrethroids such as lambda-cyhalothrin are characterised by their effectiveness at very low dose as well as their stability in light, giving residual protection of leaves from insect attack of 10 days or more. Some pyrethroid insecticides such as tefluthrin persist in soil and are used to control pests there. In all, over 30 commercial pyrethroids have been developed and they have replaced many of the older, more toxic insecticides used in public health and agriculture.

The pyrethroids all act in a similar fashion, by interfering with transmission of nerve impulses along nerve cells (Bloomqvist, 1996). They bind to a site on the sodium channel protein and prolong the opening of these channels during the action potential in nerve cells. In terms of structure and physiological activity, pyrethroids are described as either type 1 or type 2. Type 1 compounds include the natural pyrethrins, allethrin and tetramethrin, and cause hyperexcitation and convulsions in insects. Type 2 compounds, including the third- and fourth-generation pyrethroids, cause uncoordination and irregular movement in treated insects.

Pyrethroids, especially type 1, generally act very quickly – often within 1–2 min – causing a rapid knockdown of insects. The differences in behaviour of insects treated with type 1 and type 2 pyrethroids is linked to the physiological effects of these compounds on nerve cells. The effects of type 1 compounds on sodium channels may last for less than a second, whereas type 2 compounds may exert their effects for several minutes. Most pyrethroids exist as optical or geometrical isomers: compounds of the same molecular composition but with a slightly different shape. These minor changes in configuration can lead to a very great difference in their potency to insects, with some isomers often being far more toxic than others. Often it has proved impossible or prohibitively expensive to develop single-isomer products and so many pyrethroids are manufactured and sold as mixtures of isomers.

One disadvantage of the pyrethroids is that none is systemic. They are not taken up and translocated in plants. However, in addition to their stability, modern pyrethroids are effective at very low doses compared to other insecticides. For example, 5 g of a pyrethroid insecticide is sufficient to protect the same area of cereals from aphid attack as 500 g of an organophosphate (4.2.2). Also, they have a very high degree of selectivity in terms of acute toxicity: the acute toxicity of deltamethrin is about 50 000 times less to mammals than insects. Principal reasons for this lack of toxicity to warm-blooded animals are that sodium channels in mammals are much less sensitive to pyrethroids than those of insects (Vais *et al.*, 2001); the pyrethroids do not readily penetrate skin; and they are rapidly hydrolysed in mammalian tissues. Also, pyrethroids are more effective at lower temperatures, and the difference in body temperatures between mammals and insects thus allows a certain degree of selectivity. Furthermore, differences in body temperature lead to slower metabolism of pyrethroids in insects than in mammals. However, pyrethroids are not selective within insect groups, and their toxicity at low dose has on occasions caused major problems to invertebrate populations in water courses (9.6).

### 4.2.2 Organophosphorus and methylcarbamate insecticides

These compounds inhibit the passage of nerve impulses across neuronal junctions or synapses, and include some of the most toxic pesticides used in agriculture (Naumann, 1989). Although some may persist in plants and offer protection against pest attack for long periods, most neither persist in the environment to the extent of the organochlorines (4.2.4) nor bioaccumulate through food chains. The organophosphates and methylcarbamates occupied over 50% of the world market in insecticides during the mid to late twentieth century, but their market share has declined since.

The insecticidal properties of organophosphate compounds were discovered by Lange in Germany in the early 1930s and their toxicity to human beings almost immediately realised when one of his researchers was poisoned (but fortunately recovered). Some of these compounds, notably sarin and tabun, were developed as nerve gases and have been used as chemical agents in warfare and acts of terrorism.

Schrader, working for the Bayer Corporation, sought organophosphates of lower toxicity for use as insecticides, and the first organophosphate to be introduced commercially – tetraethylpyrophosphate (TEPP) – was marketed in 1943. Over 130 compounds were released between 1950 and 1970, but very few have been developed since the 1980s. Unlike the pyrethroids and organochlorines, some of these compounds are systemic, moving into plants. The entire plant may become toxic to attacking insects, and the compounds are applied less frequently when compared to non-systemic insecticides.

The popularity of organophosphate insecticides, despite their toxic nature, is linked to their low cost, and they have a very wide range of uses. The world's biggest selling insecticide in the early 1990s was chlorpyrifos, widely used in agriculture and horticulture as well as in homes, restaurants and industrial premises for ants and cockroaches. Control of sucking pests such as aphids and leafhoppers can be easily achieved with systemic compounds such as dimethoate. Pirimiphos-methyl is used widely to prevent weevil and beetle infestation of stored grain and other produce, and also in public health against fleas, bed-bugs and lice. The compound dichlorvos is highly volatile and has been used as a fumigant as well as domestically in 'fly-papers' from which it enters the atmosphere.

Organophosphates are not selective at their target site, acetylcholinesterase, but many are rapidly metabolised to non-toxic compounds in vertebrates, and this has led to the use of several for the control of ectoparasites such as fleas, lice, ticks and mites on human beings, pets and livestock.

Organophosphate insecticides occur as phosphates, where one phosphate atom is bonded to four oxygen atoms as with dichlorvos, but more commonly as phosphorothionates such as chlorpyrifos, and phosphorodithionates such as malathion, where the phosphorus atom is bonded to sulphur atoms (Figure 4.8). On uptake by insects, the P=S ('thion') group of phosphorothionates and phosphorodithionates is rapidly converted by oxidative desulphuration to the more toxic P=O ('oxon'): the process is often referred to as lethal synthesis. This conversion also takes place in light, as well as in soil, and further degradation of the 'oxon' form quickly follows in most cases: thus organophosphates

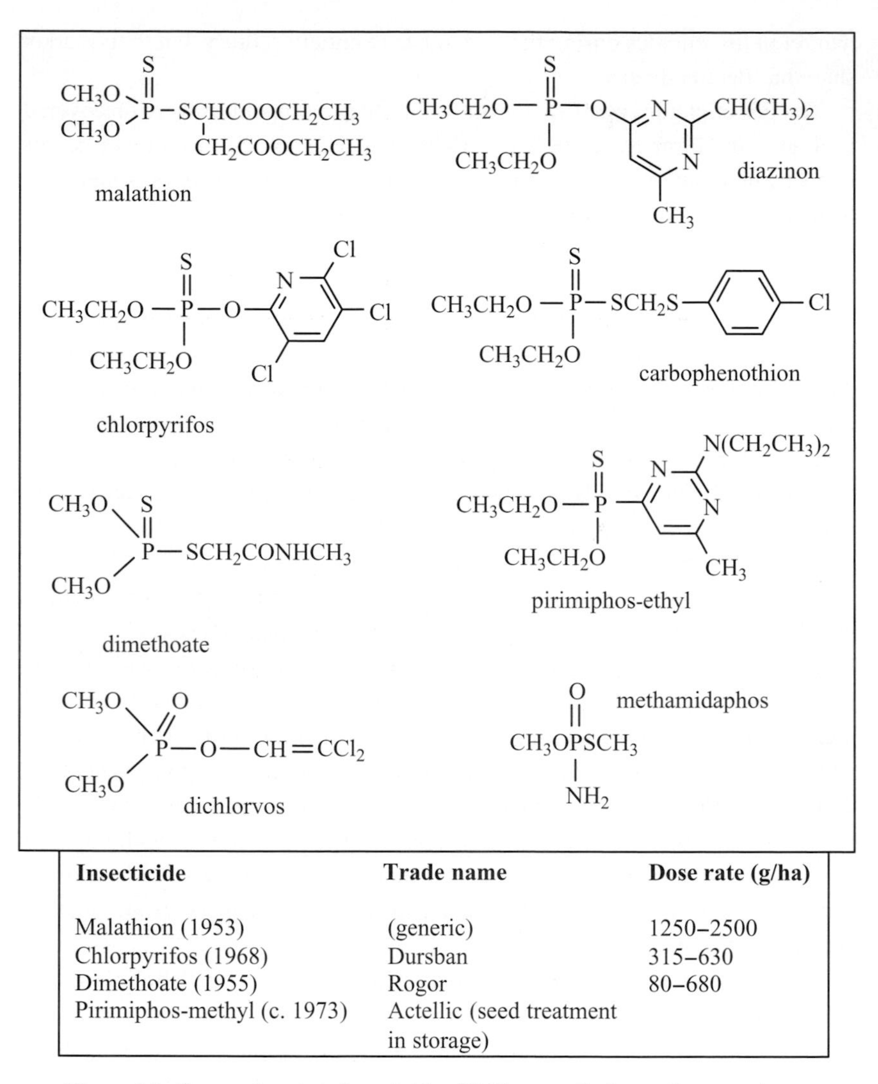


| Insecticide | Trade name | Dose rate (g/ha) |
|---|---|---|
| Malathion (1953) | (generic) | 1250–2500 |
| Chlorpyrifos (1968) | Dursban | 315–630 |
| Dimethoate (1955) | Rogor | 80–680 |
| Pirimiphos-methyl (c. 1973) | Actellic (seed treatment in storage) | |

Figure 4.8 Organophosphate insecticides. Dichlorvos, diazinon, phorate and carbophenothion are no longer approved for use in the UK.

do not accumulate in either the environment or in the tissues of human beings and other vertebrates.

In nervous tissue and at neuromuscular junctions, organophosphorus insecticides bind to acetylcholinesterase at the active site for degradation of acetylcholine, which thus accumulates. After transmission of a nerve impulse from cell to cell across a synapse, the neurotransmitter must be degraded, in

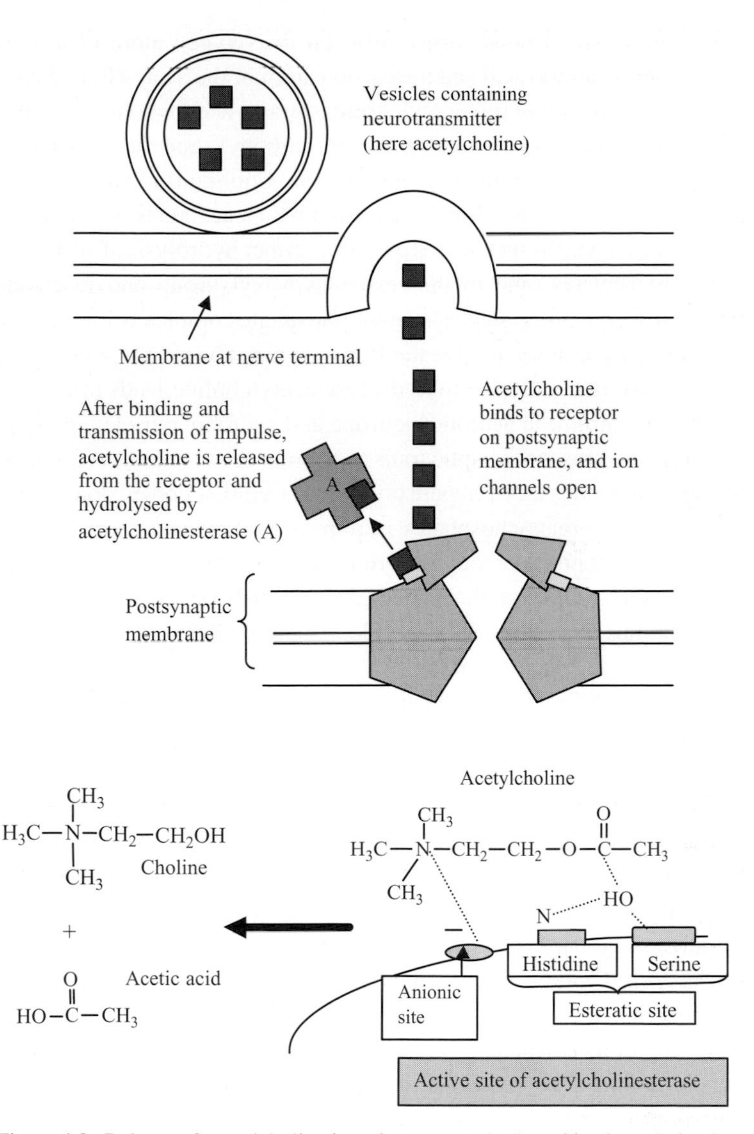


Figure 4.9 Release of acetylcholine into the synapse (top), and its degradation by acetylcholinesterase (below).

the case of acetylcholine by acetylcholinesterase, which thus clears the synaptic cleft before the next pulse of neurotransmitter is released. After liberation from its receptor, acetylcholine undergoes hydrolysis after attachment to two sites on the enzyme (Figure 4.9). The anionic site contains negative charges that attract the positively charged quaternary nitrogen present in acetylcholine.

At the esteratic site a bond forms between the oxygen atom of a hydroxyl group on a serine amino acid and the carbon atom of the acetylcholine carboxyl group. After formation of this acetylcholinesterase–acetylcholine complex, the choline oxygen breaks the bond with the carbonyl, and forms a new bond with a hydrogen atom supplied by the enzyme. Choline is released, leaving an acetylated form of acetylcholinesterase from which acetate is subsequently liberated and the enzyme is then free to perform further hydrolysis of acetylcholine.

Organophosphates bind to the serine hydroxyl group and inactivate the enzyme. An acetylcholinesterase–organophosphate complex is formed, but this is only slowly hydrolysed to give the free enzyme. The resulting lack of active acetylcholinesterase available to hydrolyse acetylcholine leads to an accumulation of acetylcholine at neurone–neurone and neurone–muscle junctions, and this severely disrupts the synaptic transmission of nerve impulses. Signs of poisoning include restlessness, hyperexcitability, tremors, convulsions and paralysis. Since most organophosphates require activation and exert their effects primarily in the central nervous system of insects, they are slower acting than insecticides such as the pyrethroids that act primarily on the peripheral nervous system, and thus have little immediate knockdown effect.

Commercial development of the methylcarbamate insecticides (Figure 4.10) began in the 1950s following extensive studies on physostigmine, a natural toxicant produced by the calabar bean (*Physostigma venenosum*). Carbaryl was the first successful synthetic methylcarbamate insecticide and was used very extensively until the late 1990s, particularly for the control of pests on fruit. This popularity was explained by its wide spectrum of activity against insects, the low acute mammalian toxicity of carbaryl and its additional side-effect of assisting fruit thinning. Indeed, for this purpose carbaryl was marketed with the imaginative trade name of 'Thinsec'. Other methylcarbamates – for example bendiocarb, used widely in public health, and pirimicarb – are also of low mammalian toxicity. Pirimicarb, remarkably among the anti-acetylcholinesterase insecticides, offers a degree of selectivity among insects in that it is very effective against aphids, but presents little risk to many other insects including bees and ladybirds.

Some methylcarbamate insecticides, notably the oxime carbamates aldicarb, methomyl and oxamyl, are toxic to mammals at low concentrations. In order to reduce health risks to those applying these chemicals, some such as aldicarb are manufactured only as granules. The oxime carbamates, like most methylcarbamates, are systemic in plants but also have a long residual activity. In addition to its effects on soil insects and, after uptake, on foliar pests, aldicarb offers very good control of soil-borne cyst nematodes, particularly those attacking potatoes. Other methylcarbamates are used additionally to control non-arthropod

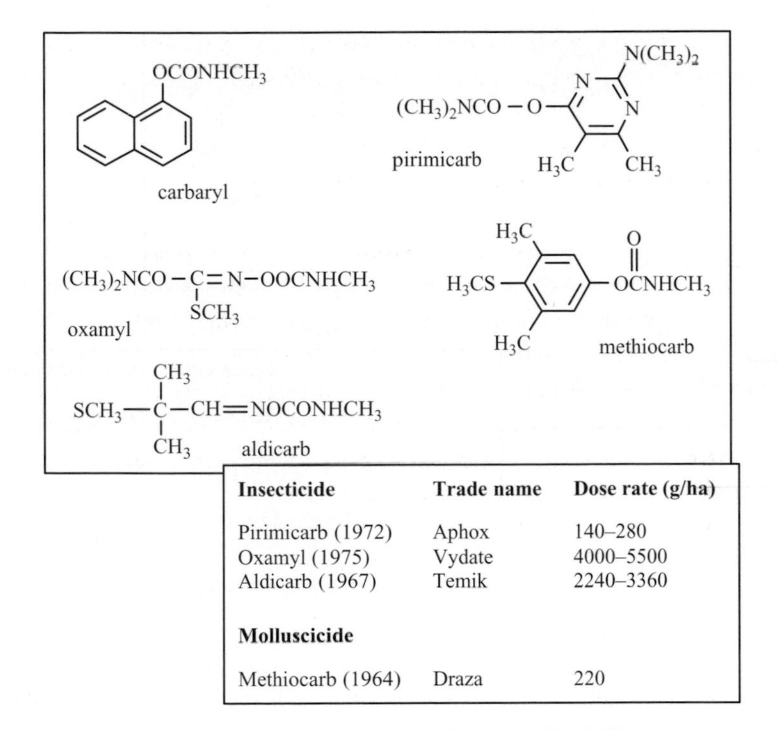


| Insecticide | Trade name | Dose rate (g/ha) |
|---|---|---|
| Pirimicarb (1972) | Aphox | 140–280 |
| Oxamyl (1975) | Vydate | 4000–5500 |
| Aldicarb (1967) | Temik | 2240–3360 |
| **Molluscicide** | | |
| Methiocarb (1964) | Draza | 220 |

Figure 4.10 Carbamate insecticides and molluscicides.

pests. For example, methiocarb is the principal active ingredient of pellets that are widely used, particularly in domestic gardens, to kill slugs and snails.

The methylcarbamates, like the organophosphates, inhibit acetylcholinesterase and the symptoms of poisoning in insects are identical. Inhibition involves binding to the same serine amino acid as with organophosphates, and the carbamylated enzyme is only slowly hydrolysed back to its active form. The carbamylated enzyme complex is less stable than that of the phosphorylated complex formed with organophosphates, and the toxic effects of carbamates in humans and other mammals can be reversed more easily than poisoning by organophosphates.

### 4.2.3 Nicotine, the neonicotinoids and spinosyns

The alkaloid nicotine occurs naturally in tobacco and probably evolved as a defence mechanism to protect plants from attack by insects. In the late nineteenth and early twentieth century, nicotine sprays and washes were used extensively for insect control in horticulture, particularly on fruit. Nicotine can be a

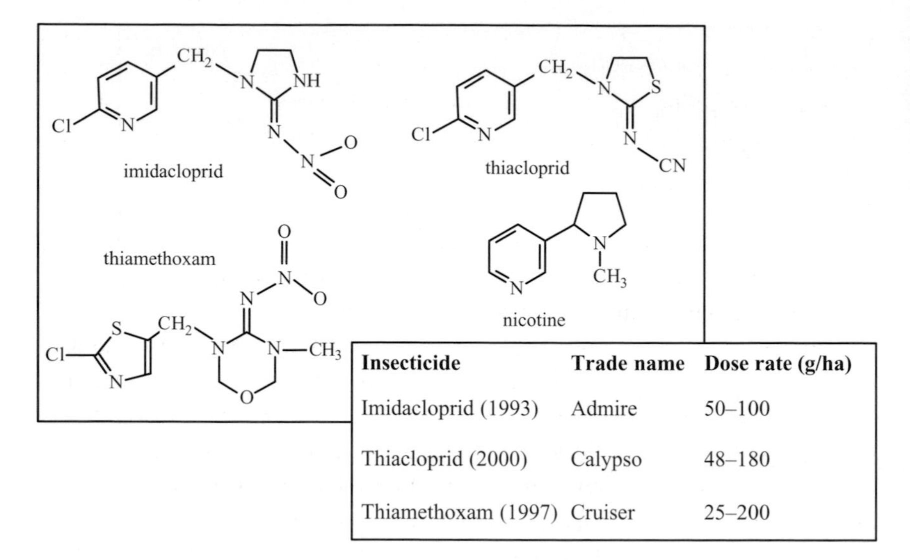


| Insecticide | Trade name | Dose rate (g/ha) |
|---|---|---|
| Imidacloprid (1993) | Admire | 50–100 |
| Thiacloprid (2000) | Calypso | 48–180 |
| Thiamethoxam (1997) | Cruiser | 25–200 |

Figure 4.11 Nicotine and neonicotinoid insecticides.

very effective insecticide in glasshouses where it is burnt in the form of shreds to release the active ingredient. Of course, access to glasshouses is restricted during such treatment. Nicotine rapidly passes through the insect cuticle and is thus an excellent contact insecticide. However, this property extends to warm-blooded animals, to which nicotine has a high acute and long-term toxicity.

Nicotine mimics the action of acetylcholine, and binds to nicotinic acetylcholine receptors at synapses in the central nervous system of insects. Nicotine is insensitive to acetylcholinesterase, and is not therefore degraded. Persistent activation of acetylcholine receptors occurs resulting in hyperexcitation, twitching, convulsions and death of the insect.

The neonicotinoids (Figure 4.11) have a similar mode of action to nicotine (Tomizawa, 2000). Introduced in the mid 1990s, the guanidine imidacloprid was, in terms of sales in 2001, the world's leading insecticide for agricultural and horticultural pest control, and other neonicotinoids such as thiacloprid, clothianidin and thiomethoxam have come to the market in the early twenty-first century. They are systemic and persistent in plants, and have low application rates and fewer non-target effects when compared to organophosphate and carbamate compounds. These properties have led to the use of imidacloprid in seed dressings, horticultural growing media as well as foliar sprays.

Neonicotinoids bind strongly to the nicotinic acetylcholine receptor in insects, but have a much lower affinity for the receptor in mammals and other vertebrates. Neonicotinoids are highly selective for the insect receptors, and with some compounds this may be due to their tertiary amine nitrogen carrying a positive charge that interacts with insect but not vertebrate nicotinic acetylcholine receptors. Whatever the mechanism of selectivity, these acetylcholine mimics are much less toxic than nicotine to human beings and other vertebrates and provide a classic example of synthetic pesticidal molecules derived from a natural product, but which have a vastly improved safety profile. Indeed, the neonicotinoids present a lower risk to health of human beings and other vertebrates than most other neurotoxic insecticides.

The spinosyn and spinosoid insecticides are structurally complex natural products derived from the soil organism *Saccharopolyspora spinosa* and also exert their effects by binding to and altering the function of the nicotinic acetylcholine receptor, although the mechanisms by which these compounds act appear to be different from that of the neonicotinoids (Sparks, Crouse and Durst, 2001). They are much less toxic to mammals than many other insecticides, although the basis of this selectivity is not yet clear. Spinosyns are potent compounds, with dose rates for the control of insect pests being close to those of pyrethroid insecticides. Spinosad, the first commercially available spinosyn, was introduced in the early 2000s.

### 4.2.4 Organochlorine compounds

These were the dominant insecticidal molecules from the 1940s to the 1960s but few organochlorine insecticides are now used for pest control in North America, Europe and the agricultural systems of wealthier nations, although they are still used for wood treatment and in public health. Organochlorines are still used in developing countries for public health, notably dichlorodiphenyltrichloroethane (DDT) for control of the malarial mosquito, and a few such as endosulfan are still employed for agricultural pest control.

The organochlorine insecticides are characterised by their stability and lipophilicity. Their chlorinated nature renders them resistant to photodegradation and breakdown by both soil microorganisms and enzyme systems of plants and animals. Most have a high octanol:water partition coefficient (8.3) and are thus readily absorbed by lipids, and this, allied to their generally slow rate of breakdown in tissues, has led to problems of bioconcentration and biomagnification in food chains. The organochlorines and especially DDT have achieved considerable notoriety, and their environmental effects have attracted much attention. It is therefore appropriate to consider their development and use.

Muller, working for the Geigy Corporation in Switzerland, was responsible for the development of dichlorodiphenyltrichloroethane or DDT. Muller set himself a series of criteria that he sought to fulfil in his search for an effective insecticide and in the light of the focus of this text on selectivity it is worth recalling these.

1. Strong toxicity to insects
2. Rapid action with insects
3. Relative harmlessness to warm-blooded animals and plants
4. Non-irritant and virtually odourless
5. Have a wide application
6. Long-lasting effect through chemical stability
7. Cheap and easy to produce in quantity.

During his work on a series of substances based on combining chlorals with hydrocarbons and phenols, Muller discovered the highly potent nature of DDT to houseflies in 1939, however the compound itself had been synthesised as far back as 1873. Although relatively slow acting when compared to standard compounds such as pyrethrum used at that time, DDT proved lethal to insects whereas recovery was sometimes experienced from treatment with pyrethrum, despite the latter's quick knockdown properties.

DDT preparations became available for use in 1941 and were extensively and successfully employed during the 1939–1945 conflict to prevent epidemics of typhus, carried by lice, and in the Far East to treat buildings where mosquitoes were prevalent. In the post war period, DDT was used extensively in areas where malaria was endemic. During the period 1945–1965, by elimination of the mosquito vector, malaria, once serious in Europe, was eradicated from most European countries and its incidence vastly reduced in other parts of the world. By the mid 1950s, Mellanby (1989) estimated that DDT had protected over 100 million people from malaria through control of its insect vector. In addition, DDT in anti-malarial campaigns has given protection against flea-borne bubonic plague; and other diseases controlled through elimination of their vectors include yellow fever, and filiariasis or river blindness. Topical applications were very effective in killing head and body lice.

The first large-scale agricultural use of DDT was on pests of fruit trees, on vines and for the control of Colorado beetle on potatoes. Very good control of a diverse range of agricultural pests was obtained – codling moths of apple and pear, fruit flies of citrus, cabbage caterpillars, flea beetles, whitefly in glasshouse crops, wireworms and many others. DDT was used very extensively on cotton pests such as bollworms, where a very high degree of control was achieved. The most important property of DDT when introduced was held to be its persistence.

This property was greatly valued in keeping buildings free of mosquitoes and cockroaches, people free from attacks by lice and bed bugs, and crops and other plants free from insect attack for weeks if not months.

By the mid 1950s, DDT was seen as a panacea for all insect problems in some parts of the world, notably the USA. Aerial spraying of DDT became common, not only for crop pests but also to control pests of pine and spruce forests, to protect elms from the insect vector of dutch elm disease, and even to remove gnats and other insects from natural waters, which irritated anglers and others seeking leisure pursuits. Elms and other species in urban areas were sprayed, often by aerial application. The consequences of this were profound; widespread and often indiscriminate spraying resulted in unwanted effects not considered during the development of DDT. Many species of insect developed resistance to DDT and its persistent nature led to the accumulation of DDT by vertebrate species with particular consequences for fish and birds (Chapter 9). Furthermore, many non-target insect species were eliminated in treated areas, influencing food webs, and in some cases leading to a resurgence or in some cases an increased incidence of pest species due to their natural predators being wiped out.

DDT readily passes through the insect cuticle and enters the blood. Indeed, lethal concentrations of DDT can be absorbed through the terminal regions of the legs of insects such as mosquitoes. The latter are very sensitive to DDT, with less than 0.001 mg per insect constituting a lethal dose. Early work established that DDT caused repeated development and propagation of action potentials in nerve cells – the process of repetitive firing. DDT is thought to directly interfere with the operation of sodium channels by binding and delaying their closure in an identical manner to that of pyrethroid insecticides (4.2.1), but other effects including interference with the transport of sodium and calcium ions may result from the action of DDT on neurones. Whatever the precise biochemical mechanism, prolonged secondary impulses or afterpotentials are induced in the neurone. These in turn induce hyperactivity and convulsions in insects, eventually leading to their death.

The low acute toxicity of DDT to mammals and other warm-blooded animals is probably due to its affinity for lipids and consequent partitioning to these in the epidermal and surface cell layers of larger animals before it can affect nerve cells. Indeed it has proved difficult to get DDT to pass through human skin. Early investigators resorted to mixing DDT with acetone and rubbing this vigorously between hands to induce nerve tremors (Mellanby, 1989). Fortunately, the experimenters suffered no long-term damage and such investigative approaches today are of course prohibited.

Dicofol is an analogue of DDT, but almost specific to mites and is still used in both agriculture and horticulture for acarine control. As with DDT, it prevents

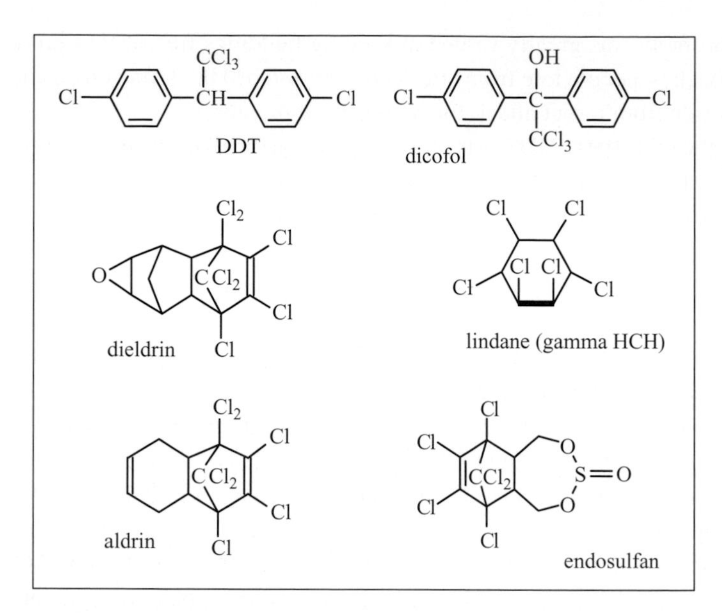


Figure 4.12 Organochlorine insecticides. Of these compounds only endosulfan was approved for use (for pest control on ornamental species) in the UK in 2005.

the closure of sodium channels and disrupts the passage of nerve impulses, but may also affect respiratory processes in mites.

Other organochlorine insecticides include hexachlorocyclohexane (HCH) and the cyclodienes. The insecticidal properties of the gamma isomer of HCH or lindane were, like those of DDT, discovered and developed during the period 1939–1945. After a concerted campaign by pressure groups, the use of lindane was prohibited in the EU from 2001. It was formerly used, primarily as a seed dressing, to control insect pests in cereals, sugar beet and other crops as well as in flea powders for cats and other domestic animals, and for timber treatment. HCH is of much the same acute toxicity to humans as DDT, and again is not readily absorbed through the skin or other external surfaces. However, it is less persistent in the environment than DDT.

By comparison, the cyclodiene group of pesticides is even more stable in the environment than DDT. These compounds include aldrin and dieldrin (Figure 4.12), which were used in seed dressings for the control of soil-borne insect pests. Persistence of the cyclodienes caused adverse effects on non-target insect species and furthermore led to biomagnification and bioaccumulation through food chains. Restrictions on their agricultural uses were imposed in many countries from the mid 1960s and they were completely proscribed in the UK by the early 1990s.

The mode of action of HCH is similar to that of the cyclodienes. These insecticides cause hyperactivity and convulsions and ultimately death of insects but unlike DDT the site of action is the synapse (Figure 4.4) where HCH and the cyclodienes antagonise the inhibitory neurotransmitter gamma-aminobutyric acid (GABA). Normally, during transmission of a nerve impulse across a synapse GABA is released from one neurone, crosses the synapse and binds to the post-synaptic membrane. The actual binding site is a receptor protein containing a chloride ion channel, and when GABA binds to this protein the channel is opened and chloride ions flow into the neurone. The function of this chloride influx is to prevent unrestricted development of nerve impulses. Cyclodienes and HCH bind to the chloride channel and block its activation by GABA, interrupting chloride ion flow through the channel. This absence of synaptic control leads to a hyperexcitation of the central nervous system. Some selectivity may be due to differences in the GABA receptor structure of insects and mammals, although the slow penetration of HCH and cyclodienes through skin and body surfaces may be at least partially responsible for differences in acute toxic effects on vertebrates when compared to insects.

### 4.2.5 Avermectins

The avermectins are a group of structurally complex natural products derived from a soil-inhabiting fungus, *Streptomyces avermitilis*, and were introduced in the mid 1980s (Lasota and Dybas, 1991). They act at very low dosage against a wide range of nematodes and insects – in some cases as low as a few grams per hectare (Figure 4.13). Avermectin B1a is the principal constituent of both abamectin and emamectin, which are used on crops, and has good contact activity against insect pests. The avermectins have limited systemic properties and will enter treated leaves. However, their principal use is as veterinary and medicinal products to control insects and especially helminthes that attack domesticated stock and human beings. The semisynthetic ivermectin is an antihelminthic and currently widely used as a drug to combat the *Onchocerca volvulus*, the nematode that causes river blindness in tropical and subtropical regions.

Avermectins, like lindane and the cyclodiene organochlorine insecticides (4.2.4), interfere with the activity of neurotransmitters in the central nervous system by increasing the flow of chloride ions from ion channels on the postsynaptic membrane. However, the precise site of action differs in that avermectins interfere with glutamate-gated chloride channels. Feeding and movement usually stop in insects and nematodes exposed to avermectins and a general paralysis may occur. However death may not occur until several days after treatment.

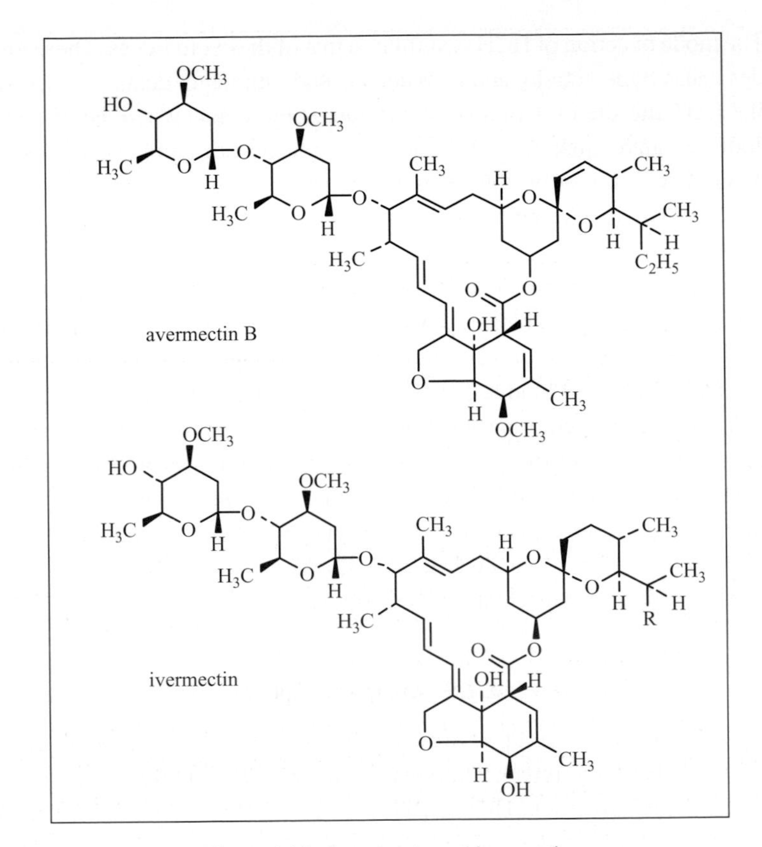


Figure 4.13 Avermectin and ivermectin.

Selectivity is evident with avermectins: glutamate-gated chloride channels predominantly occur in invertebrate nerve and muscle cells, but they do not occur in some mammalian species, and this, allied to the fact that the avermectins do not readily bind to GABA-mediated chloride channels in mammalian species, thus allows therapeutic use of avermectins such as ivermectin in treatment of helminth infestations in mammals, including humans.

### 4.2.6 Fipronil and amitraz

The phenylpyrazole compound fipronil (Figure 4.14), brought to the market in the mid 1990s, has an identical mode of action to the cyclodiene insecticides, binding to GABA-gated chloride channels. As with imidacloprid compared to nicotine, fipronil has a much better safety profile when compared to the

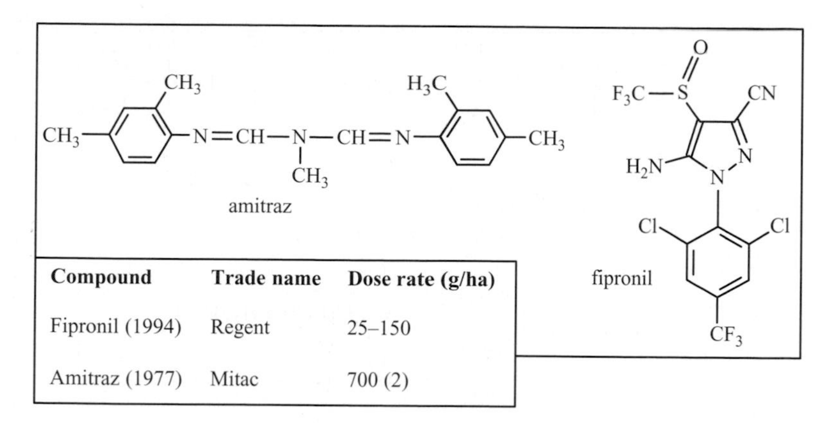


| Compound | Trade name | Dose rate (g/ha) |
|---|---|---|
| Fipronil (1994) | Regent | 25–150 |
| Amitraz (1977) | Mitac | 700 (2) |

Figure 4.14 Fipronil and amitraz.

cyclodienes. Furthermore, it has no undesirable bioaccumulative effects, and thus poses few problems to the environment. It is effective against a range of pests, and is used in agriculture, particularly as a soil insecticide, and for insect control in domestic stock, pets, houses and industrial premises.

The formamidines, of which the principal compound is amitraz (Figure 4.14), mimic the action of another neurotransmitter, octopamine. Octopamine is a peptide involved in the regulation of behaviour in insects and overstimulation of the octopamine receptor leads to tremors, convulsions and continuous flight in adult insects. In addition these compounds may act as antifeedants and suppress reproduction in insects. Amitraz is widely used against foliar pests, particularly spider mites, of cotton and fruit, as well as in public health and veterinary practice, again for the control of mites.

## 4.3 Arthropod pesticides that inhibit respiration

Some of the first synthetic organic insecticides that were released in the 1930s, such as the dinitrophenols, are effective respiratory inhibitors, but not very selective in their action. Surprisingly, some of these compounds with a poor level of selectivity, such as dinitroorthocresol (Figure 4.15), were used in some EU countries as late as the 1990s although most were withdrawn in the mid to late 1980s because of the risk they presented to human health, especially during high-volume fruit spraying. Only dinocap, which has relatively low mammalian toxicity, now remains on the UK market, being available even to

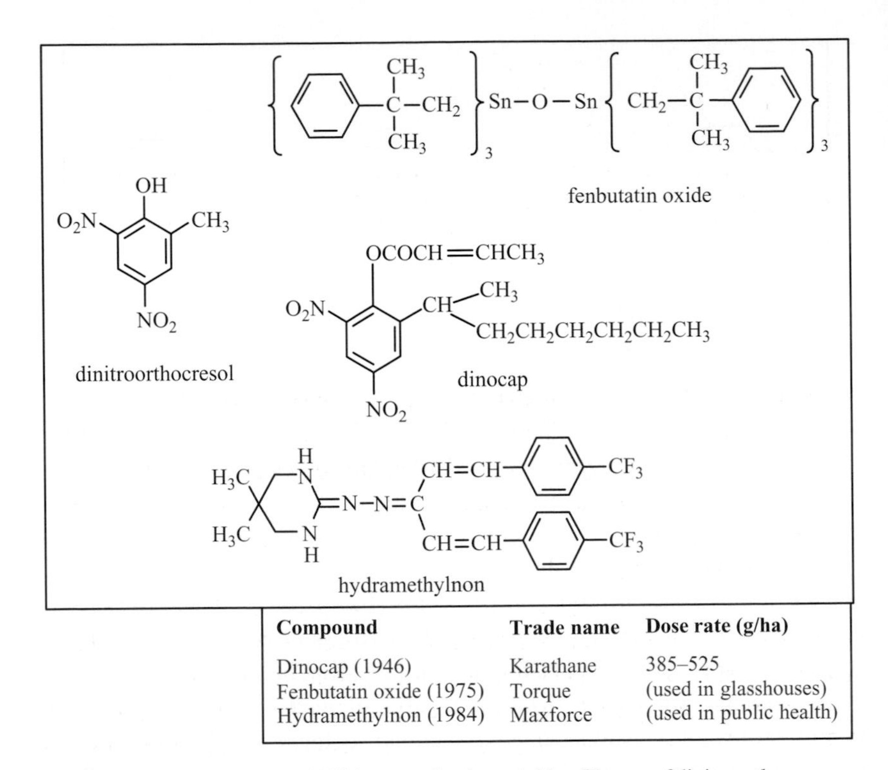


| Compound | Trade name | Dose rate (g/ha) |
|---|---|---|
| Dinocap (1946) | Karathane | 385–525 |
| Fenbutatin oxide (1975) | Torque | (used in glasshouses) |
| Hydramethylnon (1984) | Maxforce | (used in public health) |

Figure 4.15 Respiratory inhibitors used as insecticides. The use of dinitroorthocresol is now prohibited in most countries.

domestic gardeners. It gives good control of mites and is used for this purpose on fruit crops, where it has the added benefit of controlling powdery mildew fungi. The dinitrophenols inhibit and/or uncouple oxidative phosphorylation and thus prevent the formation of ATP. Selectivity, where it exists, is most probably due to metabolic breakdown in non-target organisms.

The organotin compound fenbutatin oxide (Figure 4.15) is also widely used as an acaricide. It too inhibits oxidative phosphorylation, probably at the same site as the dinitrophenols, but its selectivity, at the doses applied, to certain pest mite species is remarkable. This selectivity may be due to metabolism and detoxification in non-target species. It has low toxicity to predatory mites and insects, and has thus been widely deployed in pest management schemes utilising predators as part of integrated control strategies.

Another organotin, tributyltin (TBT; $CH_3(CH_2)_3Sn$) was widely adopted for use in marine anti-fouling paints from the early 1970s. TBT prevents the colonisation of ship's hulls by barnacles (such as *Lepas anatifera*), polychaete tubeworms as well as algae and proved very efficient in this respect, with the

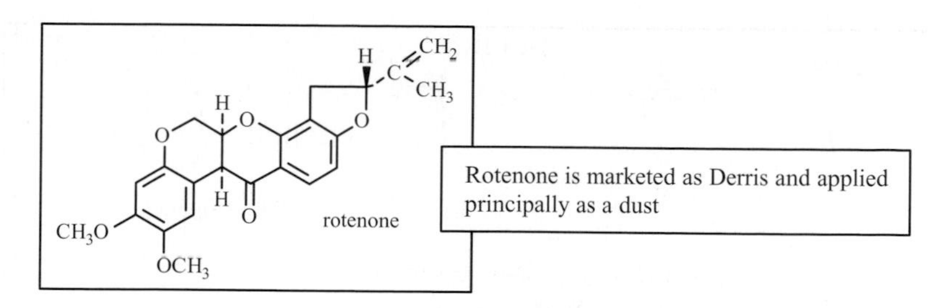


Figure 4.16 Rotenone, a natural product.

benefit of being less toxic to human beings than some earlier compounds used for this purpose. Like other organotins, TBT inhibits oxidative phosphorylation. Unfortunately, TBT slowly leaches from the hulls of ships and is acutely toxic at low concentrations to many aquatic molluscs, and even at very low levels may cause endocrine disruption in these and other aquatic invertebrate and vertebrate species (9.6).

Hydramethylnon (Figure 4.15) is another respiratory inhibitor, and is principally used in public health for control of cockroaches (*Periplaneta* and *Blatta* spp). It is a slow acting stomach poison that inhibits electron transport at the cytochrome $bc_1$ complex, in much the same way as the strobilurin fungicides (3.7.3). Careful placement in baits may prevent access to non-target species, and hydramethylnon may not be rapidly taken up, and/or detoxified by vertebrate species.

Some insecticides that inhibit respiration are of natural origin. The rotenoids, the main example of which is rotenone (Figure 4.16), are produced in the roots of two leguminous plant genera, *Derris* and *Lonchocarpus*, and as natural products are approved for use in many organic systems of growing. They have been used commercially since the 1850s. Rotenone is a piscicide as well as an insecticide, and has been used commercially to clear lakes of coarse fish prior to stocking with game fish. At the doses employed, it is relatively non-toxic to organisms that make up the fish diet, and breaks down rapidly. As an insecticide, it has been used for many years as a foliar treatment for lepidopteran larvae.

Respiratory inhibitors have continued to be developed for mite control, and compounds introduced in the 1990s include tebufenpyrad, fenazaquin and fenpyroximate (Dekeyser and Downer, 1994; Figure 4.17). These offer a much greater degree of safety to operators, and some additionally control insect pests. Rotenone and fenazaquin inhibit electron transport in mitochondria at complex 1, the NADH:ubiquinone oxidoreductase site. The subsequent disruption of energy metabolism results in gradual inactivity, paralysis and death of the mite or insect. Again, rapid breakdown in non-target organisms is the most

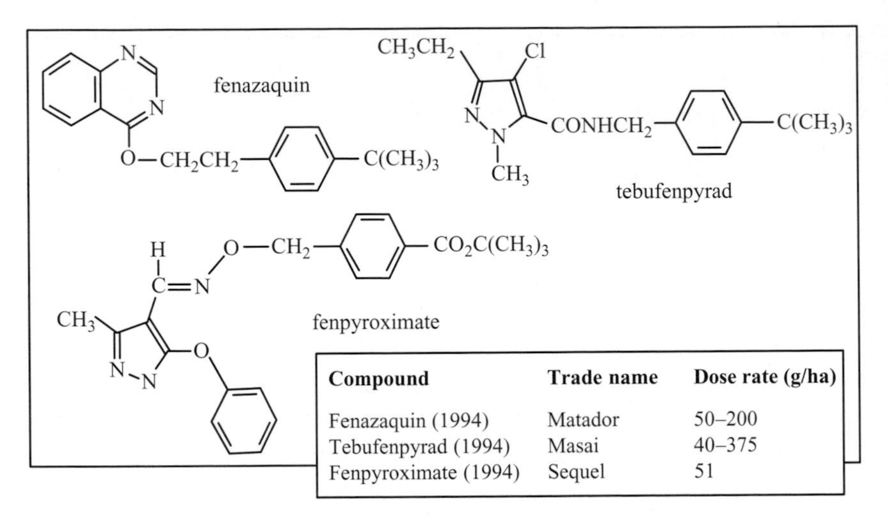


| Compound | Trade name | Dose rate (g/ha) |
|---|---|---|
| Fenazaquin (1994) | Matador | 50–200 |
| Tebufenpyrad (1994) | Masai | 40–375 |
| Fenpyroximate (1994) | Sequel | 51 |

Figure 4.17 Miticides that act by inhibiting respiration.

likely factor contributing to the selectivity of these compounds. Tebufenpyrad, fenazaquin and fenpyroximate are all rapidly metabolised in mammals.

## 4.4 Regulators of insect growth

Compounds that control insects and other invertebrate pests by interfering with the regulation of their growth are the only insecticides to exhibit true biochemical selectivity, since the systems affected are restricted to insects and other arthropods. In this respect they have much in common with the hormone herbicides. Insect growth regulators have the lowest acute toxicity to higher vertebrates of all products used to control arthropods, with compounds such as diflubenzuron, methoprene and fenoxycarb having extremely low toxicity to mammals. These insecticides include inhibitors of chitin synthesis, and compounds that may function as mimics of juvenile and moulting hormones (Dhadialla, Carlson and Le, 1998). Such processes are virtually unique to insects; although some fungi have chitin in their cell walls, it is synthesised by a different route to that in insects. As with other insecticides, the growth regulators may of course have adverse effects on non-target insects, although in some cases such as tebufenozide, a remarkable degree of specificity to single groups of insects, in this case the Lepidoptera, has been achieved.

The benzoylureas (Figure 4.18) interfere with chitin synthesis and in most cases their action follows ingestion by insects. Used in both agriculture and public health, they are very effective against coleopteran and lepidopteran larvae

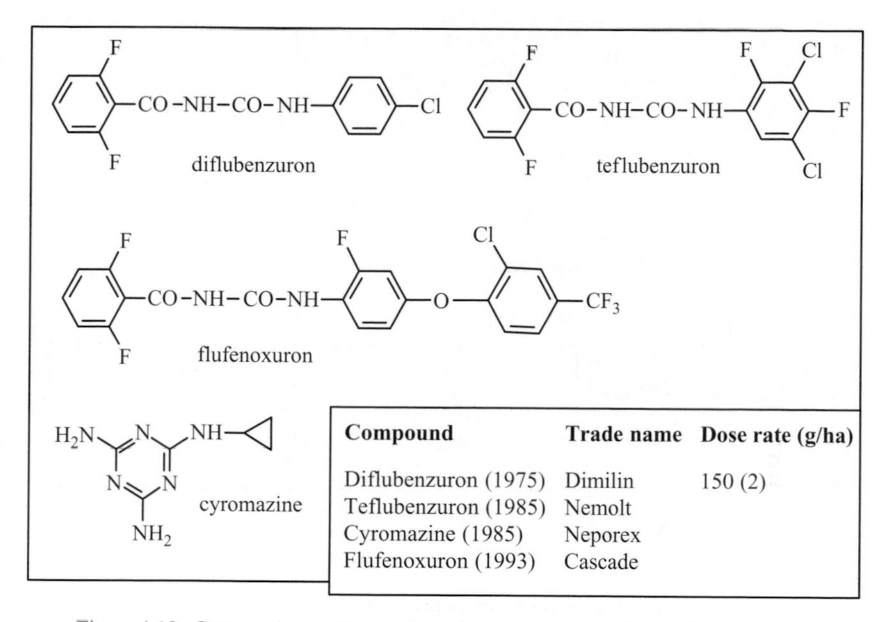


| Compound | Trade name | Dose rate (g/ha) |
|---|---|---|
| Diflubenzuron (1975) | Dimilin | 150 (2) |
| Teflubenzuron (1985) | Nemolt | |
| Cyromazine (1985) | Neporex | |
| Flufenoxuron (1993) | Cascade | |

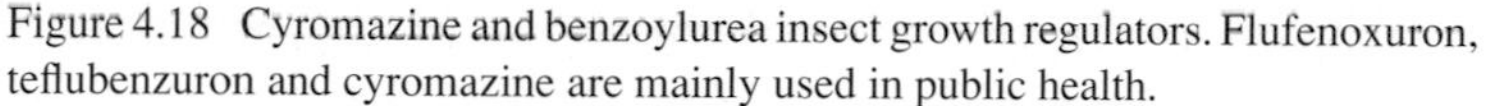
Figure 4.18 Cyromazine and benzoylurea insect growth regulators. Flufenoxuron, teflubenzuron and cyromazine are mainly used in public health.

since these are unable to resynthesise their cuticles after moulting. The cuticle in insects is composed of about 50% chitin, a polymer of *N*-acetylglucosamine, and is a vital part of the arthropod exoskeleton. The linking of *N*-acetylglucosamine units to form the long-chain chitin molecule is blocked by the benzoylureas, and the compounds are most effective if applied just before a moult. Without chitin, the cuticle becomes thin and brittle. The insect then loses its structural integrity and dies. The first chitin synthesis inhibitor to be used on a large scale was diflubenzuron, introduced in the mid 1970s. Research into more potent compounds has continued since then with the subsequent introduction of teflubenzuron and flufenoxuron.

The triazine cyromazine is also a potent inhibitor of chitin synthesis, and particularly active against fly larvae. Cyromazine is thus used for the control of dipterous leaf miners in ornamentals and vegetables in which it is systemic, and is fed to poultry to control flies in the bedding and manure of broiler sheds.

The juvenile hormone mimics include methoprene which closely resembles the sesquiterpenoid natural juvenile hormones of insects, the non-neurotoxic carbamate fenoxycarb, and the pyridine compound pyriproxyfen (Figure 4.19). These compounds are very effective if applied when juvenile hormone concentrations in insect larvae are low, typically in the final larval stage prior to

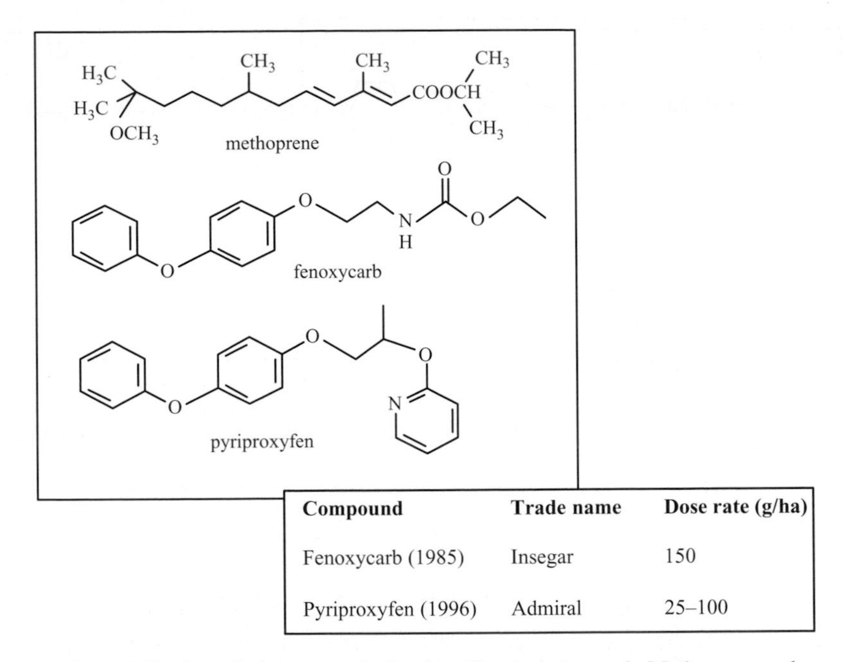


| Compound | Trade name | Dose rate (g/ha) |
|---|---|---|
| Fenoxycarb (1985) | Insegar | 150 |
| Pyriproxyfen (1996) | Admiral | 25–100 |

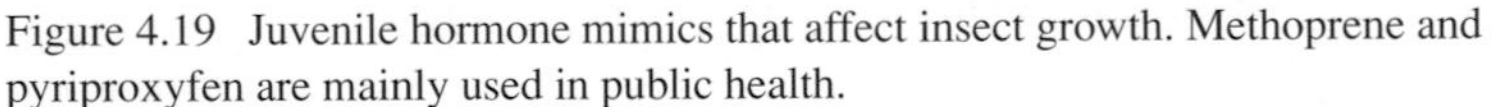

Figure 4.19 Juvenile hormone mimics that affect insect growth. Methoprene and pyriproxyfen are mainly used in public health.

pupation. They may also disrupt reproductive physiology in adult insects. Their primary use is for the control of the larval stages of flies, mosquitoes, midges and beetles. Pyriproxyfen is used to control whitefly, notably *Bemisia* sp., in cotton and tobacco.

Despite their favourable toxicological profile, the use of juvenile hormone mimics is limited in that they cannot eliminate a pre-existing pest population, and are obviously not applicable to pests that cause damage principally in the larval state.

Synthetic ecdysones, which induce moulting in insects, were developed in the 1990s. The principal compounds include tebufenozide, halofenozide and methoxyfenozide, all of which have extremely low acute toxicity to mammals and other non-target organisms, including honey bees and other beneficial arthropod species. Tebufenozide (Figure 4.20) binds to the ecdysone receptor protein of lepidopterous larvae, and acts as an ecdysteroid agonist capable of inducing lethal moults in all larval stages. Selectivity is probably due to its high binding affinity to a receptor that elicits an ecdysone or moulting response. After application feeding ceases and a premature moult ensues, causing death of the insect. Tebufenozide is widely used against the codling moth caterpillars that attack top fruit crops such as apples. The thiadazine compound

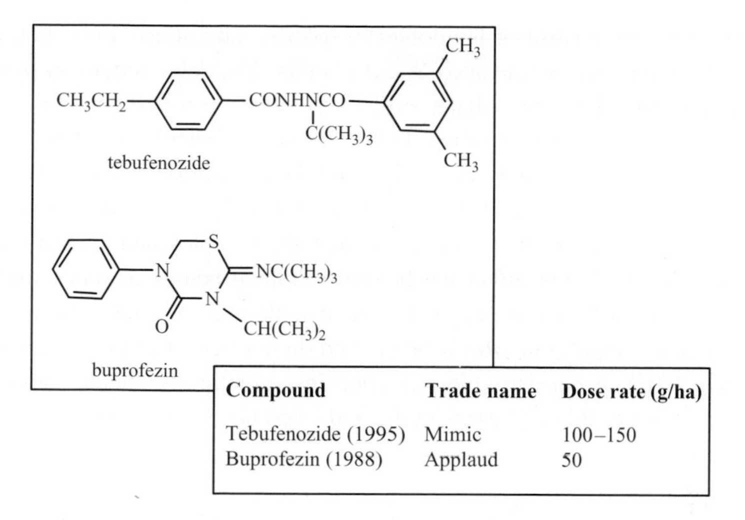


| **Compound** | **Trade name** | **Dose rate (g/ha)** |
|---|---|---|
| Tebufenozide (1995) | Mimic | 100–150 |
| Buprofezin (1988) | Applaud | 50 |

Figure 4.20 Tebufenozide and buprofezin.

buprofezin (Figure 4.20), which disrupts moulting and may also act as an insect growth regulator, is used widely for the control of whitefly of both *Bemisia* and *Trialuroides* genera. It is very effective against the larval stages of whiteflies, with little effect on non-target insect species.

## 4.5 Toxins from *Bacillus thuringiensis*

Preparations based on certain entomopathogenic fungi, bacteria and viruses have been used for the control of arthropods for many years. Most of these act through colonisation of the insect body and may be considered as biocontrol rather than chemical control agents. A few, notably those based on *Bacillus* species, owe their activity to toxic proteins, and strains of *Bacillus thuringiensis* isolated from infected caterpillars have been used for the control of lepidopteran larvae for many years. Endotoxins produced by *B. thuringiensis* are formed during sporulation of the bacterium and these are highly toxic to certain insects. The toxins act as specific gut poisons in certain species of insect: once ingested the delta-endotoxin crystals dissolve in the mid-gut of the insect. The toxins that are released then bind to the cells of the mid-gut, severely disrupting membranes leading to cell lysis. Target insects stop feeding and usually die within a few hours.

The gene sequences for endotoxin production in several strains of *B. thuringiensis* have been identified, cloned and elucidated. Gene sequences have been transferred to some crop plants including maize, cotton, rice and potatoes, primarily for the control of lepidopteran species, particularly those that penetrate deep into plant tissue such as corn borers. The delta-endotoxin product is expressed in all parts of plants, effectively acting as a permanent systemic insecticide, and is seed transmitted. These crops have been planted over millions of hectares in the USA and many other parts of the world, with the exception of EU member states where environmental and health concerns promulgated by pressure groups have so far prevented their deployment. Toxins from *B. thuringiensis* do not affect the health of human beings or other animals, and indeed are not effective against many beneficial invertebrate species. The toxins degrade rapidly in crop residues and do not persist in the environment (Burges, 2001). A considerable reduction in synthetic insecticide sprays has resulted where *B. thuringiensis*-engineered crops have been grown.

## 4.6 Selectivity, insecticide resistance and compounds with a novel mode of action

As with fungicides and herbicides, widespread adoption and use of insecticides has led to problems of resistance. Throughout the world, many species of pest insects have developed resistance to the major groups of neurotoxic insecticides, although few problems have arisen with the neonicotinoid group. Resistance has arisen most commonly in pests of intensive systems of cultivation such as field and glasshouse vegetable production, but has also posed problems for the control of vectors of viruses of arable field crops, such as the peach potato aphid *Myzus persici* in the UK and pests of intensively sprayed crops such as cotton in most countries where this crop is grown.

Resistance may result from slower penetration of insecticidal molecules into insects, and from alterations in their behaviour, but the principal causes are alterations at the site of action of the insecticide or from detoxification of the molecule. Resistance in several pest species has been associated with enhanced levels of carboxylesterases that can remove ester groups from organophosphate and carbamate insecticides. Overproduction of carboxylesterases may also contribute to pyrethroid resistance. In other cases, an alteration in the enzyme acetylcholinesterase (modified acetylcholinesterase or MACE) has led to the development of resistance in aphids to compounds such as pirimicarb.

In the 1990s, molecular studies of strains of the peach-potato aphid, as well as dipteran and lepidopteran insects resistant to pyrethroid insecticides, revealed a point mutation in the gene coding for the voltage-gated sodium channel leading to a single amino acid substitution in the protein. Pyrethroids such as deltamethrin and cypermethrin will not bind or bind only weakly to

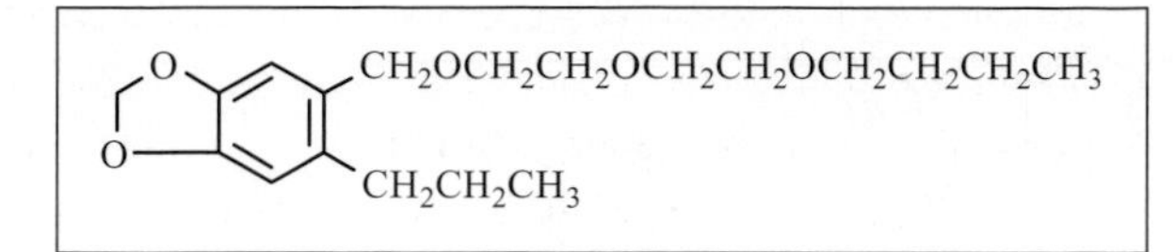


Figure 4.21 Piperonyl butoxide, added to formulations of many pyrethroid insecticides at about 100 g/ha.

such altered proteins, and are thus ineffective. This type of resistance is known as knockdown resistance or *kdr*, and also confers resistance to DDT, which has a similar mode of action to the pyrethroids. Indeed, this type of resistance may have arisen during the use of DDT and persisted through cross-resistance to pyrethroids, which were introduced shortly before the prohibition of DDT.

As with herbicide and fungicide resistance, selection pressure results in the survival and reproduction of insecticide-resistant biotypes, which may recolonise the ecological niche vacated by the control of their wild-type relatives. The problem may be made worse if the compounds applied have eliminated the natural enemies of the insect in question, since pest resurgence (of resistant individuals) may occur to greater levels than those of the original pest population.

In order to combat resistance linked to enhanced levels of detoxifying enzymes, metabolic inhibitors of these enzymes have been introduced and added to formulations. The best known of these is piperonyl butoxide (Figure 4.21) that is used widely in combination with pyrethroid insecticides. Oxidative metabolism catalysed by p450 enzymes (6.3) leads to a carbene derivative of piperonyl butoxide that forms an almost irreversible complex with the p450 enzymes themselves – thus preventing detoxification of the insecticide.

Multiple insecticide resistance has arisen in some species. Mechanisms of detoxification by enhanced carboxylesterase activity may result in resistance to organophosphate, carbamate and to some extent pyrethroid insecticides, and this has been shown in some strains of the peach-potato aphid. However, there are some instances of much greater multiple resistance. The diamond back moth, *Plutella xylostella*, is a major pest of cruciferous vegetables, particularly in the Far East. Some strains of this pest have developed resistance to insecticide groups including organophosphates, carbamates, pyrethroids, benzoylureas, avermectins and even *Bacillus thuringiensis*, and a variety of mechanisms have evolved in this species to circumvent the action of insecticidal compounds.

New compounds with novel modes of action may enable control of invertebrates that have become resistant to established molecules (Salgado, 1999).

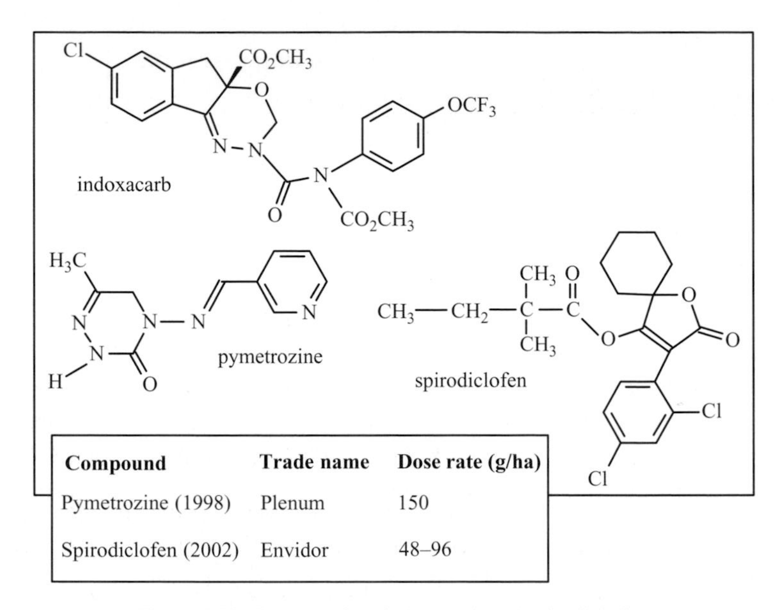


| Compound | Trade name | Dose rate (g/ha) |
|---|---|---|
| Pymetrozine (1998) | Plenum | 150 |
| Spirodiclofen (2002) | Envidor | 48–96 |

Figure 4.22 Pymetrozine, indoxacarb and spirodiclofen.

Several novel mechanisms of action have been reported among a number of insecticides brought to the market. Compounds such as pymetrozine, indoxacarb and the tetronic acids (Figure 4.22) all have modes of action that are different to those of pyrethroids, anticholinesterase compounds and those that affect nicotinic acetylcholine and GABA receptors.

Pymetrozine, a pyridine azomethine, is a synthetic antifeedant used extensively for the control of aphids and has proved particularly useful against strains resistant to other insecticides, since its mechanism of action is different to that of all other insecticide groups. Its precise mode of action and basis of selectivity have not yet been established. Indoxacarb is an ion channel blocker, affecting particularly sodium channels in nerve cells but at a different site to that inhibited by DDT and the pyrethroids. The tetronic acid derivatives spirodiclofen and spiromisafen are effective against mites, with the latter offering additional control of glasshouse whiteflies. Their mode of action may involve interference with lipid metabolism in target species. These compounds are all of very low toxicity to mammals and many other non-target organisms.

# 5

# Soil sterilants, fumigants and vertebrate poisons

## 5.1 Introduction

Most of the pesticides covered in this chapter are toxic to many organisms, and most show little or no selectivity at the biochemical level. Some are used to eradicate all weeds, pests and diseases from soil. Others are used to control mammalian pests including rats (such as *Rattus norvegicus* – the brown rat), house mice (*Mus musculus*), rabbits (*Oryctolagus cuniculus*) and moles (*Talpa europaea*). Selectivity of action is only achieved through careful placement and/or use such that these compounds do not come into contact with non-target species, and particularly people working with them. The extent to which this is achieved varies: serious illness and death have followed careless use of some of these compounds.

Soil sterilants and fumigants are often classified as general biocides rather than pesticides and kill a wide range of organisms, as do some vertebrate poisons. Other vertebrate poisons such as rodenticides may be more selective in their action, but not to warm-blooded animals, and are thus toxic at low doses to mammals and birds. However, effective antidotes to these compounds are available, reducing the risk to the health of human beings. One of the major problems with some rodenticides in the UK is their use in illegal poisoning of game birds, some of which are rare species.

## 5.2 Soil sterilants

Sterilisation of soil to remove unwanted weed seeds as well as soil-borne pests and diseases is commonly carried out with high value crops. However, it is an expensive operation, and has been superseded in some situations by other growing techniques. For example, many glasshouse crops are now raised in growing media such as mineral wool, peat or perlite, although lettuce is still

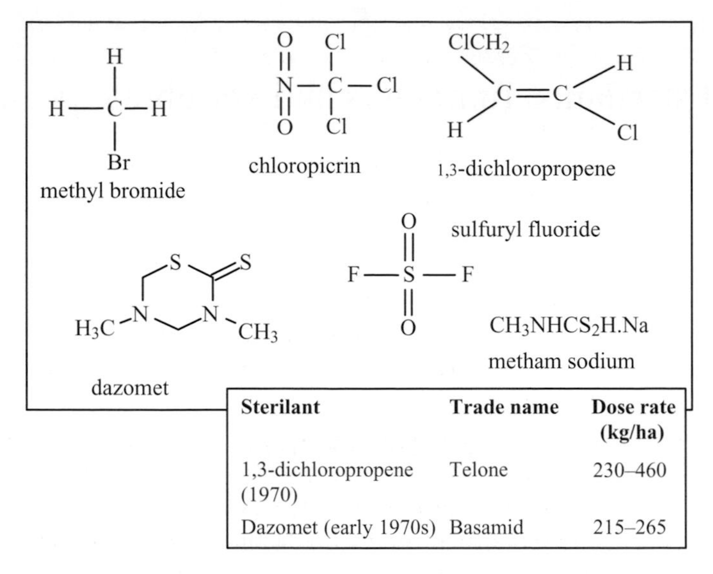


| Sterilant | Trade name | Dose rate (kg/ha) |
|---|---|---|
| 1,3-dichloropropene (1970) | Telone | 230–460 |
| Dazomet (early 1970s) | Basamid | 215–265 |

Figure 5.1 Soil sterilants.

raised in soil that is usually sterilised between crops. Soil sterilisation is still carried out for some outdoor horticultural crops such as strawberries and cut-flowers, and for these and glasshouse crops raised in soil, chemical soil sterilants are still routinely employed. The most widely used has been methyl bromide: others include 1,3-dichloropropene, chloropicrin, metham-sodium and dazomet (Figure 5.1).

Methyl bromide was formerly widely used in the USA and some other countries to clear land of weed seeds, soil-borne pests (particularly nematodes) and diseases before planting cut-flower crops, strawberries and tomatoes raised outdoors or in polythene tunnels. The USA used about 7000 tonnes in 2001. Methyl bromide is not extensively used in the UK, although the amount deployed increased from 137 tonnes to 174 tonnes from 1994 to 1998. In the developing world about 14 000 tonnes was used in 1998. Methyl bromide is supplied in pressurised canisters and is injected into soils. Here it becomes gaseous and permeates the soil: polythene sheeting is used to cover injected soil and retain the gas whilst it exerts its effects. After 7–10 days, the sheeting is removed and the gas allowed to disperse. Germination tests with quick-growing species such as cress (*Lepidium sativum*) will indicate when it is safe to plant a crop. With methyl bromide sterilisation, soil may only be out of use for about 7–10 days whereas other sterilants may take much longer to disperse.

Methyl bromide is also used as a fumigant, particularly in grain storage and for plant material in international trade. Its rapid fumigant effect and quick dispersal make it an ideal substance for this purpose.

However, methyl bromide is a general respiratory inhibitor and this, allied to its volatile nature, presents a considerable risk to human health. In countries such as the USA and UK, the risks that methyl bromide poses to health are considered to be high and thus only licensed operators are allowed to apply this chemical. Indeed, deaths following exposure to methyl bromide have been recorded in several countries. Less acutely toxic materials such as sulfuryl fluoride have been developed and latterly promoted as alternatives to methyl bromide, but concerns have arisen over fluoride residues in foodstuffs when this material is used in grain stores.

Field investigations have shown that 70–90% of methyl bromide applied to soils is released into the atmosphere and here it may cause ozone depletion, with many authorities considering methyl bromide to present a greater risk than chlorofluorocarbons to the ozone layer. For this reason, methyl bromide use for most purposes is to be phased out in Europe and the USA by 2005, and elsewhere by 2015, although human sources of the compound account for only 40–45% of the total global production of methyl bromide, with much of the rest being released from oceans. Some countries including Germany, Sweden and the Netherlands had prohibited the use of methyl bromide by 2001 (Labrada and Fornasari, 2001).

Alternatives to methyl bromide as soil sterilants include 1,3-dichloropropene, chloropicrin and formaldehyde. The latter two are toxic at low doses to human beings and other animals. Most of these compounds are applied at very high dose rates – kilograms rather than grams per hectare – in order to exert their effects (Figure 5.1). Chloropicrin or tear gas is very pungent and its smell can persist for days. 1,3-Dichloropropene is not an ozone depleter and, like methyl bromide, has a short half-life of a few hours in soil. It is widely used for control of nematodes in potatoes, tomatoes and other vegetable crops. It kills nematodes by attachment to thiol groups on enzymes and other proteins, but this action is not selective. Hence, this compound may present increased health risks to those applying the compound, and strict precautions are recommended during its use.

Metham-sodium and dazomet are applied to soil as granules and after incorporation break down to form toxic chemicals that act as general soil sterilants. Metham-sodium and dazomet decompose in soil to form methyl isocyanate, which is toxic to weed seeds, insects and soil-borne pathogens, but is not very effective against soil-borne nematodes compared with, for example, methyl bromide. The problem for growers with both metham-sodium and dazomet is that the methyl isocyanate formed takes much longer to disperse compared to fumigants. In the case of dazomet, crops may not be planted for up to 60 days after application.

## 5.3 Vertebrate poisons

Legislation in the UK allows the chemical control of rats and certain species of mice, voles, shrews and moles in addition to a few species of birds that may cause damage to crops, or are considered a public nuisance. Such species include pigeons (*Columbia livia*), rooks (*Corvus frugilegus*), jackdaws (*Corvus monedula*) and magpies (*Pica pica*). Measures to control birds and mammals in the UK often need to take into account the sensitivity frequently associated with their deployment, and sentimental affinities for furry or feathery warm-blooded animals. This image may be contrasted with the damage and suffering that can result from the activities of species such as rats. Rats consume growing crops and crop produce in storage. They excrete on food, may carry microorganisms such as those responsible for bubonic plague and Weils disease, can burrow into embankments and drains, chew through cables and invade houses and dwellings. Surveys conducted in the UK in the late 1990s showed that populations of the principal species – the brown rat – were increasing.

Chemicals used to control rats, mice and other vertebrate pests are generally toxic to other mammals and vertebrate species such as birds. The selective action of these chemicals is achieved entirely through their placement. Aluminium phosphide is a synthetic substance of high toxicity that may be used to kill moles as well as rabbits through the release of phosphine gas. Sodium cyanide may be placed into the burrows of rabbits where, in the presence of moist soil, release of the lethal gas hydrogen cyanide occurs. In the UK the naturally occurring compound strychnine, extracted and concentrated from *Strychnos* spp., was formerly used to kill moles; zinc phosphide is used in baits to control mammalian and avian pests such as squirrels, moles and rats. Animals that ingest a lethal dose of these chemicals usually die within a few hours. The risks to health from many of the above chemicals are such that the use and storage of many in the UK is governed by the Poisons Act (1972) and the deployment of these compounds is generally restricted to licensed pest operators.

Some compounds are normally formulated as baits and placed in positions accessible to the target species but which present a low risk of exposure or contact with non-target species. The narcotic alphachloralose may be mixed with grain and used to control mice and rats in buildings, as well as birds such

Table 5.1 *Toxicity of alphachloralose*

| Species | Oral $LD_{50}$ of alphachloralose (mg/kg body weight) |
|---|---|
| Rat ( *Rattus norvegicus*) | 190–300 |
| Mouse (*Mus musculus*) | 300–400 |
| Mallard (*Anas platyrhynchos*) | 42 |
| Starling (*Sturnus vulgaris*) | 75 |
| Canada goose (*Branta canadensis*) | 54 |

From Timm (1994)

as pigeons. However, this compound has been misused, notably to kill raptor birds through illegal treatment of animal carcasses that may then be ingested by hawks and other birds of prey (9.10). Alphachloralose is toxic to some species of birds at low dose (Table 5.1).

Somewhat less toxic although slower acting rodenticides (Figure 5.2) were introduced in the late 1940s, with the most widely used being the hydroxycoumarin derivative warfarin. Warfarin is an anticoagulant, effectively preventing blood clotting, and is used medicinally at specified dosages for this purpose. At low doses of around 1–10 mg per patient, it can slow down thrombin formation in blood and hence reduce the risk of clotting and arterial blockage. The concentration used in bait for pest control is around 0.005% with a lethal dose for the rat being around 1–3 mg/kg body weight.

Warfarin interferes with the regeneration of vitamin K (phylloquinone) that is involved in the synthesis of coagulant proteins such as prothrombin, essential for blood clotting. Inhibition results in a decrease in vitamin K and a consequent inability of the blood to clot. Specific inhibition of a reductase enzyme (KO reductase) that is involved in the recycling of vitamin K in the liver occurs. After ingestion, death results from extensive internal bleeding. The process is slow, and animals that have ingested warfarin generally take several days to die.

The more powerful so-called second-generation anticoagulants were introduced in the late 1970s. Most of these are also hydroxycoumarin derivatives, including bromadiolone, difenacoum and brodifacoum, and may control both warfarin-resistant and warfarin-sensitive rodents. Their potency may be linked to strong lipophilicity and thus absorption to membranes, the site of the target KO reductase enzyme (WHO, 1995). The second-generation compounds are used at similar dose rates to warfarin, at about 0.005% to 0.001% by weight of bait. Rats that have ingested these compounds die more quickly than those that have ingested warfarin, and with some species such as mice and voles, a

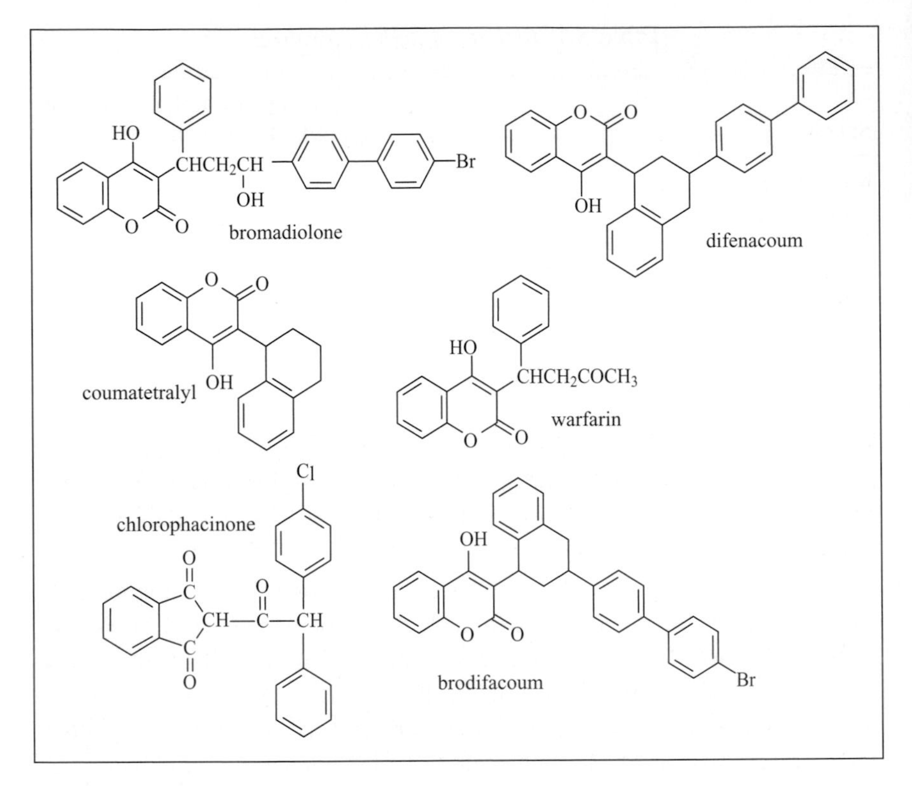


Figure 5.2 Rodenticides.

single feed may be sufficient to impart a lethal dose. Other second-generation compounds include the indanedione derivatives diphacinone and chlorophacinone, which also interfere with the regeneration of vitamin K, but in a slightly different manner to that of the hydroxycoumarin rodenticides. Although the coumarin-based rodenticides can be toxic to human beings, even at low dose, they are considered among the safest of all pesticides since, unless a very high dose is consumed, their action may be reversed by administration of vitamin K. Indeed some are marketed to the general public.

The effectiveness at very low doses of anticoagulants has led to concerns being expressed about risks to non-target vertebrates. Indeed the oral $LD_{50}$ of warfarin in terms of milligrams ingested per kilogram of body weight in species such as pigs and cats is close to that for rats. Cattle are somewhat less susceptible (Table 5.2). Studies in the UK have shown that barn owls (*Tyto alba*) may be killed by relatively low doses of second-generation rodenticides. In Finland, where second-generation rodenticides are used extensively against voles (*Microtus* and *Arvicola* spp.), placement in vole burrows markedly

Table 5.2 *Mammalian toxicity of warfarin*

| Species | Oral $LD_{50}$ of warfarin (mg/kg body weight) |
|---|---|
| Rat | 1–3 |
| Pig | 3 |
| Cat | 5–50 |
| Dog | 20–50 |
| Cattle | 200 |

From Timm (1994)

reduced the risks to non-target species. These studies also showed that birds of the corvid family (such as magpies – *Pica pica*) likely to prey/consume voles were highly tolerant of second-generation anticoagulants compared to barn owls in the UK. However, these studies also showed no ill-effects on other potential scavengers of dead (treated) voles, such as dogs and cats (Myllymaki, 1996).

Control of rodents by chemical means involves both the design and presentation of baits. The palatibility and particularly the taste and smell of the bait must be acceptable to the test species and baits should be placed in a situation inaccessible to non-target animals, particularly cats, dogs and farm stock. Bait stations have now been designed that prevent access of non-target vertebrates. One of the problems of rodent control is bait shyness, often associated with absorption of a sublethal dose and consequent avoidance of the bait. The principal method of overcoming bait shyness is to prebait with an acceptable but untreated sample of bait to familiarise animals with the baits, bait containers and bait locations before introduction of treated bait. Once a regular feeding pattern has been established, the rodents may well ingest a lethal dose at one visit at the bait location (Dennis, 1991).

Exposure at a given site for 3–6 weeks will normally control rodent infestations. However, a major problem with warfarin and other so-called first-generation anticoagulants is the development of resistance in rat populations. The resistance is expressed as a reduced affinity for warfarin to bind to the target enzyme, KO reductase. However, in many cases this resistance does not extend to second-generation rodenticides. Whereas warfarin only inhibits KO-susceptible strains of rat, brodifacoum produces a decrease in plasma prothrombin in both warfarin-susceptible and warfarin-resistant individuals (WHO, 1995).

# PART II

## Risks to Health and the Environment From Pesticides

# 6
# Pesticide toxicology

## 6.1 Introduction

Earlier chapters have indicated that some pesticidal molecules kill their target organisms quickly. In some cases, such as the pyrethroid insecticides, the death of target pests may be very fast indeed – a matter of seconds. Copper-based fungicides may kill fungal spores in minutes. Paraquat, given suitable conditions of strong sunlight, may kill plants within a day. Other pesticides may take a few days or weeks to kill their target organism, but may slow down or prevent further growth until death. Of course those involved in controlling unwanted organisms welcome this rapid action.

It is important that human beings do not experience similar short-term or acute effects from pesticidal molecules. The basis of selectivity of pesticides outlined in Chapters 3, 4 and 5 of this book and precautions taken during use of many compounds explain how their acute effects are encountered mainly by target species. However, pesticides such as the bipyridylium herbicides and some insecticides that block acetylcholinesterase activity have caused severe illness and death to human beings. The benefits of pesticides to human and animal health, crop production and storage, and in treatment of domestic and industrial premises must be set against the three million cases estimated by the World Health Organization (Harris, 2000) of acute, often life-threatening human poisonings and 100 000-plus deaths annually linked to these compounds. Most of these deaths occur in poorer countries, where inappropriate use of pesticides of high acute toxicity by poorly trained personnel has led to instances of acute poisoning and death. Deaths due to occupational exposure are far less than those due to suicides in poorer countries.

Whereas herbicides are the principal pesticides employed in the EU and North America, the much higher incidence of insect pests of crops and human beings in tropical and subtropical countries means that insecticides are often the predominantly used pesticides in these areas (Figure 1.8) and fatalities have been particularly linked to the use of organophosphate and carbamate insecticides. Also, price constraints mean that poorer countries tend to use inexpensive products and in this respect the wide use and ready availability of the bipyridilium herbicide paraquat has been frequently linked to fatalities in the developing world. Paraquat and organophosphate insecticides are the principal pesticides associated with suicide and suicide attempts in poorer countries (Eddlestone, 2000).

In addition to acute effects, much attention is now paid to the potential adverse consequences for human health resulting from long-term exposure to pesticides. In fact much of the development costs of a new pesticide are associated with investigations of possible long-term adverse health effects. Advances in analytical chemistry now mean that pesticides can be detected in very low concentrations in water and foodstuffs. The fact that pesticides occur at all has led to considerable concern among sectors of the population, particularly in affluent areas of the world where some pressure groups have been active in promoting views that residues of pesticides may present significant risks to human health no matter what the dose received.

Evaluation of risks to human health from pesticides may be judged through the principles of dose–response and exposure in exactly the same way as the effects of compounds on target organisms. The principal difference in studies on human health compared to those with most other non-target species is that direct testing is not generally an option. Toxicity data derived from test species such as rats are commonly used to judge whether adverse effects may result to the health of human beings, although some studies with human volunteers have been carried out. Epidemiological studies on the influence of pesticides on human health have been undertaken and through analysis of tissues from suicide attempts, doses of some pesticides likely to cause acute poisoning in human beings have been established. However, the direct association of longer-term ailments such as cancer with exposure to pesticides has proved rather difficult to establish due to the multiplicity of environmental influences to which human beings are exposed, as well as genetic predisposition within individuals in the population to some long-term illnesses.

Within the population as a whole, some individuals are more likely to experience greater exposure to pesticides than others (Ecobichon, 1991). Those most at risk may be workers in pesticide-manufacturing plants, distributors of pesticide products, as well as farmers and growers who handle products in the concentrated form. At a second level, in addition to those who spray pesticides, members of the public may encounter diluted compounds in spray drift or when used, in the home, workplace, on farm stock and domestic pets. Finally the population at large may ingest pesticides as residues on food, clothing or

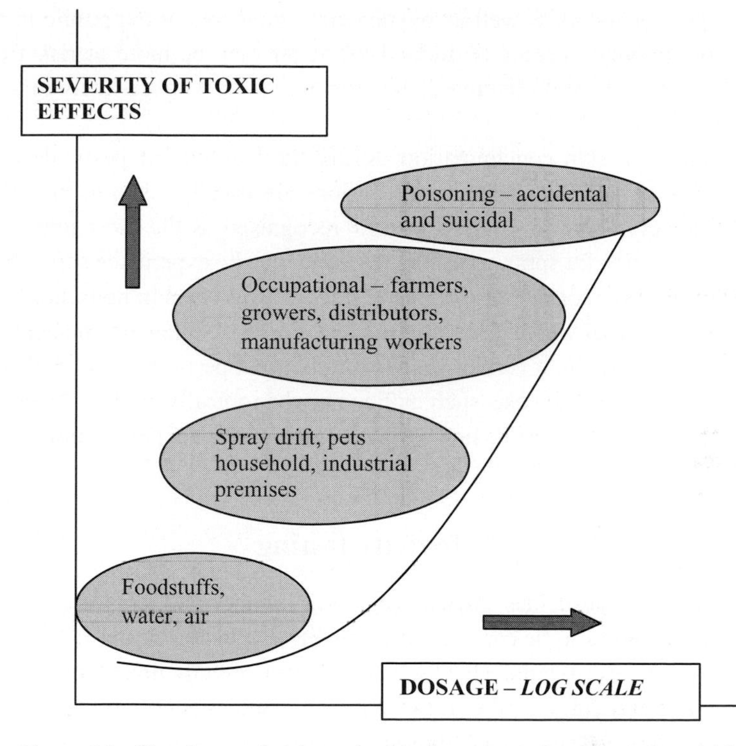


Figure 6.1 Situations and risks to health from the toxic effects of pesticides. (Adapted from Ecobichon, D. J. (1991). Toxic effects of pesticides. In: *Casarett* and *Doull's Toxicology – The Basic Science of Poisons*. 4th edn Amdur, M. O., Doull, J. and Klaasen, C. D. (eds.) (Oxford: Pergamon).

in water. The potential for exposure and relative dose encountered in these situations is broadly depicted in Figure 6.1.

The principal routes of entry of pesticides into the human body are by ingestion, inhalation and skin absorption. Marked symptoms of acute poisoning in human beings are generally associated with ingestion of pesticides in their concentrated form, usually as part of suicide attempts, especially in poorer countries of the world. However, it is pertinent to note that human beings may deliberately ingest compounds closely related to pesticides, for example, to eliminate internal fungal infections. At the other extreme, very low concentrations of pesticides may be ingested involuntarily by a large part of the global population in the form of residues in foodstuffs and drinking water.

Inhalation may occur during field spray application and within enclosed areas where pesticides are used, including not only glasshouses but also domestic and commercial premises where wood, carpets and even pets may be treated. Spray drift as noted in Chapter 8 may affect non-target organisms and these of course include spray operators as well as bystanders – members of the public in the vicinity of spraying operations. Indeed the latter may be more at risk from inhalation and skin contamination than spray operators, who are likely to wear protective clothing.

The danger of skin contamination during the handling of pesticide concentrates is obvious, and stringent precautions are usually taken to minimise contact. However, skin absorption is now recognised as the most important route of penetration for spray operators, bystanders who experience spray drift, and, albeit in lower concentrations, for residents or workers in household and commercial premises where pesticides may be absorbed following contact with treated surfaces. Again, it is relevant to point out that pesticides for control of dermatophytic fungi and pests such as lice may be topically applied to human skin and this inevitably results in a certain degree of dermal penetration.

## 6.2 Toxicity testing

Before approving pesticides for use, regulatory authorities require extensive toxicological studies to be carried out to enable evaluation of the risks that a compound may pose to human health, and these are directly linked to routes of exposure. For pesticides, studies may be summarised as acute, short-term and long-term, although there is inevitably some overlap between the categories. The toxicological tests that must be conducted for approval of a pesticide in the EU are outlined in Table 6.1 (Pesticides Safety Directorate, 2003b), and these and some of the supplementary studies that may be required are considered further in sections 6.2.1 to 6.2.3.

Medical data relating to symptoms of poisoning, and antidotes to cope with acute overdosing, must be submitted as part of registration packages. For example, first-aid measures to be used in the event of accidental exposure to concentrated form of the pesticide must be developed, particularly for eye contamination. Overall surveys of the health of personnel in manufacturing plants may be required, including details of frequency, level and duration of exposure to test compounds and any symptoms associated with such exposure.

### 6.2.1 Acute toxicity

The intrinsic acute toxicity of pesticides is of particular importance to manufacturers, distributors and spray operators who may be using or handling the pesticide in its concentrated form. The acute toxicity of a pesticide is defined as the adverse effects that occur within a short time, usually 24 h, of swallowing, inhalation or contact with eyes or skin. Test animals including rats, mice,

Table 6.1 *Toxicological studies required for pesticide approval in the European Union*

| | |
|---|---|
| **1. Acute toxicity** | |
| Acute oral – in rat | Skin irritation- usually with rabbit |
| Acute percutaneous- with rat | Eye irritation – usually with rabbit |
| Inhalation toxicity – usually with rat | Skin sensitisation – usually with guinea pig |
| **2. Short-term toxicity** | |
| 90-day dietary studies in rat or mouse | |
| 90-day studies in dog | |
| **3. Genotoxicity (mutagenicity)** | |
| i) *In vitro* studies | Bacterial assay with *Salmonella typhimurium* |
| | Gene mutation in mammalian cells |
| | Clastogenicity in mammalian cells |
| ii) *In vivo* tests | Needed *if in-vitro* tests prove positive |
| | Mammalian bone marrow cytogenetic test *or* mouse micronucleus test |
| **4. Long-term toxicity and carcinogenicity** | |
| Oral feeding to rats over 2 years | |
| **5. Reproductive toxicity** | |
| i) Multigeneration studies | Rats over two generations |
| ii) Developmental toxicity studies | Rats and rabbits |

From the UK Pesticides Safety Directorate at http://www.psd.gov.

rabbits and guinea pigs are used under carefully defined experimental conditions in studies of acute toxicity to mammals (Brown, 1980).

Acute toxicity is usually expressed as the $LD_{50}$ of a substance with respect to consumption by rats. Varying doses of the test chemical are administered orally or percutaneously (beneath the skin) to batches of test animals and the numbers that die in each batch are recorded. The $LD_{50}$ ('half the lethal dose') is the dose that is expected to kill half of the test animals in a batch. Other acute toxicity studies include investigations of skin and eye irritancy, and these are usually carried out using rabbits and guinea pigs.

It is important to note that the lower the $LD_{50}$ (i.e. the smaller the dose required to kill half of a population of test animals), the more toxic is the compound. In general, most herbicides, plant growth regulators and fungicides have high $LD_{50}$s, and thus low acute toxicity, but there are a few exceptions (Table 6.2). Paraquat, diquat and the tin-based fungicides all have relatively

Table 6.2 *Acute toxicity of pesticides expressed as $LD_{50}$s*

| Pesticide | $LD_{50}$ (mg/kg body weight – oral rat) | $LD_{50}$ (mg/kg body weight – percutaneous rat or rabbit*) | WHO toxicity class |
|---|---|---|---|
| **Herbicides** | | | |
| DNOC | 25–40 | 200–600 | Ib |
| 2,4-D | 639–764 | >1600 | II |
| Atrazine | 1869–3090 | >3100 | III |
| Isoproturon | 1826–2417 | >2000 | III |
| Metsulfuron-methyl | >5000 | >2000 | III |
| Fluazifop-butyl | 2451–3680 | >2000 | III |
| Cycloxydim | >5000 | >2000 | III |
| Carfentrazone-ethyl | >5000 | >4000 | III |
| Paraquat | 157 (30 in man) | 236–500* | II |
| **Fungicides** | | | |
| Copper oxychloride | 700–800 | >2000 | III |
| Fentin hydroxide | 110–171 | 1600 | II |
| Maneb | >5000 | >5000 | III |
| Benomyl | >5000 | >5000* | III |
| Tebuconazole | 1700–4000 | >5000 | III |
| Metalaxyl | 633 | >3100 | III |
| Azoxystrobin | >5000 | >2000 | III |
| **Insecticides** | | | |
| DDT | 113–118 | 2510 | II |
| HCH | 88–270 | 900–1000 | II |
| Dieldrin | 24–167 | 50–120 | II |
| Chlorpyrifos | 135–163 | >2000 | II |
| Malathion | 1375–2800 | 4100 | III |
| Aldicarb | 0.93 | 20 | Ia |
| Pyrethrins | 1030–2370 | >1500 | II |
| Cypermethrin | 250–4156 | >4900 | II |
| Lambda-cyhalothrin | 56–79 | 1293–1507 | II |
| Nicotine | 50–60 | 50 | Ib |
| Imidacloprid | 450 | >5000 | II |
| Thiamethoxam | 1563 | >2000 | III |
| Diflubenzuron | 4640 | >10000 | III |
| Rotenone | 132–1500 | | II |
| **Others** | | | |
| Metaldehyde | 283 | 5000 | |
| Warfarin | 186 | | Ib |
| Brodifacoum | 0.27 | 0.25–0.63 | Ia |
| Diphacinone | 2.3 | <200 | Ia |

Variations in figures are due to the procedures used to introduce the compounds to test animals as well as differences between sexes.

low $LD_{50}$s compared to those of other herbicides and fungicides and this is linked to their mode of action (2.3.2 and 3.2) – and no antidotes are available for the bipyridilium herbicides if they are ingested in high doses. Some older compounds such as the herbicides dinitroorthocresol (DNOC) and dinoseb, which posed an acute health risk due to their interference with respiratory activity in non-target species including human beings, have now been withdrawn from the market.

In contrast, a considerable number of insecticides are acutely toxic to mammals at low dose. Most insecticides act by inhibiting transmission of nerve impulses and this neurotoxicity may also be evident in non-target species that come into contact with compounds such as the organophosphate and carbamate insecticides. Several have low mammalian $LD_{50}$s (Table 6.2); in some cases such as aldicarb, methomyl and parathion very low $LD_{50}$s indeed. This high acute toxicity and their subacute toxic effects (7.3) have led to the phasing out of many of these compounds in the EU and elsewhere. Other neurotoxic insecticides such as the pyrethroids and neonicotinoids have a relatively low acute toxicity to mammals, as do most non-neurotoxic compounds used for pest control. Rodenticides are targetted at mammalian species and therefore their acute toxicity is, as may be expected, high. However, their toxic effects can be reversed (5.3).

Recognition of the potential toxicity of pesticide concentrates has led to classification systems being adopted by authorities such as the World Health Organization (WHO) and the EU with respect to the risks involved in handling concentrates, and labels that must be appended to products to reflect this. The WHO classification of pesticides by hazard is given in Table 6.3, and included for pesticides listed in Table 6.2. The EU scheme is outlined in Table 6.4 along with attendant hazard symbols.

### 6.2.2 Short-term toxicity

These studies are usually carried out to give an initial indication of possible long-term problems that may be due to bioaccumulation of test compounds, and also serve to help with the design of long-term toxicity studies. Short-term toxicity studies may include oral feeding of the compound for 28 days to rats, but, under EU regulations in place in 2001, a 90-day feeding study of both rats and dogs is necessary. In these tests, a dose–response relationship is determined, as are characteristics of any poisoning symptoms, behavioural changes, and organ- or tissue-specific pathological effects. Initial indications of the no-observed adverse effect level (NOAEL – 6.2.3) may be found within short-term toxicity studies.

Table 6.3 *The World Health Organization (WHO) classification of pesticides by hazard*

| | $LD_{50}$ rat (oral) in mg/kg body weight | | | |
|---|---|---|---|---|
| | Oral | | Dermal | |
| | Solids | Liquids | Solids | Liquids |
| 1a Extremely hazardous | 5 | <20 | <10 | <40 |
| 1b Highly hazardous | 5–50 | 20–200 | 10–100 | 40–400 |
| II Moderately hazardous | 50–500 | 200–2000 | 100–1000 | 400–4000 |
| III Slightly hazardous | >500 | >2000 | >1000 | >4000 |

From WHO Classification of Pesticides by Hazard and Guidelines to Classification 2000–2002 at http://www.who.int/pcs/docs/Classif_Pestic_2000-02.pdf.

Table 6.4 *The EU labelling scheme for pesticide toxicity*

| Classification and warning symbol | Oral | | Dermal | |
|---|---|---|---|---|
| | Solids | Liquids | Solids | Liquids |
| Very toxic | 5 or less | 25 or less | 10 or less | 50 or less |
| Toxic | 5–50 | 25–200 | 10–100 | 50–400 |
| Harmful | 50–500 | 200–2000 | 100–1000 | 400–4000 |

Figures refer to $LD_{50}$ values (rat) of compounds expressed as milligrams per kilogram of body weight

### 6.2.3 Chronic toxicity

The potential long-term adverse effects of pesticides are of importance not only to manufacturers, suppliers and spray operators, but also to consumers who may eat food or drink water containing residues of pesticides. Residues may remain on fresh produce after harvest, be present in processed foods derived from pesticide-treated crops or even in meat and dairy products from animals that have been fed with pesticide-treated feedstock. Furthermore, residues may enter drinking water supplies by run off from land to which pesticides have been applied. Indeed, most of the world's population is likely to be exposed to pesticides at some time during their life through ingestion of small, in some cases minute, concentrations of pesticides in food and drinking water. The fact that pesticides can now be detected at extremely low concentrations in food, water (and the environment) is a tribute to the art of analytical chemistry. It is the significance of these doses to public health that needs to be considered.

New pesticidal molecules now undergo exhaustive studies for potential long-term adverse effects, and those tests that are compulsory under EU law are given in Table 6.1. Studies are carried out with test animals that are fed or exposed to the compound for long periods, sometimes the entire life of the animal. The health of these test animals is monitored continuously and analysis of organs and tissues carried out post mortem. Livestock feeding studies are also undertaken, usually with cows and chickens, but occasionally with pigs, to establish the potential of residues to enter the human food chain.

A major aim of these studies of chronic toxicity is to establish that the pesticide presents no increased risk of cancer to human beings. Reproductive studies aim to identify any adverse effects on fertility and pregnancy. Allied to these are developmental toxicity (teratological) investigations to determine whether any abnormalities occur in the foetuses of test animals.

Not all chronic toxicity studies involve the use of test animals. Bacteria of the genus *Salmonella* are used to evaluate potential mutagenic effects. Changes in the DNA structure of the bacterium may occur following exposure to test chemical compounds, and thus the mutagenic potential of pesticides and other chemicals may be assessed. The use of mammalian cell cultures in toxicity testing is now routine, and the genotoxic effects of test compounds may be established by monitoring chromosome aberrations in cell cultures. Unscheduled DNA synthesis may be linked to both carcinogenesis and mutagenesis and the influence of pesticidal molecules on DNA synthesis in cultures of rat liver cells is part of modern toxicological studies.

During the mandatory tests for chronic toxicity, other information may be gathered that may assist in evaluation of long-term health risks from the test compound. Clinical observations of skin, eyes, fur, mucous membranes and effects on the respiratory, nervous and circulatory systems frequently form an integral part of chronic toxicity studies. Behavioural patterns of test animals may be monitored, including any changes that may occur in posture or gait as well as signs of hyperactivity or lethargy. Effects on the autonomic nervous system may be assessed by observing salivation, lacrimation, urination and defecation. Many regulatory authorities may require compounds to be tested for potential medium- to long-term adverse effects on the nervous system: investigations of this delayed neurotoxicity following acute exposure are required as part of registration studies in the EU. Finally, the potential of pesticides to adversely affect the mammalian immune system forms part of many current toxicological studies, and here again cell cultures such as those of mouse bone marrow are employed.

From long-term studies, the acceptable daily intake (ADI) of a pesticide is established as a primary safety standard for human beings. The ADI is considered to be the amount of pesticide residue that can be consumed every day during the life of an individual with the practical certainty that no harm will result. ADIs are set by regulatory authorities and not by the manufacturer, although toxicological data provided by the latter are often used in establishing the ADI.

ADIs for modern pesticides are largely based on information provided by studies carried out for the potential chronic effects outlined above. The dose at which no adverse effects on the health of test animals is seen is referred to as the no observed effect level (NOEL); or more commonly as the no observed adverse effect level (NOAEL). The ADIs for pesticides are generally set 100 times lower than the NOAEL, but in some cases they may be set at up to 1000 times lower. ADIs for selected pesticides, as well as maximum residue limits (MRL) values for some crop –pesticide combinations are given in Table 6.5.

Maximum residue limits (MRL values; tolerances in the USA) have been set for many crops and crop commodities, and are the maximum permitted levels of pesticide that should remain on the crop when products are used according to good agricultural practice (GAP). The setting of MRLs has not in the past been directly related to ADIs, but they are generally set such that the ADI, even with excessive consumption of food, is not exceeded. The Codex Committee on Pesticide Residues (CCPR) of the Codex Alimentarius Commission is administered by the Joint Meeting on Pesticide Residues (JMPR), a joint body of the Food and Agricultural Organization of the United Nations and the World Health Organization established in 1968, and has worked for many years to develop internationally agreed maximum limits for pesticide residues in food and feed commodities moving in international trade (FAO/WHO, 1999).

### 6.2.4 Estimates of dietary intakes of pesticides

Regulatory authorities have generally set the ADIs of pesticides at less than 1 mg/kg of body weight, but dietary intakes are generally much lower than this (7.4). Theoretical estimates of the amount of residues absorbed from dietary sources have been made, and the results of one such study for a herbicide approved for use in the UK in the late 1980s are given in Table 6.6.

Even with extreme intakes such as the theoretical consumption of as much as 65 g of peas and 63 g of Brussels sprouts every day, the total intake of residues

Table 6.5 *Parameters associated with long-term exposure to pesticides*

| Pesticide | ADI (mg/kg) | Acute RfD (mg/kg) | MRL as (mg/kg) produce for the crop listed |
|---|---|---|---|
| **Herbicides** | | | |
| 2,4-D | 0.05 | | Wheat – 0.05 |
| Atrazine | 0.005 | | For all approved crops – 0.1 |
| Isoproturon | 0.006 | | Wheat – 0.05 |
| Fluazifop | 0.005 | | Potatoes – 0.1 |
| Metsulfuron-methyl | 0.22 | | Wheat – 0.05 |
| Cinidon-ethyl | 0.03 | | Wheat – 0.1 |
| Glyphosate | 0.3 | | Wheat – 5 |
| Paraquat | 0.004 | 0.004 | Potatoes – 0.05 |
| **Fungicides** | | | |
| Fentin hydroxide | 0.0005 | | Potatoes – 0.05 |
| Maneb | 0.03 | | Potatoes – 0.05 |
| Chlorothalonil | 0.03 | | Wheat – 0.1 |
| Carbendazim | 0.03 | 0.06 | Wheat – 0.1 |
| Penconazole | 0.03 | 0.03 | Apples – 0.2 |
| Metalaxyl | 0.08 | | Potatoes – 0.2 |
| Azoxystrobin | 0.1 | | Wheat – 0.3 |
| **Insecticides** | | | |
| DDT | 0.02 | | Wheat – 0.05 |
| HCH | 0.001 | 0.06 | Wheat – 0.01 |
| Endosulfan | 0.006 | 0.02 | Wheat – 0.05 |
| Chlorpyrifos | 0.01 | 0.1 | Wheat – 0.05 |
| Malathion | 0.3 | 2 | Wheat – 8 |
| Pirimiphos-methyl | 0.03 | | Wheat – 5 |
| Aldicarb | 0.003 | 0.001 | Potatoes – 0.5 |
| Pyrethrins | 0.04 | 0.2 | Wheat – 3 |
| Cypermethrin | 0.05 | | Wheat – 0.05 |
| Lambda-cyhalothrin | 0.005 | 0.007 5 | Wheat – 0.02 |
| Fipronil | 0.0002 | 0.003 | |
| Imidacloprid | 0.06 | 0.4 | |
| Tebufenozide | 0.02 | 0.05 | |
| Diflubenzuron | 0.02 | | Mushrooms – 0.1 |
| **Others** | | | |
| Chlormequat | 0.05 | 0.05 | Wheat – 2 |
| Brodifacoum | 0.02 | | |
| Methyl bromide | 0.1 | | Lettuce – 0.05 |

Acceptable Daily Intakes (ADI) and Acute Reference Doses (ARfD's) are expressed as milligrams per kilogram of body weight and Maximum Residue Limits (MRLs) as milligrams per kilogram of produce. Derivation of Acute Reference Doses is an ongoing process. From The World Health Organisation at http://www.who.int/pcs/jmpr/sum00.pdf and the UK Pesticides Safety Directorate at http://www.pesticides.gov.uk/legislation/MRL_legislation/MRLs.xls.

Table 6.6 *Theoretical maximum daily intakes of cycloxydim*

| Commodity | Extreme (97th percentile) intake (g/day) | Highest likely residue after application (mg/kg) | Theoretical maximum daily intake, based on a consumer weighing 60 kg | | |
|---|---|---|---|---|---|
| | | | mg | mg/kg (body mass) | %ADI |
| Potatoes | 216 | 2 | 0.4 | 0.007 | 10 |
| Peas | 65 | 5 | 0.3 | 0.005 | 8 |
| Swede | 52 | 0.5 | 0.03 | 0.0004 | 0.6 |
| Brussels sprouts | 63 | 1 | 0.06 | 0.001 | 2 |
| Cabbage | 69 | 1 | 0.07 | 0.001 | 2 |
| Cauliflower | 72 | 1 | 0.07 | 0.001 | 2 |
| Rapeseed oil | 13 | <0.05 | 0.0006 | 0.00001 | 0.02 |
| Sugar | 125 | <0.05 | 0.006 | 0.0001 | 0.1 |
| Water | (2 litres) | <0.0001 | <0.0002 | 0.00003 | 0.004 |

Adapted with permission from Hignett (1991). Cycloxydim (2.5) is used to control monocotyledenous weeds in dicotyledenous crops, and hence the principal sources of residues are the latter.

of cycloxydim on produce listed in Table 6.6 is 0.7 mg per day, or expressed on a body weight basis for a consumer weighing 60 kg, 0.01 mg/kg body weight per day. The acceptable daily intake for cycloxydim is set at 0.07 mg/kg body weight per day, and thus the theoretical maximum daily intake of 0.01 mg/kg body weight per day is 14% of the ADI.

Calculations of theoretical maximum daily intakes (TMDIs) are unrealistic overestimations of actual intake, since they assume a high level of produce consumption, that all produce is treated with the pesticide in question, that residues occur at all times at the MRL, and that no losses of residues occur during processing and cooking of food. Indeed, the data in Table 6.6 may be compared with the actual situation described in Chapter 7 (7.4).

TMDIs are steadily being replaced by more realistic intake calculations (FAO, 1997), notably the international/national estimated daily intakes (IEDIs/NEDIs). These are calculated as follows

$$\text{IEDI/NEDI} = \frac{\Sigma\ \text{STMR}_\text{i} \times P_i \times F_i}{\text{Body mass}}$$

where $\text{STMR}_\text{i}$ is the supervised trials mean residue for a given food commodity; $P_\text{i}$ is the processing factor for that food commodity; and $F_\text{i}$ is the food consumption for that commodity.

The most important of the factors is the proposed adoption of STMRs instead of MRLs in predicting dietary intakes since it more accurately reflects residue levels during use. However, even NEDIs and IEDIs may be overestimations since they assume that all produce is treated, and assume (as did TMDIs) a worst-case scenario of food consumption, based on, in the case of NEDIs in the UK, consumers in the 97.5th percentile (i.e. the top 2.5% of consumers of any given food). Such consumption is unlikely to be continuous throughout the lifetime of an individual.

Within the EU and elsewhere, population subgroups such as young children and vegetarians are now considered in estimations of dietary intake of pesticides. In the UK the NEDI is expressed in terms of milligrams of pesticide per kilogram of body weight per day, with separate calculations for adults (16+ years) at 70.1 kg, schoolchildren (10–15 years) of 43.6 kg, toddlers (1.5+ years) of 14.5 kg and infants (6–12 months) at 7.5 kg. From July 2002 in the EU no infant or baby food could be sold with any pesticide residue concentration in excess of 0.01 mg/kg body weight. As seen in Table 6.5, some pesticides have an ADI of less than 0.01 mg/kg body weight and the EU has proposed that such pesticides should not be used on agricultural products intended for production of infant foods.

Further refinements in the toxicological evaluation of pesticide residues include changes in the sampling and methodology associated with analysis of residues. The variability of residues in so-called commodity units, i.e. in individual fruit, vegetables, was held to be underestimated by current sampling and extraction techniques, which involved pooling samples together prior to analysis. Consumers tend to eat produce such as apples, oranges and other fruit and vegetables singly. However, MRLs are based on composite samples where a number of commodity units are bulked and analysis is carried out on the mixed sample. Studies in the UK have shown that the degree of variation between individual commodity units within these composite samples is far greater than previously assumed (Harris and Hill, 2004). For example, a study in the 1990s revealed 30-fold differences between composite samples and individual carrots from these (Figure 6.2).

From these and other studies, the WHO and many national bodies agreed that the estimation of dietary intake should be calculated using methods appropriate to the food consumed. Where the commodity is well mixed during processing (e.g. cereals) or during consumption (cherries, grapes), available composite data are held to reflect the residue levels in the food as a whole. Where food is largely consumed as individual units (e.g. apple, jacket potatoes) the variation in residue levels between these individual units must be considered in calculations of daily intakes of pesticides. In order to take account of residues on individual

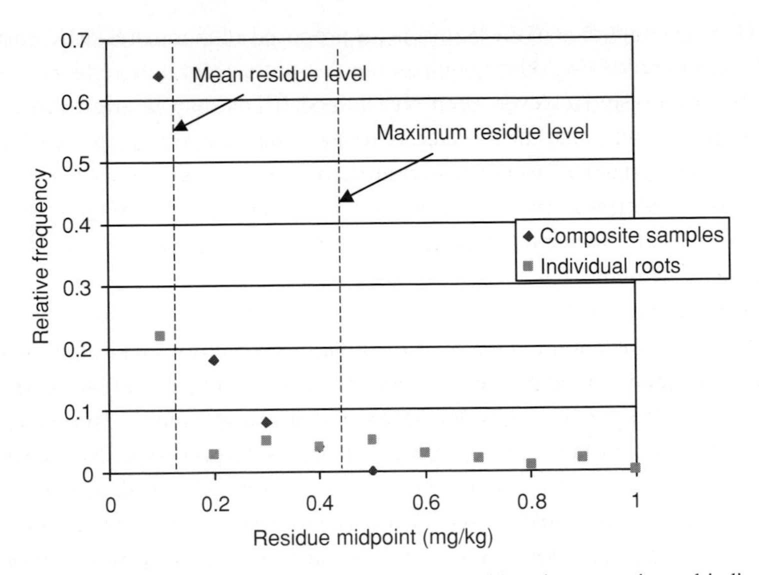


Figure 6.2 The distribution of chlorfenvinphos residues in composite and individual roots of carrots in the UK. (Adapted with permission from Crossley, S. J. 1997. Recent international activities in the methodology for dietary risk assessment. Proceedings of the IBC UK Conference, Café Royal, London, 'Pesticide residues and dietary risk assessment'.) Chlorfenvinphos is no longer approved for use on carrots in the UK.

fruit and vegetables, where occasional exposure to doses higher than the MRL may occur, the acute reference dose (acute RfD), effectively a 'short-term' ADI, has been proposed. The acute RfD is derived from short-term feeding studies to derive a no observed adverse effect level with test animals, followed as with establishment of ADIs by the application of a 100-fold safety factor. The acute RfD thus derived is the concentration of pesticide residue on a single item of fruit, vegetable or other foodstuff eaten at a single sitting which does not present an unacceptable risk to the consumer. Acute RfDs have been developed for a limited number of active ingredients and some are noted in Table 6.5.

### 6.2.5 Harmonisation of toxicity testing

The toxicity data required by regulatory authorities prior to approval of pesticides vary with different countries. For example, there are clear differences in the requirements of regulatory authorities in the UK and USA for testing the neurotoxic effects of pesticidal molecules. The Environmental Protection Agency in the USA requires a comprehensive series of neurotoxic tests. The approach in the UK is to carefully monitor the health of test animals in acute and chronic toxicity studies for any adverse neurotoxic symptoms.

The Organization for Economic Cooperation and Development (OECD) has been active since the early 1980s in coordinating procedures of toxicity testing among its member nations. Under a mutual acceptance of data (MAD) agreement, OECD countries have agreed to accept toxicity data generated in any member nation, provided the tests have been carried out to specific OECD guidelines. This harmonisation has reduced duplication of testing and thus the numbers of animals used in toxicological studies.

However, a lack of international agreement still exists on the number of tests necessary before approval of a pesticide is granted. As noted in Chapter 1, the EU has drawn up a set of Uniform Principles in an attempt to harmonise the overall procedures involved in approval of pesticides. This includes all aspects of testing such as field efficacy, environmental effects, effects on non-target species, residues as well as toxicity testing. Directive 91/414 of the EU has defined specific data requirements for toxicity testing and many of these are identical to OECD test procedures. The directive enables companies that have obtained approval/registration of a product following testing in one member state (the rapporteur state) to use the pesticide in other EU countries. A member state cannot be allowed to revoke authorisation of use on toxicological grounds, but only if it can be shown that local agricultural, plant health, climatic or environmental conditions are not compatible with the country where the product was registered. Harmonisation of MRLs for pesticides is expected to take place in the EU during 2004–2006.

## 6.3 The fate of pesticides in mammals

Studies of the fate of pesticides in mammals are often carried out at the same time as chronic toxicity studies, and serve to identify not only the mechanisms of toxicity of the pesticide but also any problems to health arising from compounds that may be formed due to metabolism. Many pesticides that enter the body at doses below the $LD_{50}$ value are transformed by metabolism into compounds that are normally less toxic and more readily excreted than the parent molecule. However, in a few cases the pesticide may be biotransformed to a more toxic substance, the outstanding example being the transformation in mammals, including human beings, of the organophosphorus insecticide parathion to the more toxic paraoxon.

Investigations of absorption, distribution in tissues, metabolism and excretion (ADME) form the basis of toxicological studies for drugs, antibiotics, food preservatives, flavourings as well as pesticides. Indeed, the metabolism of these so-called xenobiotics is much better known than that of most naturally occurring substances. ADME studies include determination of the rate and extent of absorption by different routes and at different doses, movement of the compound into body organs and tissues, its metabolism or accumulation there, and finally the mode and rate of excretion. With pesticides, investigations of bioaccumulation may be additionally required if the compound has a long half-life in test species, in order to identify potential food-chain problems. Most studies are carried out using radiolabelled pesticides, which enable the compound and its metabolites to be traced in the tissues, organs and excretory products of test animals.

Studies of pesticide metabolism are carried out not only with test species such as rats, but also with animals likely to enter the food chain such as cattle, pigs, sheep and chickens. Livestock may feed on pasture or compounded foodstuffs that may contain residues, and indeed may be treated directly with pesticides to protect them from or cure fungal or parasitic infections. Consequently, risks that may occur to human health from consumption of meat, eggs, cheese and milk need to be assessed.

Exaggerated doses are used in metabolism studies, but there is a need to recognise that at high dose levels saturation of processes such as absorption and metabolism may occur. In practice, such saturation may happen following exposure to pesticide concentrates, but is not likely to occur from residue concentrations in food and water. In oral studies, a dose of around 10 mg of compound per kilogram of body weight of the test animal is often used for metabolism studies. This is considerably higher than dietary doses (Chapter 7), but allows the detection and characterisation of metabolites.

Extrapolation from results obtained in toxicity tests with animal species to establish safety margins for exposure of human beings has been criticised on the grounds that the metabolism of chemicals may differ in different species. Differences do occur in some cases, but metabolic routes leading to in most cases detoxication and/or excretion do appear to share common features in many species. Indeed the degradation of a pesticide in plants, its breakdown in soil by microorganisms, and detoxication of the compound in animal tissues may be accomplished in some cases by identical mechanisms.

Metabolism of pesticides in human beings has been investigated. A considerable body of data was gathered in the 1950s, 1960s and 1970s from studies with volunteers, but, in view of ethical considerations, such experimentation became unpopular, and in some countries prohibited by law. Sadly, some data are available for pesticide metabolism in human beings from studies of suicide and attempted suicide.

The metabolism of foreign chemicals of either natural or synthetic origin is biphasic (Millburn, 1995; Figure 6.3). Most pesticides are not very

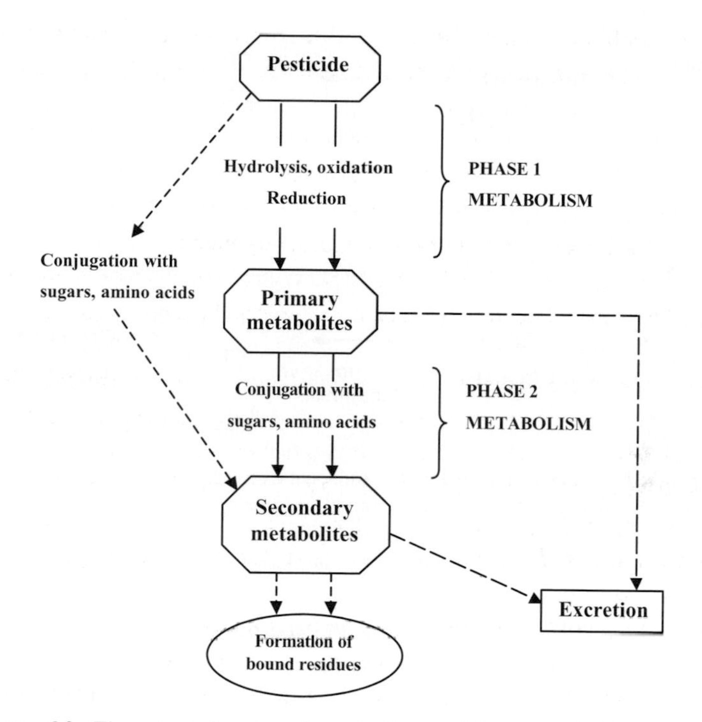


Figure 6.3 The principal routes of metabolism and fate of pesticides in living systems.

water-soluble, and the primary or phase 1 reactions in metabolism serve to render substances more reactive and water-soluble. This reactivity allows the conjugation reactions of phase 2 metabolism, where the metabolites produced in phase 1 form complexes with common cellular constituents such as simple sugars or amino acids. These complexes may be readily excreted in urine or bile. Some pesticides may be directly conjugated, some such as atrazine excreted after primary metabolism and a few such as the (highly water-soluble) hormone herbicide 2,4-dichlorophenoxyacetic acid (2,4-D) may be excreted unchanged.

Hydrolases and oxygenases are the predominant enzymes involved in the primary metabolism of pesticides, but other reactions including reduction, dechlorination, hydroxylation, epoxidation and dealkylation may occur. Many of these enzymes have a low substrate specificity, which means they can metabolise a range of pesticidal (and other) molecules.

Of primary interest are those molecules of high mammalian toxicity, mainly insecticides, and many studies of the metabolic fate of these compounds have been undertaken. Many insecticides contain ester, amide and phosphate linkages and these may be hydrolysed. Organophosphate, carbamate and pyrethroid insecticides, dithiocarbamate fungicides, as well as urea and

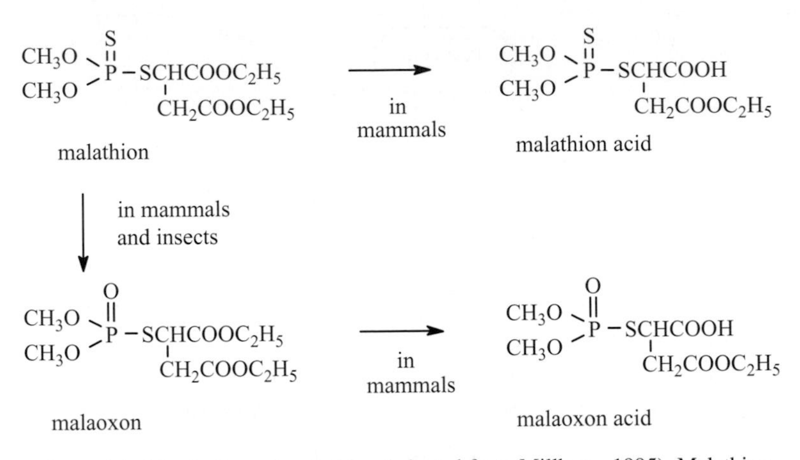


Figure 6.4 Metabolism of malathion (adapted from Millburn, 1995). Malathion is hydrolysed to malathion acid in mammals, and oxidised to malaoxon in insects and mammals, but only the latter readily hydrolyse malaoxon to its acid. Insects do not readily detoxify malaoxon and are thus killed.

carbamate herbicides may all be metabolised by hydrolytic enzymes. These processes of hydrolysis and oxidation usually lead to compounds of lower toxicity in mammals. For example, the organophosphorus compound malathion is rapidly hydrolysed and oxidised in mammals (Figure 6.4). The metabolites produced (malathion and malaoxon monoacids) are of low toxicity and readily excreted.

Oxidation by monooxygenase enzymes, especially those from subcellular microsomes, is of major importance in the detoxication of pesticides. The smooth endoplasmic reticulum in the liver is rich in these enzymes and one of the principal functions of this organ is to detoxify potential toxins of both natural and synthetic origin. The activity of monooxygenase enzymes involves the separation of atoms in an $O_2$ molecule and the placement of one of the atoms in an appropriate substrate. The other atom forms part of a water molecule:

$$R - H + O_2 + [2H] \rightarrow R.OH + H_2O.$$

The activity of microsomal monooxygenases is associated with a haem protein – cytochrome p450 – that couples electron flow to the reduction of the $O_2$ molecule. Microsomal monooxygenases are partially responsible for the rapid detoxication of pyrethroid insecticides (both natural and synthetic) in mammals and this explains the lack of acute toxic effects of these compounds to mammalian species. Many insects (other than those resistant to pyrethroids, which often have enhanced levels of monooxygenase activity) cannot or only slowly detoxify pyrethroids and are thus killed. In fact, the activity of pyrethroid insecticides may be enhanced by inclusion of inhibitors of monooxygenase

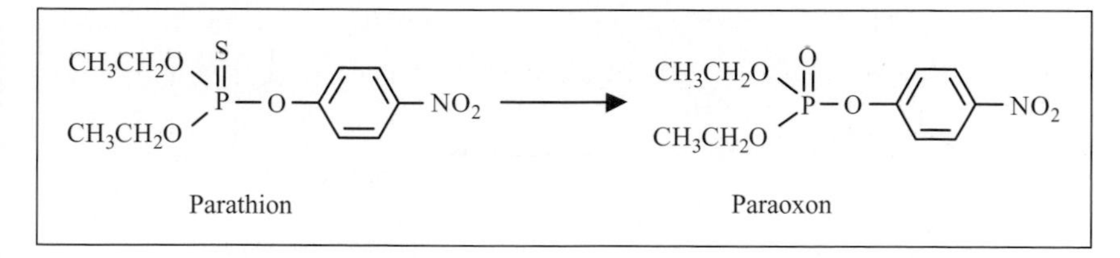


Figure 6.5 Oxidative metabolism of parathion to paraoxon.

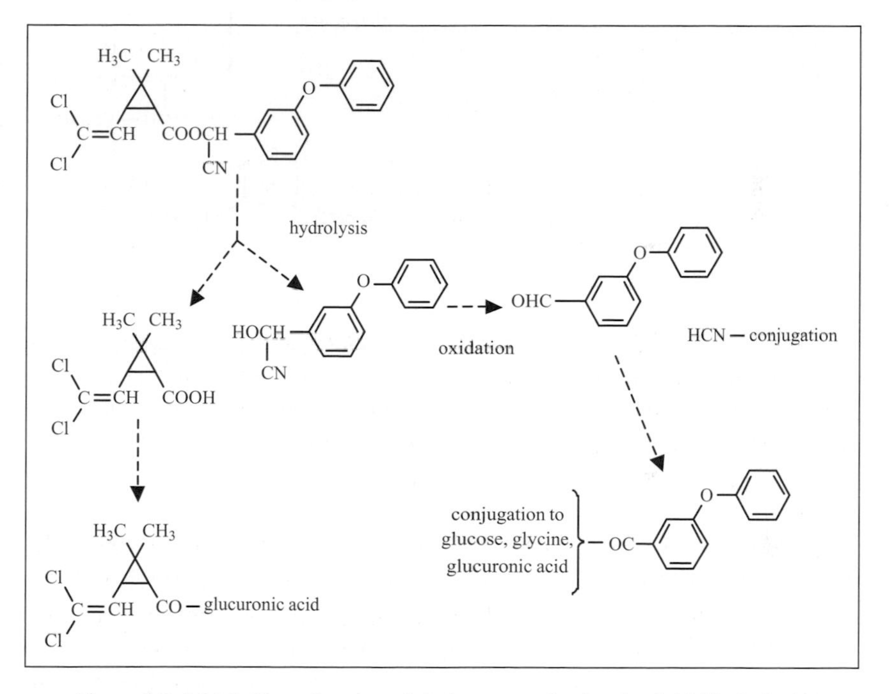


Figure 6.6 Metabolism of cypermethrin in mammals, showing initial hydrolysis, oxidation of the alcohol metabolite with release of cyanide and 3-phenoxybenzoic acid – both subsequently conjugated, as is the acid metabolite (right side of the diagram). Adapted with permission from Milburn (1995).

activity such as piperonyl butoxide. Monooxygenases are also involved in the detoxication of some organophosphorus and carbamate insecticides, but in one case – that of the organophosphate parathion – their activity results in the formation of a metabolite, paraoxon, that is more toxic than the parent molecule (Figure 6.5).

The organophosphoros, carbamate and especially pyrethroid insecticides may be rapidly metabolised in the human body at sublethal doses, but this is not the case with the organochlorine insecticides. The biotransformation and

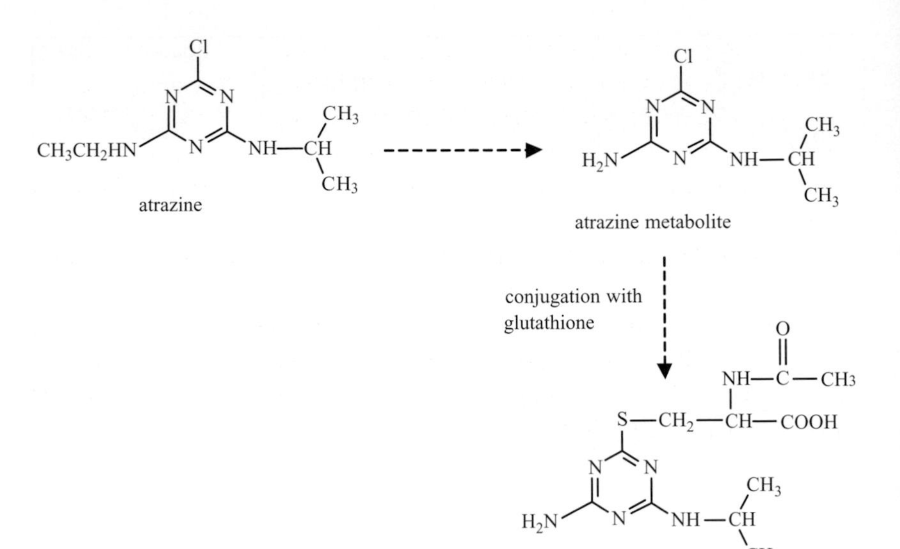


Figure 6.7 The principal route of degradation of atrazine in rat. Adapted with permission from Timchalk, C., Dryzga, M. D., Langvardt, P. W., Kastl, P. E., and Osborne, D. W. (1990). Determination of the effect of tridiphane on the pharmacokinetics of [$^{14}$C]-atrazine following oral administration to male Fischer 344 rats. *Toxicology*, **61**, 27–40. The conjugated metabolite is excreted in the urine.

excretion of these pesticides is slow because of their complex chlorinated ring structure, the same properties which contribute towards their persistence in the environment (Chapter 8). Elimination of DDT from the body occurs very slowly, at roughly 1% of the total quantity stored in the body per day.

Phase 1 metabolism of pesticides usually exposes chemical groups such as hydroxyl, carboxyl, amino and sulphydryl groups that may conjugate with small, often highly water-soluble molecules in the cell. Increased water solubility enhances the prospect of excretion of the molecule. Conjugation may involve sugars such as glucose, where the conjugate produced is a glucoside, and sugar derivatives such as glucuronic acid where the product is a glucuronide (Millburn, 1995). Glucuronides are the main conjugates in mammals and the enzymes responsible for linking glucuronic acid molecules to pesticide metabolites are glucuronyl transferases. Many herbicides, fungicides and insecticides are converted to glucuronic acid conjugates after primary metabolism in the liver. In some cases direct conjugation of a pesticide without prior (phase 1) metabolism may occur. The tripeptide glutathione often acts in this way, being conjugated to pesticidal molecules by glutathione-*S*-transferases.

The step-by-step metabolism of many pesticides is well documented. For example, the pyrethroid insecticide cypermethrin is metabolised initially by hydrolysis to an acid and an alcohol derivative which are both subsequently conjugated, principally to glucuronic acid, and excreted (Figure 6.6). The rapid nature of these reactions means that the acute toxicity of cypermethrin is low in mammals. Atrazine, frequently detected in drinking water supplies in the 1980s and 1990s, is metabolised in mammals by *N*-dealkylation and glutathione conjugation (Figure 6.7) and excreted in urine.

Finally, some pesticides may be excreted unchanged from mammals. 2,4-D is excreted rapidly (in 1–2 days) and unchanged by human beings, as are other hormone herbicides such as (4-chloro-2-methylphenoxy) acetic acid (MCPA) and picloram (Roberts, 1998).

# 7
# Pesticides and human health

## 7.1 Introduction

In the nineteenth century, when pesticides were first used on a large scale in Europe and the USA, compounds of high mammalian toxicity containing for example cyanide and arsenic were widely employed. The use of such materials and organic compounds of high toxicity such as dinitroorthocresol (DNOC) continued well into the twentieth century. Diagrams and photographs from that era show little use of protective clothing or care in handling and storage of pesticides (Figure 7.1), and risks of exposure were undoubtedly very high. It is then not surprising that ill health and fatalities occurred among those involved with the manufacture and application of pesticides. Indeed, deaths of spray operators were linked to the use of compounds such as DNOC as late as the 1950s in the UK, and into the 1970s in some other parts of Europe.

In occupational use, pesticides present a much higher degree of risk to workers and others in developing, often poorer, countries than in developed, wealthier countries and some of the reasons for this are summarised in Table 7.1. Additional risks are posed by the much greater use of insecticides including organophosphates, carbamates (Thompson and Richardson, 2004) and some organochlorines such as endosulfan in warmer climates.

## 7.2 Risks to health from pesticide concentrates

The most pronounced adverse acute effects to human health from pesticides come from exposure to manufactured products since these normally contain high concentrations of the active ingredient. The proportion of active ingredient in pesticide concentrates may be as high as 80% of the product, and exposure to even small doses for a short period of time may result in adverse effects.

Figure 7.1 Handling and spraying of pesticides in the 1940s and 1950s (photographs by kind permission of Mr. T. H. Robinson).

Table 7.1 *Areas of concern leading to problems with pesticides in developing countries*

- High rates of illiteracy linked to complex label instructions
- Labels may be in an unfamiliar language
- Lack of access to clean water for washing
- Eating, drinking and smoking during pesticide use
- Absence of medical facilities including access to antidotes
- Lack of training in pesticides use
- Shortage of technical and cultural controls to minimise hazards
- Cost of protective clothing
- Problems of wearing protective clothing in hot, humid climates
- Mixing of toxic concentrates by hand
- Inappropriate use of pesticides
- Lack of awareness of application rates, re-entry periods and harvest intervals
- Reuse of concentrate containers for food or water storage

Many of these are considered by Dinham (1995) and Dinham (2003)

The potential for the concentrated forms of pesticides to cause acute harm is now well recognised and precautions have been widely implemented to minimise risks of contact, inhalation and ingestion. Consistent improvements in safety procedures and their adoption by manufacturers, distributors and spray operators in developed countries have reduced the potential for pesticides in their concentrated form to adversely affect the health of personnel handling these substances. The regulations in force in the UK appertaining to storage, preparation for use and application of pesticides are among the most stringent in the world. Personnel directly involved in the storage, distribution and sale of pesticides on a commercial scale must possess an appropriate certificate of the British Agrochemical Standards Inspection Scheme (BASIS). The adherence to exposure limits for those involved in manufacturing and formulating and in some cases applying products is mandatory in many countries including the UK, where they are defined as occupational exposure limits (OELs). These in turn are expressed as either maximum exposure limits (MELs) or occupational exposure standards (OESs). The MEL is the maximum concentration of an airborne substance to which an employee may be exposed by inhalation. An OES is defined as the concentration of an airborne substance to which employees may be exposed by inhalation every day of their working life without harm. Similar safety criteria have been developed in other countries; for example, in the USA threshold limit values (TLVs) are the equivalent of OELs in the

Table 7.2 *Occupational exposure standards of some pesticides in the UK*

| Pesticide | Long-term exposure (8 h time-weighted average) | Short-term exposure (15 min) |
|---|---|---|
| *Insecticides* | | |
| Endosulfan | 0.1 | 0.3 |
| Chlorpyrifos | 0.2 | 0.6 |
| Diazinon | 0.1 | 0.3 |
| Nicotine | 0.5 | 1.5 |
| Pyrethrins | 5 | 10 |
| *Fungicides* | | |
| Benomyl | 10 | 15 |
| Captan | 5 | 15 |
| *Herbicides* | | |
| Diuron | 1 | 1 |
| Picloram | 10 | 20 |
| *Rodenticide* | | |
| Warfarin | 0.1 | 0.3 |

The units are milligrams of active ingredient per cubic metre of air.
From Document EH40/2002, UK Health and Safety Executive

UK (Nigg, 1998). In the UK, OESs have been developed for many industrial chemicals including about 50 pesticides and some of these are listed in Table 7.2. In the UK from 2005, MELs and OESs were merged to form a single exposure standard – the Workplace Exposure Limit (WEL).

The introduction in the UK of the Control of Pesticides Regulations (1986) as part of the Food and Environmental Protection Act (1985) addressed the safe use of concentrated forms of pesticides by those involved in their application. Spray operators and others who apply pesticides on a commercial basis must now hold Certificates of Competence in Pesticide Application. The basic certificate is awarded for successful completion of the foundation module, Pesticides Application 1 (PA1), which is concerned with the safe use of pesticides. Those who apply pesticides must comply with the guidelines given in the UK's Code of Practice for the safe use of pesticides on farms and other premises. Those sections of the code relating to human health that form an integral part of PA1 are outlined below.

| STATUTORY CONDITIONS RELATING TO USE | |
|---|---|
| FOR USE ONLY AS AN AGRICULTURAL FUNGICIDE | |
| Crops: | Potatoes |
| Maximum Individual Dose: | 0.463 kg product /ha |
| Maximum Number of Treatments: | 3 per crop |
| Latest Time Of Application: | 7 days before harvest |
| Operator Protection: | Engineering control of operator exposure must be used where reasonably practicable in addition to the following personal protective equipment. See PRECAUTIONS (marked *)- However, engineering controls may replace personal protective equipment if a COSHH assessment shows they provide an equal or higher standard or protection. |
| Environmental Protection: | See PRECAUTIONS (marked †) |
| Other Specific Restrictions: | See PRECAUTIONS (marked 0) |
| READ ALL PRECAUTIONS BEFORE USE | MAFF 07305 |

Figure 7.2 An example of a statutory box on a UK pesticide label (by kind permission of Bayer Crop Science).

### 7.2.1 Compliance with instructions on the product label

A legal obligation exists in the UK to follow label instructions, especially those contained in the so-called statutory box (Figure 7.2). A great deal of information may be given on product labels and in recent years considerable improvements in label design and layout have taken place. Internationally recognised hazard symbols are now used on labels and product information is generally clear and unambiguous.

### 7.2.2 Formulation and container design

The development of containers with wide necks and handles that are easily gripped has reduced the chances of spillage during the addition of pesticide concentrate to spray tanks. Refillable, returnable containers are now available which can be attached to spray tanks and from which pesticide concentrates can be added to the tanks without the need for the operator to open the container. Advances in technology include the development of water-soluble sachets and bags of powder formulations. Provided these are kept dry and handled carefully, there is little chance of contamination during storage or addition to spray tanks.

### 7.2.3 The use of protective clothing

The use of protective clothing lowers the risk of contact with pesticide concentrates during addition to spray tanks or other application equipment. The minimum level of protective clothing that must be worn when handling pesticide

Figure 7.3 Protective clothing and application of pesticides with a hand-held mini-boom sprayer. Photograph by kind permission of Joe Martin, Agrisearch UK.

concentrates in the UK is a coverall, face shield, solvent-resistant gloves and wellington boots (Figure 7.3). Further protective equipment such as solvent-proof aprons, goggles and respirators may be required when using products of high toxicity.

### 7.2.4 The use of appropriate measuring procedures

Risks of contamination by pesticide concentrates are reduced if they are dispensed in a suitable place and in a safe manner. For example, powder formulations of pesticides should be measured out in still conditions, preferably indoors, and the mixing of powders with a little water to form a slurry prior to addition to the spray tank reduces risks from dust arising from the powder. Careful pouring of liquid formulations of pesticides will avoid spillages.

Safe and secure storage of pesticides is now mandatory under the Control of Pesticides Regulations in the UK and this, coupled with the measures listed above, has reduced risks to health of operators, and also to the environment from pesticides in their concentrated form.

Precautions to avoid contact with pesticide concentrates during occupational use form part of recommendations in many other developed countries including the USA, Canada and member states of the EU, and as in the UK these may be reinforced by strict safety laws. The 1992 Worker Protection Standard (WPS) in the USA includes requirements for protection during application, adherence to label instructions, pesticide training and effective provision for decontamination by pesticides. Of particular note in the USAWPS is strong emphasis on the exclusion of workers from areas that have been treated with certain pesticides, particularly insecticides. As with the Food and Environmental Protection Act (1985) in the UK, the United States Environmental Protection Agency (USEPA) through its Worker Protection Standard and The Pest Management Regulatory Agency (PMRA) in Canada have developed a common examination on core principles for all pesticide applicators.

In the light of the precautions described above, risks to the health of personnel involved in the manufacture, distribution and application of pesticides are low in the UK and other developed countries. Deaths due to acute poisoning by pesticides in the UK are rare and in agricultural practice have averaged less than one per year since 1950. Most of these fatalities have occurred from fumigant pesticides in grain stores. In the USA the number of deaths attributed to acute occupational pesticide poisoning has progressively declined over the last 40 years.

Far more deaths have been associated with suicide attempts. For instance the National Poisons Unit in the UK recorded 1080 suicide attempts involving paraquat (with 286 proving fatal) between 1980 and 1985, just prior to the introduction of the Control of Pesticides Regulations. However, data accumulated since then continue to implicate paraquat in suicide attempts with 34 deaths between 1990 and 1993 resulting from ingestion of this compound, notwithstanding the strong emetic added to the formulation by its manufacturers. Ingestion of as little as 3 g may cause death (Marrs, 2004). Suicide attempts, particularly with organophosphate insecticides and rodenticides, are also the major cause of death associated with the pesticide concentrates in states of the USA such as California (Anon, 2003a).

Suicides are major causes of death in many poorer countries. Mortality resulting from deliberate consumption of pesticides is much greater than accidental or occupational exposure (Eddleston, 2000). Within the developing world, the World Health Organization has attributed about two-thirds of the 100 000 pesticide-associated deaths to ingestion of concentrates of organophosphate and carbamate insecticides, as well as paraquat. The comprehensive survey of Eddleston indicates the importance of pesticides in suicide attempts in developing countries (Table 7.3). The large number of suicides reported for 'Other

Table 7.3 *Pesticide-related poisonings from results of 96 hospital-based surveys of acute poisonings in the developing world*

| Region and number of studies undertaken | Number of patients poisoned | Percentage of poisoning attempts |
|---|---|---|
| Sub-Saharan Africa (18) | 1 729 | 13 |
| Middle Eastern crescent (15) | 955 | 11 |
| India (6) | 1 777 | 59 |
| China (1) | 42 | 9 |
| Other parts of Asia and islands (30) | 170 798 | 55 |
| Latin America and Caribbean (6) | 575 | 27 |

Adapted from Eddleston (2000)

parts of Asia and islands' evident in the table is due to the inclusion of two major studies from Sri Lanka.

In Sri Lanka, over 100 000 deaths over the period 1983–1998 have been attributed to suicide associated with consumption of pesticide concentrates. Easy access to pesticides, which can be bought 'over the counter' has been a primary reason for the high suicide rate in Sri Lanka, at 40 per 100 000 in 1995 compared to 11 per 100 000 in the UK (Eddlestone, Rezvi-Sheriff and Hawton, 1998). An epidemiological study from the agricultural district of Galle in Sri Lanka revealed a poisoning incidence of 75 per 100 000 of the population, with a corresponding high death rate of 22 per 100 000. Pesticides, especially organophosphates, carbamates and paraquat, were responsible for around 60% of poisonings. Figures from other countries suggest similar patterns of pesticide poisoning. In 1995 in China, around 48 000 cases of pesticide poisoning were reported with 3204 deaths, many of which were due to deliberate ingestion. In a national suicide survey conducted in China in 2001, of 250 000 suicide attempts, two-thirds were found to be due to consumption of pesticides, particularly rodenticides (Phillips, 2002).

Potential long-term adverse effects to health resulting from occupational exposure to pesticide concentrates have been studied by monitoring the health of workers in pesticide-manufacturing plants. Three studies in the USA of workers potentially exposed to phenoxy herbicides containing the contaminant dioxin have revealed that although exposure was high, no adverse mortality rates were evident. The health records of workers from 1952 to 1981 at a Colorado pesticide plant manufacturing compounds such as aldrin, dieldrin, dichlorvos and dibromochloropropane revealed mortality rates close to the predicted pattern. A detailed study of the health records of over 5000 workers from a company in

Table 7.4 *Mortality and incidence of cancer in workers from a manufacturing plant*

| | Number of deaths | Number with cancer |
|---|---|---|
| Expected (based on population average) | 1223 | 314 |
| Recorded among workers | 1039 | 297 |

Data from Coggon *et al.* (1986)

the UK manufacturing the herbicide MCPA, conducted over 20 years at a time when health and safety procedures at work were much less stringent than now, revealed no adverse health effects. Overall mortality was less than that of the national population (Table 7.4), as was mortality from cancer. However, few studies have been carried out on the long-term health of workers in manufacturing plants in less developed parts of the world, where pesticide manufacturing is being increasingly concentrated.

Although few effects on mortality of workers have been shown, in some cases other long-term adverse effects have been associated with occupational exposure to pesticides (10.3). In the USA, the nematicide dibromochloropropane, although not linked to premature death, has been linked with male infertility among workers in manufacturing plants and those spraying the compound (Lahdetie, 1995). This pesticide, now withdrawn in the USA, has never been approved for use in the UK.

## 7.3 Risks to health from diluted pesticides at or following application

Although the actual dose applied to crops and in other situations is usually a much lower concentration than that of the manufactured product, many instances of ill-health resulting from exposure to diluted concentrates have been documented in spray operators and other pesticide applicators, as well as in people who have been contaminated by spray drift, have come into contact with treated crops, surfaces or pets, or who have entered areas treated with pesticides before the applied compounds have had sufficient time to disperse or degrade (Stephenson and Ritcey, 1998). As noted earlier (6.1), some of these people may be more at risk than spray operators, who may be wearing protective clothing. In some instances, particularly in developing countries, deaths

Table 7.5 *Examples of acute poisonings due to pesticide ingestion*

| Case history | Outcome |
|---|---|
| Iraq 1971. Wheat and barley seed treated with a methylmercury fungicide and intended for sowing was consumed | Estimates of 6000 poisonings with 400 deaths |
| California and Oregon, USA 1985. Marketing of watermelons treated with the carbamate insecticide aldicarb within the prescribed harvest interval | 1175 people suffered acute ill effects of vomiting, nausea, diarrhoea, dizziness |
| Sudan 1991. Consumption of bread made with maize flour containing endosulfan | 350 people poisoned; 35 deaths |
| Northern Ireland and Eire 1992. Marketing of cucumbers treated with aldicarb, which was not approved for use in this crop | About 30 people suffered ill effects |
| Cuzco, Peru 1999. Mixing of parathion with milk powder, and consumption by children | 24 children died |
| Benin 1999. Inappropriate use of endosulfan – a cotton insecticide – on foodstuffs and food crops | An estimated 75 poisonings with 37 deaths |

Data from Ferrer and Cabral (1989): Conway and Pretty (1991) and various issues of *Pesticides News*

have occurred, but more commonly acute toxic effects have been felt, from which some contaminated individuals have often recovered, but others have experienced long-term illness.

Some of the most serious adverse effects have occurred through consumption of produce shortly after treatment with pesticides or by mixing pesticides directly with foodstuffs, and some of these are outlined in Table 7.5. In developing nations, local epidemics of poisoning have occurred and, as Table 7.5 shows, are still happening. In the Borgou province of the West African state of Benin in 1999, inappropriate use of the organochlorine insecticide endosulfan on foodstuffs and food crops led to at least 37 people losing their lives.

Contamination of skin and inhalation of organophosphate insecticides in spraying operations has caused death as well as chronic illness, particularly in developing countries. The Pesticides Action Network has collected information on the health and environmental impact of pesticides from several countries in poorer parts of the world, and many case histories are documented in Dinham (1995) and Harris (2000). Applications of organophosphate insecticides have been associated with adverse health effects and sometimes death in South America (Paraguay), Asia (the Cordilleran region of the Philippines) and Africa (Kenya and Senegal). Case studies in the districts of Tegal and Brebes in Indonesia revealed adverse effects in spray operators of fatigue (in 60% of operators), muscle stiffness (54%), dry throat (30%), dizziness (21%), blurred vision (15%), and stinging eyes (15%) – all occurring within a few hours of spraying. The principal pesticides used in this vegetable-growing region included organophosphates such as triazophos, methamidaphos and chlorpyrifos as well as carbamates such as methomyl. Similar comprehensive studies in part of the Cordilleran growing region of the Philippines revealed complaints of dry itchy skin in over 70% of spray personnel, itchy eyes and blurred vision (66%), abdominal pain (50%), dizziness/nausea (42%) and fatigue (39%). Here, the principal pesticides used included the organophosphate insecticides triazophos and methadimaphos. In Senegal, deaths were reported among sprayers using mixtures of carbofuran (the most likely toxicant), benomyl and thiram.

Comparisons between the precautions taken by those handling and applying pesticides in the UK and their general lack in developing countries explain the high risk that diluted concentrates of pesticides pose to spray operators in poorer countries. Table 7.6 offers some comparisons between the requirements of the foundation module for pesticides application in the UK, and the situation in most developing countries as described in many issues of *Pesticides News*, the FAO website and in Harris (2000).

Aerial spraying of pesticides has caused health problems to local inhabitants in some countries. A survey conducted in the Peraya district of Kerala in South India monitored the health of 4000 families in an area sprayed with copper oxychloride, endosulfan with mancozeb, and carbaryl to control pests and disease of cashew nuts (Usha, 2000). Between 600 and 800 families reported adverse effects ranging from acute skin, eye and nasal irritation as well as headaches and dizziness to chronic conditions including tiredness and eczema. Infertility in male members of the population as well as gynaecological problems were also reported.

Sublethal effects from diluted concentrate have also been seen in developed countries and again these have often involved organophosphate and carbamate insecticides. Problems of spray drift affecting bystanders have been reported in developed countries, and indeed proved to be a major focus of attention at the open meeting of the Advisory Committee for Pesticides in the UK in 2003, where a number of case histories were presented (Anon, 2003a). In this respect, a major report of the Royal Commission on Environmental Pollution in the UK (2005) has firmly concluded that bystanders and those living in the vicinity of regularly sprayed areas may suffer not only acute but also long-term

Table 7.6 *Comparisons between the knowledge required under the Food and Environmental Protection Act (FEPA) in the UK and provided through the National Proficiency Test Councils PA1 examination, with typical situations extant in developing countries*

| Pesticides Application 1 syllabus from the National Proficiency Test Council in the UK | Typical situations and practices in developing countries (after Dinham, 1995; Harris, 2000) |
|---|---|
| 1. Legislation | Legislation is actually extant in many countries, but rarely complied with or indeed known at farm level |
| 2. Label interpretation | Many labels are not in the language of the country; spray operators may be illiterate |
| 3. Sprayer design and safety | Hand-held sprayers are most common rather than tractor-mounted; sprayers are often in poor condition |
| 4. Protective clothing | Protective clothing may not be worn due to availability and discomfort in hot conditions |
| 5. Safe and secure storage of pesticides | Pesticides are rarely securely stored and may be readily available from untrained distributors |
| 6. Mixing and pouring | Water is often taken direct from local streams/rivers; mixing is not always accurate due to lack of measuring devices |
| 7. Accidental contamination | Frequent in developing countries, due to lack of protective clothing |
| 8. Record keeping | Records are sporadically kept; most often to comply with the requirements of multiple retailers importing produce from developing countries |
| 9. Health records | Few records are kept |
| 10. Environmental considerations | In many areas little regard is given to environmental contamination |

health problems. The RCEP report felt that the latter chronic conditions were vastly underreported. Table 7.7 shows a comparison of the concentrations of pesticide as prepared for spraying with $LD_{50}$ values of compounds commonly used in the UK, and illustrates the potential risks to health from contamination that might arise from spraying operations. Although the figures are not exactly comparable, it is likely that the concentration of pesticide in spray-strength

Table 7.7 *Concentrations of pesticides prepared to spray-strength compared to $LD_{50}$ values. Data for spray strength have been derived from appropriate product manuals*

| Pesticide | Highest recommended concentration in spray tank as mg/l | $LD_{50}$ values: rat oral; mg/kg |
|---|---|---|
| Epoxiconazole | 625 | >5000 |
| Chlormequat | 7330 | 800–900 |
| Isoproturon | 7000 | 1826–2417 |
| Cypermethrin | 125 | 250–4000 |
| Fenpropimorph | 3375 | >1400 |
| Glyphosate | 2400 | >5000 |
| Azoxystrobin | 1250 | >5000 |
| Kresoxim-methyl | 500 | >5000 |
| Tebuconazole | 1250 | 1700–4000 |
| Mecoprop | 2700 | 930–1166 |

form – to which bystanders may be exposed – is sometimes considerably higher than the $LD_{50}$. Formulation constituents other than the active ingredient may also be present at relatively high concentrations at spray strength.

Immersion of sheep in baths or troughs of diluted insecticide to kill external parasites such as ticks, lice and especially scab is a common practice in the UK and some other countries. In the early years of the last century, toxic compounds containing compounds of arsenic and phenol were used for sheep dipping. These were superseded after 1948 by organochlorine insecticides, notably HCH (lindane) and dieldrin. Their use was so successful that, by 1952, sheep scab had been eliminated from flocks on the UK mainland. However, the persistence of dieldrin led to its prohibition as a sheep dip constituent in the UK from 1965, and the use of HCH was gradually phased out in the mid 1980s. These were replaced in the late 1960s and 1970s by less persistent organophosphate insecticides, primarily diazinon, chlorfenvinphos and propetamphos. Sheep scab returned to England in 1972, probably on imported animals. Compulsory nationwide dipping was reintroduced in 1976, having been considered unnecessary during the period of use of organochlorines. This situation lasted until 1992, when mandatory dipping was scrapped and a policy was introduced requiring farmers to dip sheep if scab was found in their flocks (Beesley, 1994).

The procedures of dipping animals under stress in large volumes of pesticides, frequently in warm weather, presents a much higher risk of worker contamination in the UK than almost any other method of applying pesticides.

Splashing associated with the dipping, and proximity to and handling of dipped animals dramatically increase the chances of inhalation and skin contact, requiring rigorous attention to the use of protective measures.

The advent of organophosphate use in sheep dipping in the UK led to an increase in reports of ill-health among farmers, shepherds and others involved in the dipping process. From 1985 until 1992 the Veterinary Medicines Inspectorate in the UK received over 300 reports of suspected adverse reactions to sheep dips. The Pesticides Exposure Group of Sufferers (PEGS) estimates that over 3000 people in the UK may be suffering chronic ill health from sheep dipping operations, and a study carried out at the Institute of Occupational Health in the UK during the late 1990s indicated more frequent symptoms of neuropathy among sheep dippers compared to workers in an unrelated industry (Anon, 1999). The problems of ill health among those involved in sheep dipping led to the withdrawal of organophosphate sheep dips in the UK from 1997, and their replacement by pyrethroid insecticides, notably cypermethrin. Paradoxically this has led to environmental problems, particularly for aquatic invertebrates, to which pyrethroids are more toxic than organophosphates (9.6).

Problems of adverse health due to insecticides have been linked to re-entering treated crops and other areas before the pesticide has dissipated. As with sheep dips in the UK, many of these problems have arisen following the change in use from organochlorine to organophosphate compounds for pest control. Many well-documented incidents of ill health and even death due to re-entry into treated crops come from the cotton and fruit-growing areas of the USA. In California, particular problems were experienced with parathion (which on citrus foliage was transformed to the even more toxic and persistent paraoxon), as well as other organophosphates such as dimethoate and azinphos-methyl. Adverse effects were exactly the same as those associated with exposure to organophosphate sheep dips in the UK, viz headaches, nausea, vomiting, dizziness and long-term changes in behaviour, and are now recognised as the medical condition known as organophosphorus-induced delayed polyneuropathy (OPIDP). The effects have been linked to depressed acetylcholine sterase activity in susceptible individuals, and in some cases the neurological effects appear irreversible. Increasingly stringent precautions in California have led to a steady reduction in the numbers of illnesses and injuries associated with pesticides after spraying in fields – from 373 in 1987 to 231 in 1995 and then less than 100 after 1999 with only 45 incidents reported in 2001 (Anon, 2003b).

Notwithstanding major advances in biological control in the last decade of the twentieth century and their adoption in glasshouses, fumigants and other pesticides are still used widely to control pests and diseases. The intensive management systems needed to maintain production of high-quality fruit and vegetables often entails the routine use of pesticides. The general soil sterilant methyl bromide is an extremely toxic substance that has caused serious health effects and occasionally death. It is due to be phased out worldwide on both toxicological and environmental grounds by 2015 (5.2). However, relatively little data are available for pesticides and occupational health among glasshouse workers, and this also extends to personnel in other non-crop situations where pesticides are routinely applied such as to amenity areas, golf greens and other sports surfaces.

Exposure to diluted pesticides also occurs in homes, offices and other enclosed premises. Treatment of buildings with fungicides to prevent decay of timber, and with insecticides to prevent or control pests of public health interest such as cockroaches may result in occupants experiencing ill health from the compounds used. Whereas pesticides used in outdoor situations may be degraded by the action of ultraviolet light in the air and on surfaces, and broken down by microorganisms in soil (8.5), rates of dissipation inside buildings may be very much slower. Timbers treated with the insecticide HCH (lindane) and the fungicide propiconazole may be guaranteed free of pest and disease attack for 10 or 20 years, illustrating the longevity of the compounds inside buildings. Carpets treated with the pyrethroid insecticide permethrin are commonly guaranteed for several years from insect attack.

Occasional instances of adverse health effects have been reported in those applying pesticides to homes, offices and industrial premises, but reports of illness in occupants have also been recorded. Exposure to the organochlorine lindane has been a particular cause of concern, and contributed to the decision to withdraw this compound for use in the EU from 2001.

In the UK the Health and Safety Executive (HSE) investigate incidents of contamination by pesticides during or shortly after their use, and the number of incidents during the 1990s that alleged ill health due to pesticide contamination is shown in Figure 7.4. The Pesticide Incidents Appraisal Panel considers all of these cases and in most years fewer than 25% are confirmed as, or likely to be, due to pesticide contamination. The majority of confirmed cases have been associated with crop spraying of insecticides, and again organophosphates feature prominently.

Again, the situation in countries such as the UK may be contrasted with that in other parts of the world. Data from Thailand of occupational instances of pesticide poisoning from 1987 up to 1994 are shown in Figure 7.5: the instances are much higher than in the UK.

Long-term effects of exposure to pesticides among spray operators have also been intensively scrutinised. Soft tissue cancers and lymphomas are major

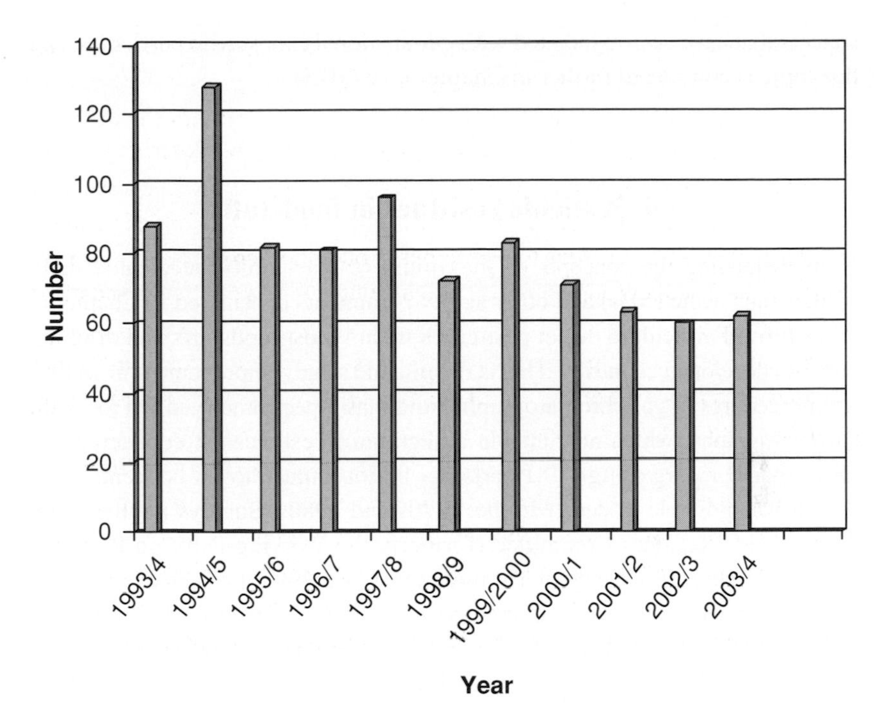


Figure 7.4 Numbers of alleged ill health incidents reported to the Pesticides Incidents Appraisal Panel of the UK Health and Safety Executive 1993–2004. Data from http://www.hse.gov.uk/fod/pir0304.pdf.

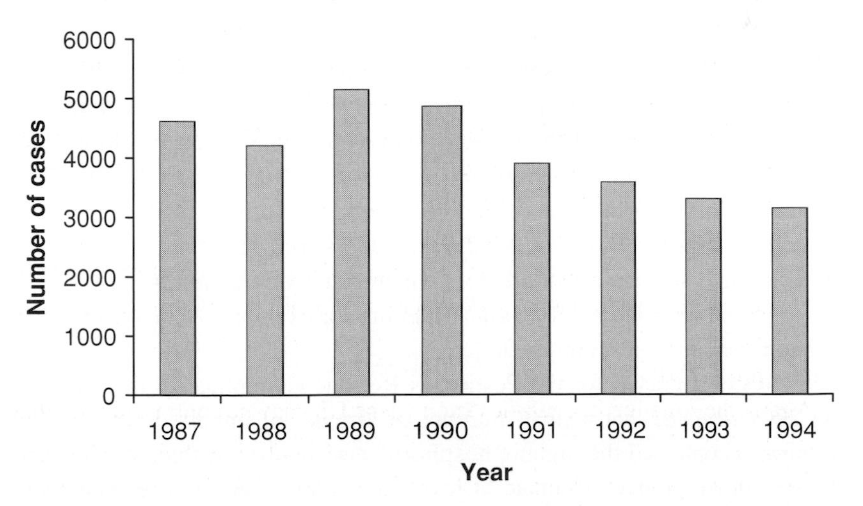


Figure 7.5 Occupational poisoning cases due to pesticides in Thailand (from the Epidemiological Division, Ministry of Health quoted by Harris, J. (2000), *Chemical Pesticide Markets, Health Risks and Residues.* CABI, Wallingford).

cancers that have been associated with occupational exposure to pesticides, and this topic is considered further in Chapter (10) (10.3).

## 7.4 Pesticide residues in foodstuffs

Notwithstanding the concepts of maximum residue limits, acceptable daily intakes (and acute RfDs) and other safety parameters considered in Chapter 6, it has proved difficult to detect pesticides within most foodstuffs and drinking water in developed countries. This is despite the steady improvement in analytical procedures of gas chromatography, and mass spectrometry allied to liquid chromatography, which now enable detection of pesticides at concentrations lower than 0.1 μ/kg or μg/l (0.1 parts per billion); much lower concentrations than it was possible to detect in the 1970s and 1980s. Surveys published by the Pesticide Residues Committee (formerly the Working Party on Pesticide Residues) in the UK have regularly shown that over 70% of produce from retail outlets such as supermarkets contained no residues, with over 25% showing residues below the MRL and less than 1% of produce exhibiting residues in excess of the MRL (Figure 7.6). In the 2000–2003 UK surveys, very few pesticides were detected above the permitted maximum residue levels in any of the staple foods – none in bread and milk with fewer than 1% above the limit in potatoes.

The small numbers of foodstuffs in which pesticides have been detected in excess of permitted maximum residue levels have been targetted for action. Particular examples from the UK have included imported nectarines, peaches, pears and celery as well as home-grown lettuce and carrots (Table 7.8. and Figure 7.7 respectively). Residues of the plant growth regulatory substance chlormequat regularly occurred on pears imported during the 1990s to the UK from Belgium and Holland, where the compound was applied primarily to improve the appearance of pears. Both this and the regular occurrence of fungicide residues on celery and spinach imported from Spain resulted in action being taken in their respective countries of origin. Belgium, for example, suspended the use of chlormequat on pears in 1999.

Surveys conducted for the Pesticides Residue Committee in the UK during 2001 revealed that a small number of samples of imported peaches and nectarines contained the organophosphorus insecticides, methamidophos and its breakdown product acephate at levels above the maximum residue level. Imported peaches and nectarines were therefore targeted for analysis in 2003. Of 90 samples, residues of methamidaphos and acephate were detected in five with a (marginal) maximum residue limit exceedance in only one sample.

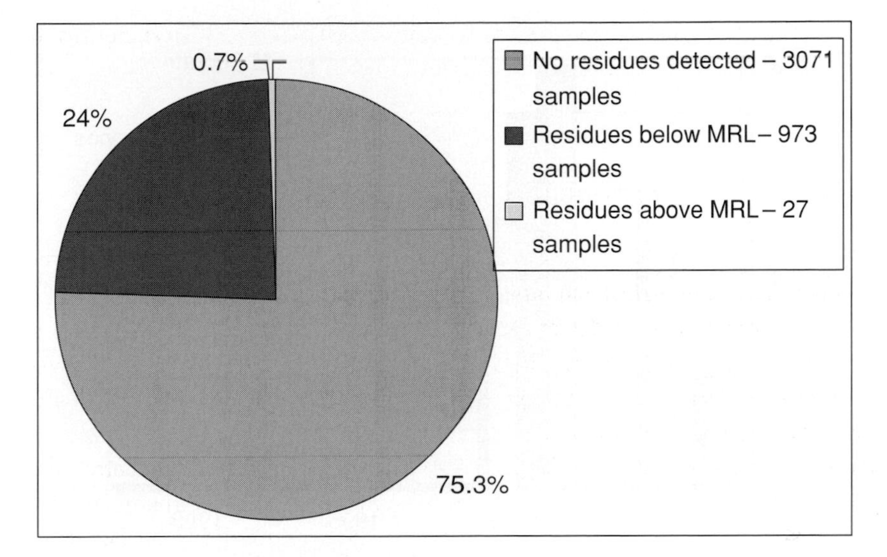


Figure 7.6 A summary of the UK Pesticide Residues Committee findings for 2003. Samples were analysed for over 170 000 pesticide/commodity combinations. From http://www.pesticides.gov.uk/committee/PRC/PRC_annual_rep_2003.pdf.

Table 7.8 *Pesticide/foodstuff combinations exceeding maximum residue limits in the UK in 2002 and 2003*

| Foodstuff | Number of samples analysed | Number of samples containing residues | Number of samples containing residues above MRL | Compounds detected above MRL |
|---|---|---|---|---|
| Lettuce 2002 | 108 | 41 (37%) | 3 (3%) | Bromide |
| Lettuce 2003 | 107 | 42 (39%) | 5 (5%) | Bromide |
| Potatoes 2002 | 241 | 93 (39%) | 3 (1%) | Chlorpropham |
| Potatoes 2003 | 144 | 44 (31%) | 1 (0.5%) | Chlorpropham |
| Grapes 2002 | 72 | 43 (59%) | 0 | |
| Grapes 2003 | 72 | 45 (63%) | 4 (6%) | Methomyl |

Adapted from http://www.pesticides.gov.uk/uploadedfiles/Web_Assets/PRC/PRC_annual_rep_2003.pdf

Residues of organophosphorus pesticides were monitored extensively in UK-grown carrots during the 1990s. Organophosphates replaced persistent organochlorine compounds such as $\gamma$HCH in the 1980s for control of carrot fly (*Psila rosae*), the major pest of the crop. In the early 1990s, samples of UK carrots were found to contain organophosphates, notably triazophos and chlorvenfinphos, sometimes above the maximum residue limit (Figure 7.7). Indeed, in 1992, almost one-quarter of carrots analysed in the UK for the Working

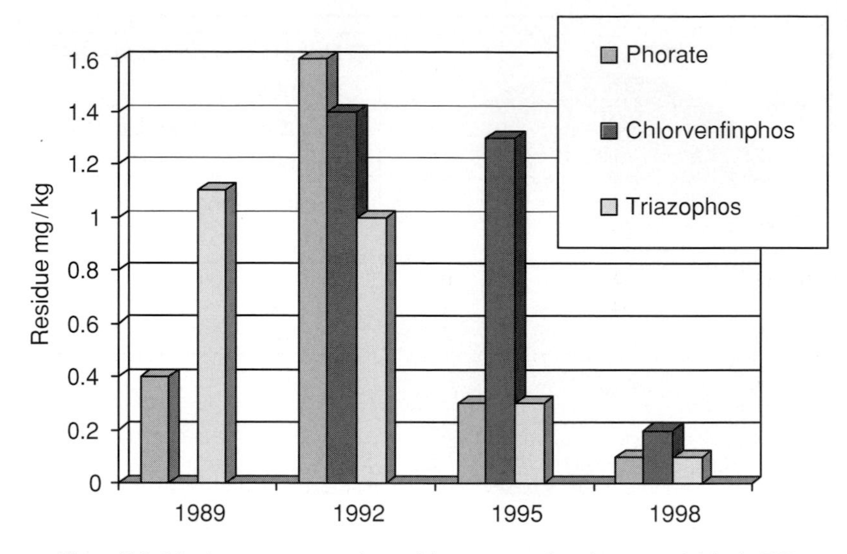


Figure 7.7 Maximum concentrations of three organophosphorus pesticides in UK carrots 1989–1998. Adapted from http://www.pesticides.gov.uk/prc.asp?id=789.

Party on Pesticide Residues had triazophos residues in excess of the maximum residue limit. This prompted action in the UK and in 1999 no home-grown carrots exceeded maximum residue limits for organophosphorus insecticides – however, this was due in part to a revision in the value upwards from 0.3 mg/kg to 3 mg/kg for triazophos! Nevertheless, the area of UK carrots treated with organophosphorus compounds declined markedly in the late 1990s and since triazophos was not supported for review in the EU, problems associated with its persistence in crops no longer exist. Indeed in 2003 only 1 sample out of 132 examined was found to contain traces of an organophosphorus insecticide and this was well below the maximum residue limit.

The occurrence of fungicide residues in lettuce is a recurring problem in the UK linked to growth in winter when conditions favour the development of downy mildew (*Bremia lactucae*) and grey mould (*Botrytis cinerea*) diseases in the crop. Regular high humidity in glasshouses at this time has led to growers using several fungicide sprays in efforts to control these diseases. This problem has been going on for many years: in 1979–1980 pesticide residues on lettuce were often found to exceed maximum residue limits in force at that time (Conway and Pretty, 1991). Additionally, the use of methyl bromide (5.2) as a soil sterilant prior to planting lettuce has led to levels of inorganic bromide in this crop exceeding maximum residue limits, sometimes by a factor of 10 or more.

Table 7.9 *Pesticide residues in the surveillance programme for UK lettuce during the late 1990s/early 2000s*

| Year | Number of lettuces analysed | Numbers containing pesticide residues | Numbers exceeding MRLs | Pesticides detected with no approval for use on lettuce in UK |
|---|---|---|---|---|
| 1997 | 17 | 6 | 1 | Tolclofos-methyl |
| 1998 | 25 | 15 | 8 | Procymidone, oxadixyl, furalaxyl |
| 1999 | 72 | 40 | 8 | Oxadixyl |
| 2000 | 35 | 24 | 0 | |
| 2001 | 180 | 54 | 2 | Azoxystrobin |
| 2002 | 108 | 41 | 3 | Vinclozolin |

Adapted from http://www.pesticides.gov.uk/prc.asp?id = 959

The withdrawal of methyl bromide should eliminate such problems of bromide contamination.

Some fungicides not approved in the UK for use on lettuce, such as the dicarboximide vinclozolin, were detected in samples taken during the early 1990s and approved compounds such as iprodione were found in some lettuce heads at concentrations above the maximum residue limit. Action was taken by MAFF in England through an enforcement programme including sampling of lettuce on grower holdings. This resulted in the prosecution of some growers under the Food and Environmental Protection Act with a consequent reduction in residue problems. However, breaches of the regulations were still occurring in the early 2000s, and residues of pesticides, albeit in most cases below the maximum residue limit, were still routinely found then in UK grown lettuce (Table 7.9). Targetting of crops such as lettuce where residues of pesticides are likely to occur is now normal practice in the UK and several other countries in surveys for pesticide residues.

Analysis of many foodstuffs has revealed residues of more than one pesticide, and concerns over the effects of multiple residues in foodstuffs on the health of consumers were addressed in the late 1990s. Residues of pesticides with an identical mode of action, such as organophosphate insecticides, may be at least additive in their effects on human beings (and other non-target organisms). In other cases, the possibility of interactions occurring between multiple residues with adverse consequences for consumers has been highlighted. The Food Standards Agency in the UK set up a Working Group on the Risk Assessments of Mixtures of Pesticides (WIGRAMP) to investigate potential problems arising from exposure to mixtures of pesticides in foodstuffs. Multiple residues have been especially associated with fruit and vegetable crops such as apples, oranges and lettuce (Reynolds and Hill, 2002; Table 7.10).

Table 7.10 *Pesticide residues in a range of foodstuffs available in the UK in 2001*

| Foodstuff | Number of samples analysed | Number containing residues | Number containing multiple residues | Number with residues above MRL |
|---|---|---|---|---|
| Apples | 252 | 68 (27%) | 29 (12%) | None |
| Cucumber | 72 | 42 (56%) | 12 (17%) | 3 (4%) |
| Potatoes | 239 | 80 (33%) | 18 (9%) | 3 (1%) |
| Strawberries | 179 | 115 (64%) | 68 (38%) | 4 (2%) |
| Tomatoes | 144 | 26 (18%) | 8 (6%) | None |

Adapted from http://www.pesticides.gov.uk/prc.asp?id = 796

In addition to fruit, vegetables and staple dietary foods such as cereals and potatoes, animal products have also been surveyed for the presence of residues. Most surveys conducted since the late 1990s have shown that pesticide residues are detected in extremely low amounts in animal products. Of 707 such samples analysed in the UK in 2003, no pesticide residues could be found in 99% with residues in the other 1% below maximum residue limits (Anon, 2004a).

Targetting samples for pesticide analysis since 2000 in the UK has involved sampling manufactured products as well as food for toddlers and infants, whose health has been considered to be more at risk from residues. In the 2001 round of analyses, infant foods were found to have a low percentage of residues (5% of 154 samples) but processed foods such as crisps were found to have residues in 25–45% of samples examined (Anon, 2002a). In 2003 no residues were found in any infant foods (Anon, 2004a).

Surveys of foodstuffs for pesticide residues in other parts of the developed world reveal results similar to those in the UK. Data submitted to the European Commission show similar results to those of the UK, with a few countries showing higher maximum residue limit exceedances (Table 7.11). Many of these breaches are linked to imported food, particularly of fruit and vegetables from outside the EU, and sampling in many countries is aimed at produce thought likely to contain residues. Indeed, the European Commission now operates a Rapid Alert system with respect to toxins in foodstuffs marketed in the EU. When a member state finds that produce contains residues in excess of the maximum residue limits, or detects a pesticide that is not permitted for use on the product, other member states are immediately informed of the nature and origin of the foodstuff in question.

In the USA the California Department of Pesticide Regulation (CDPR) has targetted the crops and produce most likely to contain pesticide residues. Their

Table 7.11 *Pesticide residue analysis in Europe 1999*

| Country | Number of samples | Percentage with no detectable residues | Percentage with residues | Percentage exceeding Maximum Residue Limits |
|---|---|---|---|---|
| Austria | 546 | 49 | 40 | 11 |
| France | 4 553 | 46 | 33 | 8 |
| Germany | 13 000 | 51.3 | 39.7 | 3.1 |
| Italy | 7 802 | 68.4 | 30.3 | 1.3 |
| Netherlands | 1 500 | 58 | 39 | 3 |
| Spain | 2 898 | 57.4 | 39.2 | 3.4 |

Data from European Commission. Annex to Monitoring of Pesticide Residues in Products of Plant Origin in the European Union, Norway and Iceland 1999. SANCO/397/01 (available at http://europa.eu.int/comm/food/fs/inspections/fnaoi/reports/annual_eu/ann397_en.pdf
In some cases, residues were determined with no MRL for the commodity in question

Table 7.12 *Pesticide residue analysis in some subtropical countries*

| Country | Number of samples | Percentage with no detectable residues | Percentage with residues | Percentage exceeding maximum residue limits |
|---|---|---|---|---|
| Costa Rica 1993 | | 45 | 55 | 11 |
| Pakistan early 1990s | 550 | 61 | 39 | 14 |
| Karachi (Sindh) area within the above study | 250 | | 37 | 18 |
| Egypt 1997 | 2318 | 81.5 | 18.5 | 1.9 |

Quoted in Harris (2000); Dogheim *et al.* (2002)

multi-residue screens are some of the most comprehensive in the world. They routinely look for more than 300 pesticides and their metabolites, and the CDPR test more than 12 000 samples of 160 different kinds of fresh fruit and vegetables. As in the UK studies, no residues are detected in about 70% of samples and, in those with residues, only 0.9% breached legal limits (Wells, 1994). Most of these were pesticides not approved for use on the commodity, and were probably the result of spray drift.

As with acute toxicity, the situation is very different in poorer countries (Table 7.12). Few comprehensive surveys exist of pesticide residues in foodstuffs in the developing world. The technology and expertise required are expensive and many countries in poorer parts of the globe do not have the resources to finance such facilities.

Conway and Pretty (1991) review some early data that show concentrations of pesticides well in excess of maximum residue limits on foodstuffs, quoting examples of vegetables in Indonesia containing very high levels of organophosphates. Very high levels of organochlorines including DDT and lindane were reported in fish, cow's milk and beef in developing countries. Much of the residue originated from spraying cotton in the vicinity of the animals as well as near water-courses. Further instances from the 1990s are reported by Harris (2000). In Costa Rica during 1993, pesticide residues were found in 55% of food samples with 11% exceeding maximum residue limits. In India, the Indian Council of Medical Research reported in 1993 that of 2205 cow and buffalo milk samples, lindane was detected in 85% of samples with 41% exceeding tolerance limits.

As part of a study on tomato production in Brazil (Zavatti and Abakerli, 1996) pesticide residues were found to exceed Brazilian maximum residue limits. Of 32 samples analysed in a multi-residue programme, 8 contained methamidaphos above the maximum residue limit and 25 contained ethylene thiourea, a metabolite of the fungicide maneb, above the maximum residue limit for this substance. In addition, nine samples contained endosulfan, which is not registered for use on tomatoes in Brazil.

The work of the World Health Organization through its Joint Meeting on Pesticide Residues (JMPR) Committee that defines maximum residue limits for foodstuff/pesticide combinations may be important in restricting residues in foods, but surveillance data to monitor practice in situ are lacking.

## 7.5 Pesticide residues in drinking water

Legislation has been introduced in some countries with respect to the maximum admissible concentration (MAC) of pesticides in natural waters used as sources of drinking water. For example, the EC Drinking Water Directive of 1980 set standards for pesticides (and other chemicals) in drinking water of 0.1 μg/l for an individual pesticide. These standards coincided approximately to detection limits for pesticides at the time the legislation was introduced, and were not based on any evaluation of toxicological data: in the EU all pesticides are treated as equal with respect to their toxicological properties under the Drinking Water Directive. Furthermore, compliance with the regulations may depend on the number of pesticides assayed by drinking water suppliers.

A more pragmatic approach has been adopted by the Environmental Protection Agency (EPA) Office of Water in the United States. The EPA has the authority to set drinking water standards under the Safe Drinking Water Act (SDWA), and the 1996 amendments to the Act require the EPA to set standards using a scientific approach that includes evaluation of occurrence of potential

Table 7.13 *Standards for selected pesticides in drinking water defined by the Environmental Protection Agency in the USA*

| Pesticide | MCLG | MCL |
|---|---|---|
| Atrazine | 3 | 3 |
| 2,4-D | 70 | 70 |
| Dibromochloropropane | 0 | 0.2 |
| Glyphosate | 700 | 700 |
| $\gamma$HCH (lindane) | 0.2 | 0.2 |

MCLG is the maximum contaminant level goal; MCL the maximum contaminant level; and the figures are in μg/l. Figures from USEPA website at http://www.epa.gov/safewater/standard/setting.html

contaminants in the environment, human exposure and risks of adverse effects in the general population. After consideration of health effects, the EPA establishes a maximum contaminant level goal (MCLG), defined as the maximum concentration of a contaminant in drinking water at which no known or anticipated adverse effect on the health of persons would occur. As with ADIs for pesticides in food, the EPA MCLGs consider risks to subpopulations such as infants, children and the elderly.

Once the MCLG has been established, the EPA enforces a maximum contaminant level (MCL): the maximum permissible level of a contaminant in a public water system. The MCL is set as close as possible to the MCLG, and is based on current detection technology. Some examples of these toxicologically derived values for pesticides in US drinking water are given in Table 7.13.

In practice, pesticides have been regularly found in drinking water supplies in Europe above the MAC as defined by the EC Drinking Water Directive. The implementation of the Water Supply (Water Quality) Regulations of 1989 led to a great increase in the number of assays carried out by water companies in the UK. About 130 of the 450 pesticides used in the UK were assayed in water samples in 1989. From 540 000 determinations made in that year about 2% – around 11 000 – exceeded the EC limit. The pesticides most commonly encountered were atrazine, simazine, isoproturon, chlorotoluron and mecoprop. A surprising aspect of these figures in many parts of the UK was the common presence of the herbicides atrazine and simazine, even in rural areas where these compounds had not been extensively employed. Residues of these compounds are now known to have come from railway tracks, and the track beds of closed

Table 7.14 *Pesticides in drinking water supplied by four water companies in England in 1993 and 1998*

| | Water company | | | |
|---|---|---|---|---|
| Pesticide | Anglian Water | North-West Water | Severn Trent Water | Thames Water |
| Atrazine | 80 (0) | 33 (0) | 45 (0) | 5263 (10) |
| Simazine | 32 (0) | 1 (0) | 7 (0) | 3652 (1) |
| Diuron | 0 (0) | 1 (0) | 32 (0) | 3242 (0) |
| Isoproturon | 33 (0) | 1 (0) | 12 (0) | 2860 (0) |
| Mecoprop | 13 (0) | 5 (0) | 157 (0) | 44 (0) |

Figures refer to the number of samples exceeding 0.1 μg/l for individual pesticides. 1998 values are in parentheses. Data from White and Pinkstone (1995). The occurrence of pesticides in drinking water. In: *Pesticide Movement to Water. BCPC Monograph 62.* Pp. 263–268. BCPC Publications. Farnham, UK: Annual Reports of the UK Drinking Water Inspectorate 1995 onwards at http://www.dwi.gov.uk/reports.shtm

railways, where formerly atrazine and simazine were used extensively in the 1950s and 1960s for weed control. Isoproturon, chlorotoluron and mecoprop are some of the commonest pesticides applied for weed control in cereal crops.

Breaches of the EC limits declined from a high of 3% in 1992, to 2% in 1993, to 1.2% in 1994 and to 0.8% in 1995. In 1994, of 1 112 269 tests undertaken, about 13 000 exceeded the EC limit. Atrazine breaches dropped from 12 728 in 1992 to 1724 in 1994, and those of diuron from 4647 to 1302 over the same period. However, because of these recurring breaches of EU limits, water companies in the UK invested heavily in treatment processes such as those based on activated charcoal that could remove traces of pesticides from drinking water derived from surface sources such as rivers and reservoirs. By the late 1990s this resulted in a dramatic reduction in residues above the EU limit, with only 11 in 1998 (Table 7.14).

These figures refer to drinking water provided by suppliers from treated sources, but samples taken from groundwater sources such as private wells and boreholes may also contain residues of pesticides (Beitz *et al.*, 1994). In many countries, wells and boreholes form the principal source of drinking water (Table 7.15) and concern has been expressed at the levels of pesticides that have been detected in some cases. Studies undertaken in the USA in Pennsylvania, Maryland and Delaware in the early/mid 1980s found concentrations of some

Table 7.15 *Relative proportions of drinking water sources in various countries*

| Country | Groundwater % | Surface water % |
|---|---|---|
| Belgium | 67 | 33 |
| France | 65 | 35 |
| Great Britain | 28 | 72 |
| The Netherlands | 70 | 30 |
| Spain | 22 | 78 |

Data from Roberts (1999)

soil-applied herbicides in groundwater to exceed 50 μg/l beneath maize crops. Atrazine and simazine, used extensively for weed control in maize, were the principal compounds detected (Anon, 1990).

The National Water Quality Assessment Program (NAWQA) was established in the USA in 1991 and, like the Drinking Water Inspectorate in the UK, produces annual reports that include pesticide occurrence. During 1991–2001 pesticides were detected in over 50% of shallow wells analysed by NAWQA in the USA (Kolpin and Martin, 2002). However, less than 1% exceeded USEPA drinking water standards. The principal compounds detected were the triazine herbicides atrazine and simazine, the former occasionally at over 1 μg/l, a concentration that does not exceed the recommended USEPA Health Advisory (Table 7.14), but which is ten times the EU limit for pesticides in drinking water.

It is notable that most compounds detected in groundwater supplies are soil-applied pesticides such as the triazine herbicides, and these may continue to leach through the soil profile and into aquifers over many years. Thus, the presence of pesticide residues in drinking water derived from wells and boreholes may continue for many years to come.

## 7.6 Pesticides in human tissues

Because of their bioaccumulative properties, the organochlorine insecticides have been intensively studied in human tissue. Analyses in some parts of the world have shown very high levels of organochlorines in human tissues. In the early 1970s, DDT in the fat of human beings in many parts of the world reached concentrations of 10 000–20 000 μg/kg. Studies listed by Conway and Pretty (1991) reported values of around 18 000 μg/kg in Central Europe and up to 10 000 μg/kg in the USA. In India and Pakistan, concentrations above

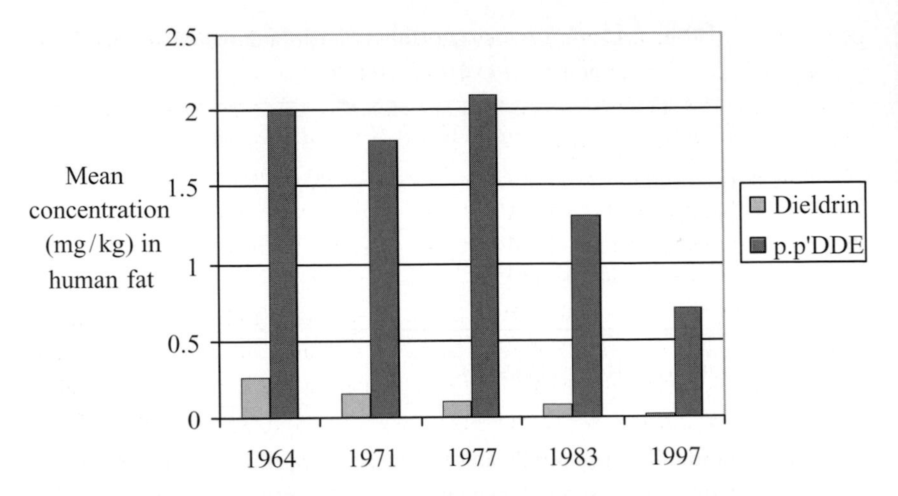


Figure 7.8 Concentrations of organochlorine pesticides in human fat. Data from the Seventh Annual Report of the UK Committee on Carcinogenicity of Chemicals, Food, Consumer Products and the Environment at http://www.advisorybodies.doh.gov.uk/coc/1999ar.pdf. Figures given are mean values for two or three years preceding the date given.

Table 7.16 *Residue levels of DDT in human milk. Values given are means expressed as μg total DDT compounds per kilogram in milk fat*

| Country/ region | Date | DDT μg/kg milk fat | Date | DDT μg/kg milk fat | Date | DDT μg/kg milk fat |
|---|---|---|---|---|---|---|
| UK | 1965 | 5 040 | 1979–1980 | 1150 | | |
| USA | 1962 | 9 250 | 1970–1971 | 4250 | 1989 | 550 |
| Canada | 1968 | 5 500 | 1978 | 1400 | 1986 | 385 |
| Sweden | 1972 | 2 930 | 1981 | 1060 | 1992 | 283 |
| Mexico | 1976 | 13 180 | 1980 | 6750 | 1997 | 3975 |
| India | 1980 | 4 800 | 1984 | 6000 | 1994 | 1200 |
| Hong Kong | 1985 | 13 800 | 1992 | 2171 | 2001 | 390 |
| Japan | 1979 | 925 | | | 1999 | 345 |

From data quoted in Conway and Pretty (1991); Smith (1999); Wong *et al.* (2002)

20 000 μg/kg were regularly recorded in the 1960s and 1970s, with some extreme cases of individuals having 150 000 μg/kg (0.15 g/kg) DDT in their tissues. It is perhaps a testament to the low risk to human health of DDT that few instances of ill health were reported in individuals carrying such body-burdens of this pesticide. Since its withdrawal from use in many countries, levels of DDT and its metabolites in fatty tissue have declined worldwide. This has also

occurred with other organochlorines such as dieldrin and chlordane that have also been detected in human tissues (Figure 7.8).

Much concern has been expressed about the levels of organochlorine insecticides in breast milk, which is lipid-rich and in which organochlorines may accumulate. In human breast milk in the 1960s and 1970s, concentrations of 5000–10 000 μg of DDT per kilogram of milk fat were not uncommon, but these have subsequently declined in many countries following the prohibition of DDT for most uses, with levels usually below 1000 μg of DDT per kilogram of milk fat (Table 7.16; Smith, 1999). However, much higher concentrations have been occasionally detected where DDT is still used in vector control programmes for malaria, and it is likely that infants in these areas may receive doses above the WHO acceptable daily intake for infants of 20 μg of DDT per kilogram of body weight per day.

# 8
# Pesticides in the environment

## 8.1 Introduction

The dose and extent to which non-target organisms are exposed to pesticides is associated with the release, distribution and fate of these compounds in the environment. Selectivity, in addition to being conferred by the mode and biochemical site of action of pesticides, may also be influenced by the delivery of the compound to the target organism. The most desirable method of delivery is directly to the problem pest, disease or weed; here entry to the environment may be localised and effects on non-target species kept to a minimum. Examples of such targetted applications mainly come from human health and veterinary practice, where individual patients or animals may be treated. In such situations, cast skin or hair, washings and excreta may be the only source of pesticides entering the environment.

Careful positioning of pesticides may ensure selectivity to the target organisms. Application of compounds to containers in which high-value ornamental plant species are growing is one such example. Closed systems of growing with recycling of irrigation water are becoming accepted in glasshouse complexes in Europe, and pesticides applied in the irrigation water of hydroponically raised crops are not likely to enter the external environment. Rodent pests in catering and domestic premises, farms and elsewhere are usually controlled by careful positioning of poisoned baits, such that these are not accessible to non-target species. However, careless placement of these baits, many of which have a high mammalian toxicity, has on occasions resulted in deaths to pets and domesticated stock.

Seed treatment also involves the localised applications of pesticides, and is regarded as an efficient method of application, giving a much smaller environmental input than for example spraying. Seed treatments offer very low inputs of fungicides and insecticides that can offer control of soil pests and diseases, and in some cases foliar attacks due to the systemicity of the compound. Paradoxically however, some of the most serious effects on non-target species, particularly birds, have resulted from consumption of insecticide-treated grain. Some seed-applied systemic organophosphates and carbamates have even led to adverse effects from consumption of foliage into which the insecticides have passed.

A degree of selectivity can be achieved with pesticides that are specifically applied to soil in that above-ground species may not be affected. This is especially important with some of the carbamate and organophosphate insecticides that are toxic at low dose to surface dwelling and other fauna. Of course, soil-applied pesticides pose the greatest risk to contamination of groundwater beneath the subsoil.

## 8.2 Spraying

Most pesticides are introduced into the environment in a diluted form of the concentrate as sprays in agricultural, horticultural and amenity situations to control weeds, pests and diseases. Spraying and treatment of households and commercial premises for insect and rodent control is another means by which these compounds are introduced into the environment, as is timber treatment, where fungicides are frequently used. In grain stores, aircraft, glasshouses and some domestic situations pesticides may be deployed as mists, fogs or as fumigants (Matthews, 1992).

Most spraying in the UK is carried out on arable crops. Tractor-mounted or towed sprayers deliver diluted concentrate of the pesticide from tanks on the rear of the sprayer through a series of nozzles on a horizontal boom usually about 50 cm above ground level (Figure 8.1). Hydraulic pressure applied through the tank drives the dilute concentrate through the nozzles producing a range of droplet sizes. The droplet size can be altered by adjusting the hydraulic pressure and by varying nozzle size (Figure 8.2). Nozzles with a small orifice produce a fine spray of small droplet size (less than 100 μm diameter), with those of larger orifice size producing correspondingly larger droplets.

Other systems include hand-held sprayers (Figure 7.3) that are used not only in agricultural and horticultural practice but also in amenity situations to treat hard surfaces such as roads and pavements as well as timber and other surfaces inside buildings. Air-assisted sprayers, often referred to as mist blowers, may be used to gain good coverage of fruit trees and bushes. Aerial spraying of pesticides is now rare in the UK, but common in some other parts of the world.

Figure 8.1 Applying pesticides from a tractor-mounted spray boom.

Figure 8.2 Flat fan (upper) and cone (lower) nozzles.

Agricultural spraying, although the most convenient way to apply pesticides, is not very efficient in that much of the applied product fails to find its intended target. As little as 1% of pesticide sprayed on a crop may reach the target weed, pest or disease (Graham-Bryce, 1977). The remainder stays either on the crop, or may enter the air, soil or natural waters. Spray drift and the volatilisation of chemicals already applied may result in the pesticide occurring in areas away from the selected site of application. Locally, compounds may enter water courses or, in the case of herbicides, damage neighbouring crops and vegetation. Control of drift is one of the most important factors in limiting the non-target effects of pesticides both to members of the public as well as surrounding fauna and flora.

In the UK and many other developed countries, strict procedures frequently backed by legislation have been introduced to minimise off-target effects resulting from the drift of pesticides. Codes of conduct have been drawn up in response to the introduction of the Control of Pesticides Regulations (1986) as part of the UK Food and Environmental Act of 1985. The codes outline the principal factors that influence spray drift, such as wind speed, nozzle size and pressure, that must be taken into account when considering whether to spray.

Drift is influenced by the droplet sizes produced during passage through spray nozzles. According to the size and shape of their apertures, nozzles may produce fine, medium or coarse sprays giving droplet size diameters respectively of between 100 μm and 200 μm, between 200 μm and 300 μm, and over 300 μm. Whereas sprays consisting largely of fine droplets may give good coverage of foliage and other treated surfaces, they also present the greatest risk of drift. Coarser droplets may not pose so great a risk of drift, but may not give such good coverage and require higher application rates, leading to increased risks of soil and possibly groundwater contamination.

The principal types of nozzle employed in pesticide spraying are cone, flat-fan and deflector (Figure 8.2). Cone nozzles have a round orifice, behind which is a so-called swirl plate that rotates during spraying and produces a cone-shaped pattern of spray droplets. Cone nozzles are often used to give good spray coverage when used to treat shrub or tree foliage as well as timber and other materials within buildings where drift may be of little concern. They are the most popular type of nozzle on sprayers used by hobby (amateur) gardeners but, particularly in inexpert hands, may be the principal source of drift in domestic gardens and allotments. In crop spraying the favoured nozzle is the flat fan with an elliptical orifice that produces a sheet of spray. Different flat fan nozzles serve to deliver fine, medium or coarse sprays. Deflector or floodjet nozzles were formerly widely used in the UK by municipal authorities to ensure good coverage and kill of unwanted vegetation on roads and footpaths.

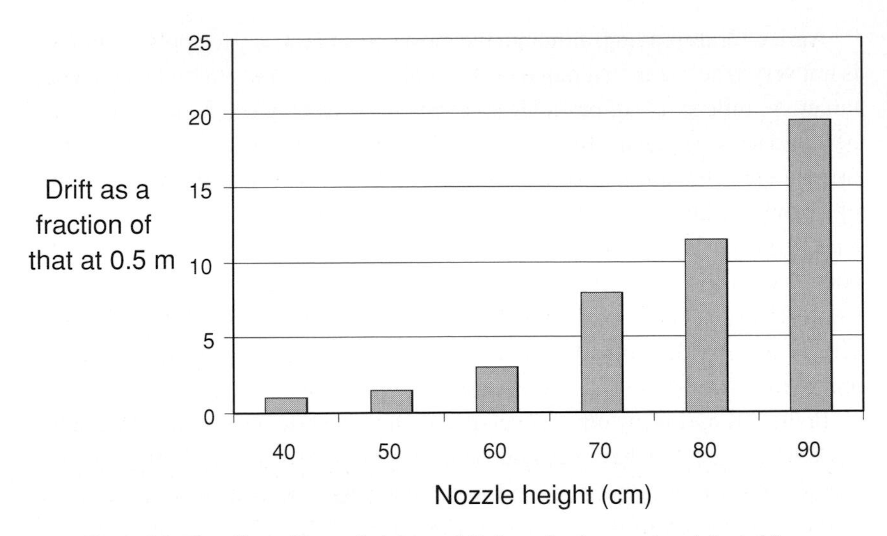


Figure 8.3 The effect of boom height on drift from flat fan nozzles. Adapted from Miller, P. C. H. (1999). Factors affecting the risk of drift into field boundaries. *Proceedings of the Brighton Crop Protection Conference, Weeds*, 439–446.

Improvements in nozzle design and use have centred on features to reduce drift (Taylor *et al.*, 1999). Low drift nozzles that produce coarser droplets at low flow rates by the inclusion of an orifice plate have been developed. Air inducing nozzles, where air is drawn into the liquid stream producing droplets with air bubbles, have been introduced. Nozzle selection charts are now available and these indicate preferred nozzle choices that minimise drift without compromising efficiency.

Drift is further influenced by the operation of spraying apparatus, and especially the position of the spray boom. Generally the higher the boom above the crop or treated surface, the greater is the potential for drift (Figure 8.3). Spraying as close as practicable to the target crop or other area to be sprayed will reduce drift, and in agricultural applications modern tractor-mounted sprayers have sophisticated technology to prevent lateral and vertical movements of spray booms, but which may also allow low frequency movement in order that the ground profile may be followed (Miller, 1999, 2003).

Weather conditions, and particularly the strength and direction of wind, may have a major effect on drift from agricultural and horticultural crop spraying. In the UK and many other countries, clear guidance is given on wind speed and spraying (Table 8.1). Slight breezes blowing away from susceptible crops or vegetation, particularly with herbicides, are considered the safest conditions in which to spray. Still conditions, particularly on warm days, are not considered

Table 8.1 *Wind speed as a guide to spraying in the UK*

| Approximate air speed at boom height | Beaufort scale of wind speed (at a height of 10 m) | Description | Visible signs | Spraying |
|---|---|---|---|---|
| Less than 2 km/h | Force 0 | Calm | Smoke rises vertically | Use only medium or coarse sprays |
| 2–3.2 km/h | Force 1 | Light air | Smoke drifts | Acceptable spraying conditions |
| 3.2–6.5 km/h | Force 2 | Light breeze | Leaves rustle; wind felt on face | Ideal spraying conditions |
| 6.5–9.6 km/h | Force 3 | Gentle breeze | Leaves and twigs in constant motion | Increased risk of spray drift |
| 9.6–14.5 km/h | Force 4 | Moderate breeze | Small branches move; dust and loose paper are raised | Spraying inadvisable |

ideal for spraying since convection currents formed in such conditions may carry fine spray droplets. In reality, because of the limited number of days when spraying may be possible, most spray operators would apply pesticides, particularly insecticides and fungicides, on calm days, but may elect to use medium or coarse nozzles for this purpose.

The distance between sprayed areas and field boundaries can influence effects arising from spray drift and, in this respect, unsprayed field margins or buffer zones now form part of good agricultural practice in many countries. Such zones reduce the likelihood of pesticides entering ditches, streams and other watercourses that often border fields. The nature of the field boundary itself may influence drift: grass/wild flower strips established around sprayed areas have been shown to ameliorate the effects of drift by up to 50% compared to bare stubble. The width and height of trees and shrubs at field margins may have a pronounced effect on drift: high thick boundary hedgerows will clearly restrict spray movement. However, in the UK and elsewhere in Europe, removal of field boundary hedgerows to create larger fields that can be cultivated and sprayed more efficiently formed a common part of agricultural practice under production-driven EU agricultural support programmes in the 1970s and 1980s. Thus many field margins, particularly in predominantly arable areas, may have no hedgerow boundary with, in many cases, a simple open ditch into which field drains run.

Several countries in the EU now require assessment of the risk to water courses to be evaluated prior to spraying. Spray operators in the UK must now carry out a LERAP (local environment risk assessment for pesticides) assessment before applying pesticides and maintain records of this assessment. For many products a buffer zone of 1 m from the water course must be maintained, but for many insecticides, which may affect aquatic invertebrates, the buffer zone may be extended to 5 m.

## 8.3 Physical and chemical properties of pesticides

Globally, pesticides that are volatile and resistant to degradation by ultraviolet light may be transported vast distances after spraying. The discovery of organochlorine compounds in polar ice and snow (Gregor, 1990), thousands of miles from the nearest site of application, is evidence of such long distance movement. The fate and distribution of pesticides in the atmosphere, as well as soil and water, is closely associated with the physical and chemical properties of these molecules.

The stability of pesticides in terms of resistance to photolysis, chemical hydrolysis and microbial degradation is to a large extent a function of their chemical structure. The variation in ring structures and different types of chemical bonding between these in organic molecules largely determine their stability. Of particular note is the stability of chlorinated ring structures. This stability, initially viewed as a positive attribute in the case of organochlorine insecticides and offering for example long season insect control, is now seen in environmental terms as a great disadvantage, resulting in persistence and detrimental effects on non-target organisms. Resistance to metabolic breakdown in the tissues of non-target species has also allowed these compounds to accumulate in food chains (Chapter 9). Some other pesticides, such as triazine and some urea herbicides, certain triazole fungicides and a few modern insecticides, such as tefluthrin and imidacloprid, are stable for months in the soil environment. However, unlike the organochlorines, none of these compounds appears to bioaccumulate in food chains.

The vapour pressure of pesticides (Table 8.2) is an important physicochemical property. Compounds with a high vapour pressure are generally volatile and may readily enter the atmosphere. Volatility may present problems with respect to pesticide use in enclosed situations such as homes, shops and restaurants but the volatile nature of some compounds is actually exploited for pest and

Table 8.2 *Vapour pressure and volatility of selected pesticides*

| | Pesticide | Vapour pressure (mPa at 20°C) |
|---|---|---|
| Non-volatile | Glyphosate | negligible |
| | Cypermethrin | 0.002 |
| | Benomyl | 0.005 |
| | Chlorothalonil | 0.076 |
| Low volatility | Diclofop-methyl | 0.25 |
| | Mecoprop (acid formulation) | 0.31 |
| Moderate volatility | Captan | 1.3 |
| | Clopyralid | 1.33 |
| | Malathion | 5.3 |
| | $\gamma$HCH | 5.6 |
| Volatile | Trifluralin | 9.5 |
| | 2,4-D (acid formulation) | 11 |
| Highly volatile | 2,4-D (methyl ester) | 13.33 |
| | Dichlorvos | 2100 |
| | Nicotine | 5650 |

disease control in grain, vegetable and fruit stores and also for soil application. Methyl bromide has been widely employed as a soil sterilant and fumigant and the organophosphate dichlorvos has been used in fly papers to control flying insects. The manner in which a compound is formulated may also affect its volatility. The ester formulations of phenoxyacetic hormone herbicides are far more volatile than the corresponding acid formulations and the risk of spray drift and movement of pesticides into non-target areas of the environment are obviously much greater with the ester formulations.

The presence and longevity of pesticides in the environment are also influenced by the degree of ionisation of these molecules, whether they are acidic or basic in nature or indeed non-ionic. For example, the highly basic quaternary N-containing pesticides such as paraquat ionise completely in aqueous solutions as cations, and these are readily adsorbed to mineral particles in soil. Other compounds such as the hormone herbicides may dissociate in soil water to acidic forms and may become bound to organic matter.

Another physicochemical property of pesticides is the octanol:water partition coefficient ($K_{OW}$), which, along with the organic carbon partition coefficient $K_{OC}$, is used in environmental chemistry for calculations and predictions of systemicity, lipophilicity, structure–activity relationships, bioaccumulation and

Table 8.3 *Octanol:water partition coefficients of selected pesticides*

| Pesticide | Octanol:water partition coefficient (log $K_{OW}$) | | |
|---|---|---|---|
| Cypermethrin | 6.6 | | |
| Aldrin | 5.2 | ↑ | Lipophilic |
| Diflufenican | 4.9 | | |
| Triadimefon | 3.1 | | |
| Atrazine | 2.5 | | |
| 2,4-D | 2.5 | ↓ | Hydrophilic |
| Metalaxyl | 1.7 | | |
| Thifensulfuron | 1.6 | | |
| Dimethoate | 0.7 | | |

bioconcentration. The coefficient is the relative amount of a chemical moving into the water or octanol phase when shaken in a mixture of these and is commonly expressed in logarithmic form as log $K_{OW}$. Some values for pesticides are given in Table 8.3.

In general, compounds with a value of more than 4 are sparingly water soluble, and readily absorbed by lipids. Such pesticides are not generally systemic, or readily mobile in soils. Some, notably the organochlorine insecticides, have high log $K_{OW}$ values and because of their high lipid solubility, allied to a slow rate of degradation, can persist in the environment and bioaccumulate in food chains. Conversely, compounds with a log $K_{OW}$ of less than 2 are generally water soluble and systemic in plants, but may pose environmental risks since they have the potential to leach from soil into watercourses.

The solubility of pesticides is related to the octanol:water partition coefficient, and is clearly an important property since those compounds that are readily soluble in water may be more likely to leach than those of low water solubility. Some examples of water solubility are given in Table 8.4. However, solubility should not be considered alone. The rate of degradation of pesticides in soil and water must also be considered since some compounds, for example the herbicides 2,4-D and thifensulfuron-methyl that are highly water-soluble, may be rapidly degraded in soil.

In addition to the active ingredient, pesticide formulations usually contain substances that may improve shelf-life, dispersal of the active ingredient in the spray tank, spread and adherence to the leaf surface as well as uptake into plants and/or the target organism. Many pesticides are rather insoluble and thus are formulated to achieve good dispersal in the spray tank largely by mixture with

Table 8.4 *Leaching potential as defined by mobility class and water solubility of pesticides*

| Mobility class defined by the USEPA | Water solubility (mg/l) | Examples (solubility in water as mg/l) |
|---|---|---|
| Very high | $>10^6$–3000 | Aldicarb – 4930<br>Thifensulfuron – 6250 |
| High | 3000–300 | Fenpropidin – 530<br>2,4-D – 311 |
| Medium | 300–30 | Malathion – 145<br>Isoproturon – 65 |
| Low | 30–2 | Dimethoate – 23<br>Atrazine – 22 |
| Slight | 2–0.5 | Fenoxaprop – 0.9<br>Imidacloprid – 0.6 |
| Immobile | <0.5 | DDT – 0.0023<br>Cypermethrin – 0.0004 |

Adapted from Sieber (1988)

solvents or surface-active agents. The latter include a range of materials such as sodium lauryl suphate, octylphenol and polyoxyethylene derivatives. Such wetters and surfactants may also be added separately to spray tanks as adjuvants, chemicals that serve to enhance the performance of pesticides. Adjuvants may allow use of lower doses of the active ingredient, and in some cases enhance the selectivity of the pesticide.

Some pesticides are used as dry powders or dusts with the active ingredient coated on to a powdered kaolinite clay. Such formulations are used for spot treatments in domestic and industrial premises, and sometimes in glasshouses, but are rarely used in the field – the potential for drift of dust formulations is obvious. Wettable powders possess wetting and dispersing agents as a small percentage of their formulation, and these generally present few risks to spray operators or the environment.

Many pesticides were formerly formulated as emulsifiable concentrates. Here pesticides are dissolved in solvents such as cyclohexanone or xylene, and an emulsifier added to ensure the production of a stable emulsion with water in the spray tank. Solvents such as cyclohexanone and especially xylene may present long-term risks to the health of operators, and there is little information about the fate and effects of these formulation additives in the environment. For these reasons emulsifiable concentrates have been steadily replaced by other

Table 8.5 *Calculated tropospheric lifetimes of selected pesticides*

| Pesticide | Lifetime |
|---|---|
| Malathion | 3 h |
| Dieldrin | 1.1 day |
| $\gamma$HCH | 7 days |

Adapted from Atkinson, Kwok and Arey (1992).

formulations such as suspension concentrates, where the active ingredient is mixed with finely milled particles of powdered clay along with a dispersing agent. Further formulation developments that offer greater safety to operators as well as more efficient targetting of pesticides include microencapsulation. Here, the active ingredient is embedded in a polymer or contained in a polymeric shell; such formulations are often used with insecticides and insect growth regulators because of their volatility. Granular formulations of pesticides present a lower risk of operator contamination compared to dusts and sprays, and are often incorporated into soil, reducing potential non-target effects to, for example, ground-dwelling arthropods in the case of insecticides.

## 8.4 Pesticides in the atmosphere

The principal means by which pesticides enter the atmosphere is volatilisation from droplets produced during spraying operations. However, even after application, pesticides may enter the atmosphere by volatilisation from crop and soil surfaces, especially if these are moist. In some cases, losses can be as much as 90% of the applied dose; within 7 days of application in the case of the herbicide trifluralin and the insecticide $\gamma$HCH (Spencer and Cliath, 1990).

After volatilisation, pesticides may undergo transformations in the atmosphere. Some organophosphorus insecticides and many herbicides undergo photolysis and participate in chemical reactions with hydroxyl ($OH^-$) and nitrate ($NO_3$) radicals as well as ozone ($O_3$) and may have lifetimes of a few hours or less. Others, notably the organochlorine insecticides, have much longer gas-phase lifetimes, often several days, and thus may be transported considerable distances in the atmosphere (Table 8.5).

The most common pesticides detected in the atmosphere are the organochlorine insecticides including DDT, its breakdown products and isomers of HCH.

Table 8.6 *Pesticides detected in air and rainwater in Brimstone, Oxfordshire 1990–1991*

| Pesticide | Concentration in air ($ng/m^3$) | Peak concentration in rainfall ($\mu g/l$) |
|---|---|---|
| γHCH | 0.041–1.50 | 0.5 |
| Simazine | 0.014–0.29 | 0.23 |
| Isoproturon | 0.005–0.15 | 0.13 |

Adapted from Harris *et al.* (1992)

Extensive surveys of the incidence of organochlorine insecticides in various parts of the globe have been carried out. A comprehensive survey by Tatsukawa, Yamaguchi, Kawano, Kannan and Tanabe (1990) found concentrations of γHCH of up to 15 $ng/m^3$ in the atmosphere. The highest concentrations were found in localities close to industrially developed areas but organochlorine insecticides have been found in the Antarctic and Southern Atlantic areas, albeit at extremely low atmospheric concentrations of around 22 $pg/m^3$, for γHCH, and DDE and DDT at up to 400 $pg/m^3$.

Pesticides in the atmosphere may be deposited by rainfall (Dubus *et al.*, 1998). Studies have shown the presence of pesticides, and not just organochlorines, in rainfall in several parts of the world. In the UK and other parts of Europe atrazine, simazine, isoproturon and γHCH have been detected in the air and rain water (Table 8.6).

## 8.5 Pesticides in soils

Pesticides form a very small proportion of the organic components entering soils. A multitude of chemicals enter soils as the breakdown products of dead plants, animals and microorganisms and these ultimately form part of the humus fraction of soils – as indeed do the end products of many pesticides (Weber, 1994).

Processes determining the fate of pesticides in soil include volatilisation, adsorption to soil particles and organic matter, photochemical decomposition, plant uptake, chemical and microbial breakdown and leaching. Photochemical decomposition, as well as occurring in the atmosphere, may take place on soil surfaces. Some soil-applied pesticides such as the herbicide trifluralin are photolabile and inactivation and photodegradation of such compounds may occur on exposed soil surfaces.

Table 8.7 *$K_d$ values and leaching potential of selected pesticides*

| Pesticide | $K_d$ value in loam soil containing 3% organic matter | Distance pesticide leached if rain exceeds evapotranspiration by 25 cm and assuming no degradation |
|---|---|---|
| 2,4-D | 1 | Much leaching of pesticides into subsoil |
| Simazine | 1–4 | |
| Diuron | 10 | Much pesticide leached into soil but peak concentration in top 20 cm |
| Lindane | 50 | |
| Parathion | 200 | Small amount of leaching with peak concentration in top 5 cm of soil |
| DDT | 10 000 | No significant leaching |
| Paraquat | 100 000+ | |

Partially adapted from Riley (1991)

With the exception of seed treatments, a very small proportion of pesticides applied to crops is actually absorbed by the target organisms. This may be around 1–2% of the applied dose of soil herbicides, and much less in the case of soil-applied insecticides and fungicides. Some of the low-dose herbicides, for example the sulphonylureas, and insecticides such as the avermectins may be exceptions, where correspondingly more of the applied dose is taken up by the target species. However, uptake by target species is not a major mechanism of pesticide dissipation from soils (Riley and Eagle, 1990).

Once pesticides enter soil, they may exist within the solid, gas and water constituents. This distribution between the various components is commonly expressed as partition coefficients. The most important of these is the soil:water adsorption coefficient $K_d$ which is:

$$\frac{\text{Concentration of pesticide adsorbed to soil particles}}{\text{Concentration of pesticide in soil water}}$$

The soil adsorption coefficient is usually measured as:

$$\frac{\text{Quantity of pesticide (mg) in soil (g)}}{\text{Quantity of pesticide (mg) in water (ml)}}$$

High $K_d$ values mean that a chemical is likely to be strongly adsorbed on to soil particles, and thus will not tend to leach. Table 8.7 gives $K_d$ values of selected pesticides, along with comments on the likelihood of their leaching. In general, pesticides existing as anions at neutral soil pH values, such as the herbicide 2,4-D, are weakly adsorbed. Non-ionic or uncharged molecules such as the triazine and urea herbicides (e.g. atrazine, diuron) may be adsorbed on to organic matter by low energy bonds. With most pesticides, a good correlation is evident between the adsorption coefficient and the organic matter content of soils. The higher the organic content of soil, the greater is the binding of many pesticides and this may lead to some soil-applied pesticides being ineffective when applied to peats, peaty and organic soils. As noted earlier, the partition coefficient between organic carbon and soil water – $K_{oc}$ – is often used to express the adsorption of a pesticide to organic matter, and a good correlation exists between this coefficient, the soil:water partition coefficient $K_d$ and the octanol:water partition coefficient $K_{ow}$.

Although binding to organic matter is the principal mechanism of adsorption for most pesticides in soil, some may bind additionally to clay mineral particles. The clay fraction of soils is negatively charged and positively charged molecules such as paraquat are strongly bound to negative sites on clay particles, and may only be released by treatment with strong acids. No leaching occurs, and for all practical purposes such compounds are inactivated on contact with soil.

## 8.6 Chemical and microbial degradation of pesticides in soils

Most organic chemicals of natural or synthetic origin may be degraded or metabolised in soil by microorganisms. The annual turnover of biodegradable material in soil is estimated on a global basis at $2 \times 10^{11}$ tonnes. Much of this material is hydrocarbons and carbon-based polymers of natural origin, primarily the decomposing remains of plants animals and microorganisms. Global annual production of synthetic organic chemicals has been estimated at $2 \times 10^8$ tonnes – about 0.1% of organic matter in the biosphere; pesticides account for about $2.5 \times 10^6$ tonnes per year – about 0.000 001% of synthetic organic chemicals and thus form a miniscule proportion of the biodegradable material referred to above.

Removal of pesticides from soil is not considered complete until the parent compound has been completely mineralised to $CO_2$, methane, inorganic nutrients and humus compounds. Mineralisation is dependent mainly on microbial transformation of pesticides with some chemical hydrolysis. Microbial transformations occur most readily where pesticidal molecules have some similarities to those already present in nature. For example, the breakdown of hormone-based herbicides such as 2,4-D is rapid in soils, negating to some extent the high leaching potential of these compounds. By contrast, some chemicals may contain functional groups and structures rarely found in naturally occurring substances; these may be degraded slowly and thus persist in the environment.

Examples include some sulphonate- and halogen-containing pesticides such as the sulphonylurea herbicides and particularly the organochlorine insecticides.

Degradation of pesticides in soil is carried out primarily by bacteria, notably species of *Pseudomonas*. Extracellular enzymes such as esterases, hydroxylases and hydrolases produced by soil microorganisms which break down components of organic matter are also involved in the degradation of pesticides. In many cases the transformations carried out are plasmid mediated. Plasmids are extrachromosomal pieces of DNA and contain genes that are involved in pathways of catabolism for naturally occurring substances, and it is likely that such pathways may become adapted to deal with novel synthetic organic substances such as pesticides entering soil.

Knowledge of the precise pathways and rates of degradation of pesticides is required by regulatory authorities in most countries as a condition of approval for use. Pesticide degradation is usually determined with radiolabelled compounds under controlled conditions, and lysimeters (Figure 8.7) are commonly used. The molecules may be labelled in different parts so that the fates of individual components of the molecule can be followed. The breakdown of the herbicide 2,4-D by soil microorganisms has been extensively studied, and Figure 8.4 shows the principal routes and steps involving complete degradation of both the aromatic ring and side chain.

The degradative routes of many pesticides in soil (as well as in plants and mammals) have been published in the definitive reference works of Roberts and his co-authors. (Roberts, 1998, 1999). Most compounds are transformed to so-called intermediate metabolites. For example, the herbicide bentazone is degraded microbially to the metabolites 6-OH bentazone, 8-OH bentazone and anthranilic acid isopropylamide. Isopropylamide is a natural product which occurs during the metabolism of citric acid by some microorganisms. Other pesticides may undergo similar fates in soil. Insecticides such as chlorpyrifos are broken down, in this case to the metabolite 3,5,6-trichloropyridin-2-ol, and this may be further degraded. The insecticide imidacloprid is soil applied and slowly degrades to the metabolite 6-chloronicotinic acid, some of which may be bound within soil organic matter but most of which is transformed to $CO_2$ (Figure 8.5). In some cases, pesticide degradation in soil may be so rapid that intermediates have not been detected, for example for the herbicide clopyralid.

Pesticide metabolites often become bound to the organic matter fraction in soils, forming part of the humus fraction, from which they may only be released by strong acids/alkalis. Often about half of the carbon present in pesticidal molecules becomes bound in this way, with the remainder being liberated as $CO_2$. The actual amounts of these bound metabolites or residues within soil organic matter are very small. An application of a herbicide at 0.5 kg/ha may

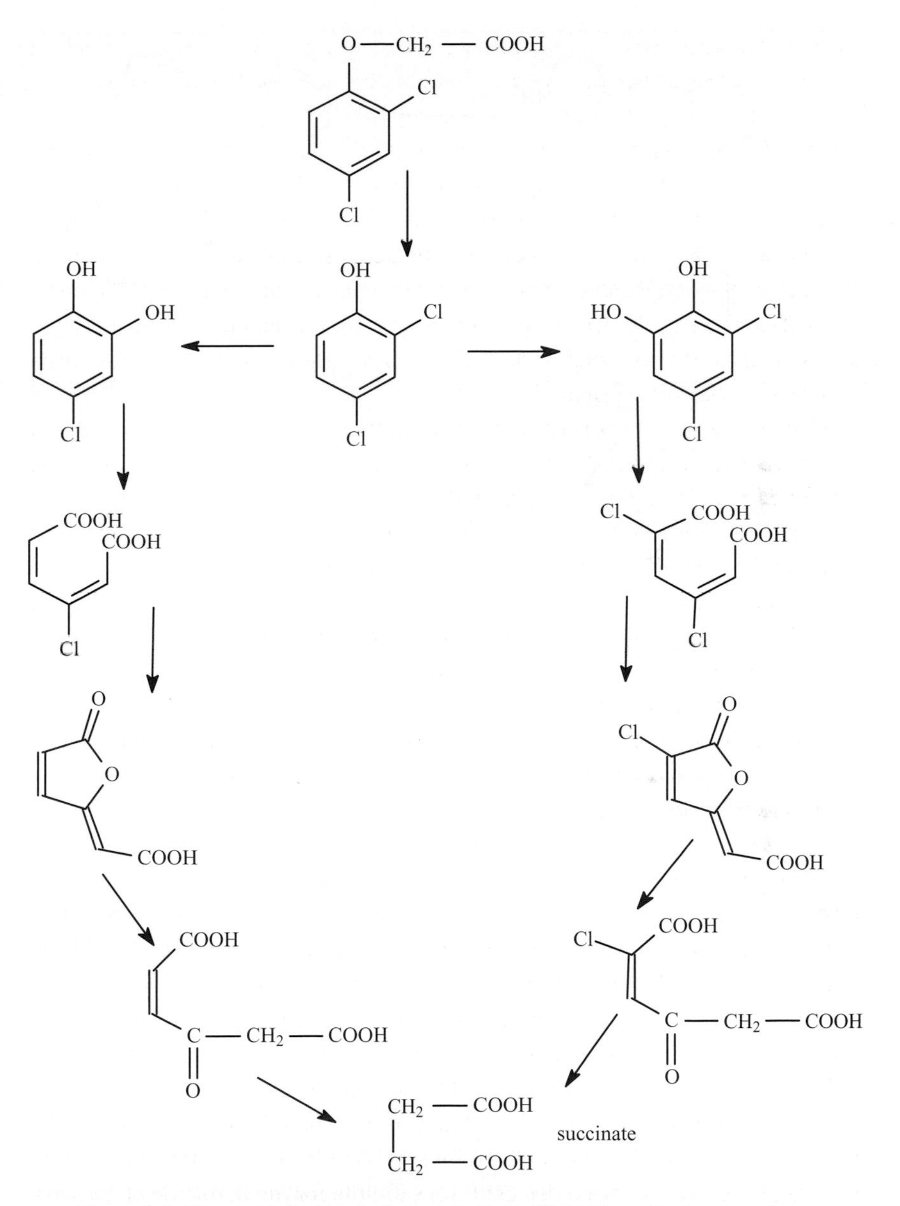


Figure 8.4 Structural changes in the metabolism of 2,4-D by soil microorganisms. Adapted from Roberts (1998). *Metabolic Pathways of Agrochemicals.* Part One. *Herbicides and Plant Growth Regulators*. London: Royal Society of Chemistry.

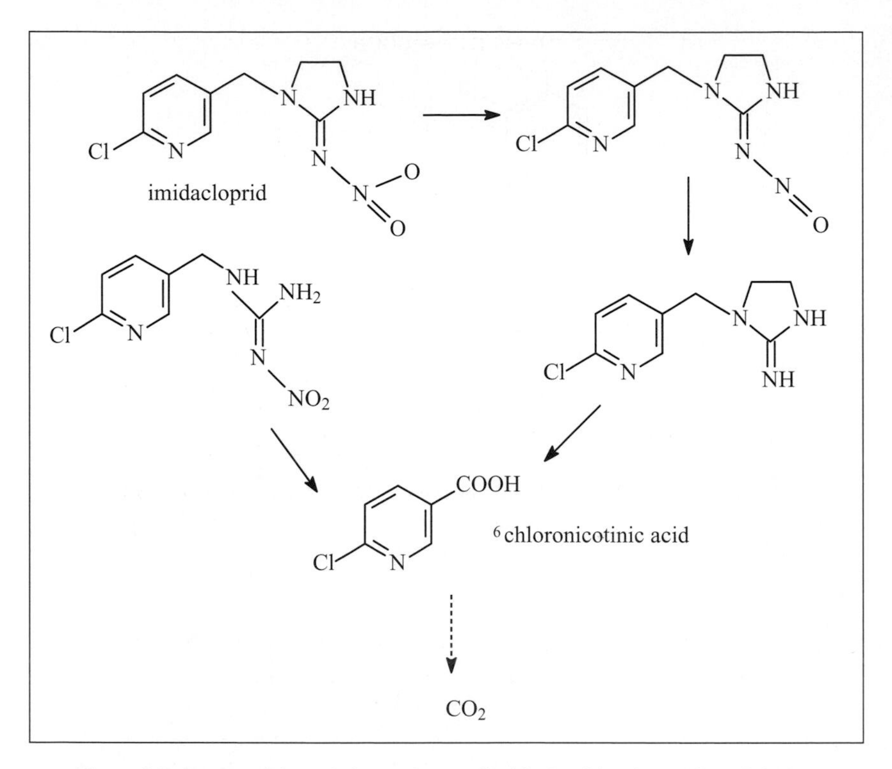


Figure 8.5 Proposed degradation pathway of imidacloprid. Adapted from Scholtz, and Spiteller (1992).

leave bound residues at 0.1 mg/kg in the top 20 cm of soil. This may be compared to the normal range of 1–5% (10 000–50 000 mg/kg) total organic matter in soil. The bound residues are probably slowly degraded with the rest of the humus fraction in soils (Riley and Eagle, 1990). Many widely used compounds such as glyphosate, imidacloprid (Figures 8.5 and 8.6) and azoxystrobin break down and release $CO_2$ as the principal degradation product.

The rate of degradation is often expressed as the half life of the pesticidal molecule in soil, and this varies according to the different compounds. Persistence is of course for farmers and growers a desirable quality with, for example, herbicides and insecticides applied to soils and other areas for long season weed and insect control respectively. However, persistence itself can be detrimental to following crops in the case of herbicides (9.4.1), and persistence in many cases increases the chances of pesticides and their metabolites leaching from soil. The half lives of some common soil-applied and other pesticides are given in Table 8.8.

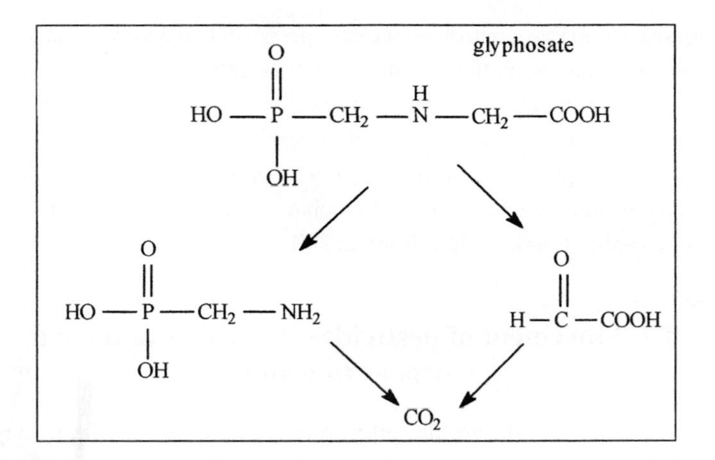


Figure 8.6 Degradation of glyphosate in soil. Adapted from Roberts (1998) *Metabolic Pathways of Agrochemicals*. Part One. *Herbicides and Plant Growth Regulators*. London: Royal Society of Chemistry.

Table 8.8 *Half lives of selected pesticides in soil*

| Pesticide | Half life in soil (days), unless otherwise stated) |
|---|---|
| Benomyl | 19 h |
| Glyphosate | 3 |
| 2,4-D | 7 |
| Thifensulfuron methyl | 6–12 |
| Azoxystrobin | 7–28 |
| Isoproturon | 6–28 |
| Atrazine | 30–50 |
| Aldicarb | 60–90 |
| Diuron | 90–180 |
| Picloram | 30–330 |
| DDT | 2–15 years |

Paradoxically, rapid degradation of pesticides in soil can pose problems for farmers and growers. Repeated application of the same pesticide can result in the establishment of a population of soil microorganisms capable of rapidly breaking down the pesticide once applied (Riley and Eagle, 1990). This phenomenon of accelerated or enhanced degradation is of course advantageous for foliar-acting pesticides, which do not exert their activity through the soil. It is however disadvantageous with respect to pesticides applied to soil, which must persist to enable control of weeds, pests and diseases in soil. In fact some pesticides may be ineffective on soils where enhanced degradation occurs. Accelerated degradation of the soil-applied insecticides aldicarb and carbofuran has resulted in unsatisfactory control of nematode and insect pests in potato and sugar beet respectively. In the UK, poor control of white rot (*Sclerotium cepivorum*) of onion crops has been associated with enhanced degradation of the dicarboximide fungicide iprodione in soil.

## 8.7 Movement of pesticides from soils and hard surfaces to water

Leaching through the soil profile, often referred to as matrix flow, may ultimately lead to the passage of pesticides and their metabolites into groundwater. Leaching is influenced by the quantity and frequency of precipitation, the composition of the soil, as well as the physicochemical properties of pesticides noted earlier (8.3). Leaching of molecules through soil may be monitored using lysimeters (Figure 8.7). These are cylinders, usually metal and commonly

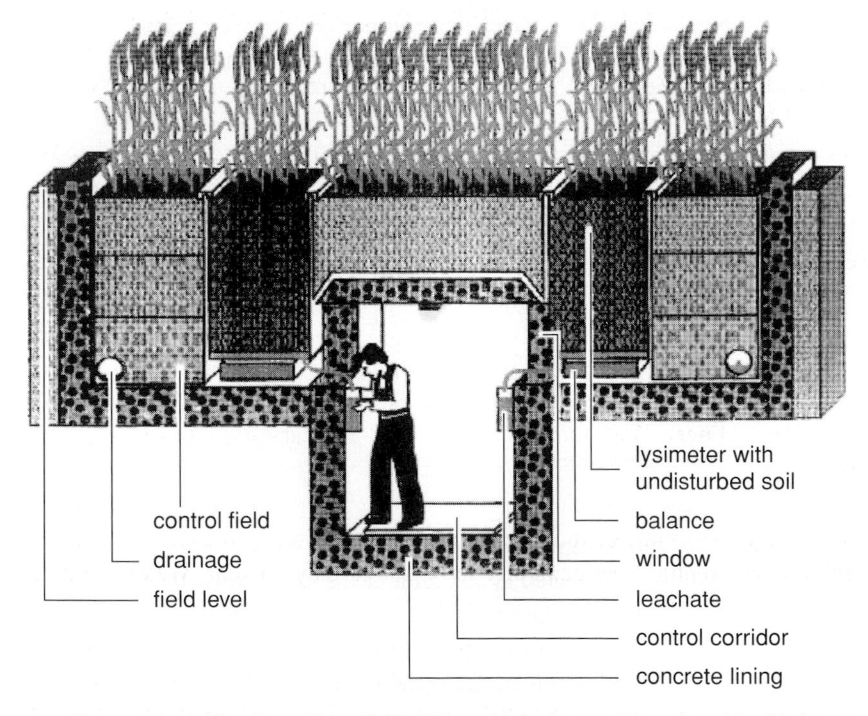


Figure 8.7 A diagrammatic representation of a lysimeter. Reproduced by kind permission of Bayer Crop Science.

1 m deep and 0.8 m in diameter, which are inserted into or filled with soil. Lysimeters are often sited in growing crops, allowing monitoring of pesticide leaching in conditions as close as possible to normal field situations. Indeed, crops can be sown and raised within the lysimeters. Radiolabelled pesticide may be applied to the soil in the lysimeter. Removing the lysimeter complete with soil and extracting the labelled pesticide at different depths may determine its fate. Lysimeters allow the study of leaching to be undertaken with different soils of varying characteristics at the same experimental site.

Although leaching may occur more quickly in, for example, sandy soils, the flow of groundwater into first the unsaturated and then the saturated layers (aquifers) beneath soils is generally slow and dissipation of pesticides may occur during such matrix flow. During movement towards aquifers, pesticide degradation may occur, albeit at a slower rate than in soils. Indeed experiments on chalk aquifers in southern England have shown a decline in concentration of herbicides such as atrazine and isoproturon with depth (Chilton *et al.*, 1995). In the lower horizons of soils, the low organic matter and air content relative to surface layers ensures a much lower level of microbial activity, and thus lower potential for pesticide degradation.

The potential for pesticides applied under normal agricultural practice to leach through the soil and reach groundwater was initially considered as negligible, but pesticides have been detected in borehole and well water in both the USA and Europe (Bewick, 1994). In the late 1970s the detection of the nematicide dibromochloropropane (DBCP) in California and of the insecticide/nematicide aldicarb in a domestic well at Long Island, New York were the first indications in the USA that pesticides were moving from soils into aquifers. Atrazine, regularly applied to maize crops for weed control in the USA, began to appear in drinking water supplies during the 1980s, sometimes at concentrations above 1 μg/l (Richards and Baker, 1990). In the UK, herbicides such as isoproturon, widely applied to cereal crops, have been regularly and widely detected in both natural and drinking waters during the last two decades at concentrations above the EU limit for pesticides in drinking water of 0.1 μg/l.

Pesticides may enter watercourses more rapidly through cracks and macropores in soils; routes of movement often termed preferential flow. Soils that exhibit preferential flow generally have a high clay content. Water normally moves through the matrix of clay soils at a slow rate due to the small pore diameter of clay particles, but, when dried out, cracks in the soil may swiftly conduct water to the drainage system. Field drainage systems installed to alleviate problems of waterlogging in clay soils may in turn aid the movement of pesticides to watercourses. Comprehensive studies at Brimstone Farm in Oxfordshire have shown the importance of preferential flow in the movement of the herbicide isoproturon from clay soils into drainage channels and streams, with over 70% of the leached pesticide travelling to the drainage system by this route (Beulke, Brown and Dubus, 1998). Finally, a few pesticides may actually be directly applied to natural waters, usually for weed control.

Considerable seasonal variations occur in the levels of pesticides in run-off. Rapid but transient increases in pesticide concentrations may occur in surface run-off following periods of heavy rainfall. Measurements in the UK have shown the presence of isoproturon in run-off water from clay soils at levels of around 100 μg/l, compared to 30–50 μg/l in field drainage. These concentrations were seen in run-off and drainage systems following the first appreciable rainfall after application of the herbicide in autumn (Williams *et al.*, 1991). Indeed, run-off of pesticides from agricultural soils in the UK is most pronounced in the autumn, when herbicides such as isoproturon have been applied for control of weeds in cereal crops. At other times of the year, pesticide concentrations in run-off may be much lower.

Risks to the health of non-target species may obviously be greatest where pesticides in their manufactured, concentrated form are released at a single point into the environment and this point-source pollution can have devastating effects. Such events are fortunately rare, but contamination of natural waters by local spillages of concentrated pesticides still occurs in developed countries and is common in some underdeveloped areas of the world. Point-source pollution principally arises from contamination of hard surfaces such as concrete, gravel or asphalt where loading, filling and mixing of concentrates in spray tanks is carried out. Spillages may be readily washed into drains; whereas the amount of pesticide passing to ground and surface waters such as rivers and streams is about 0.1–2% of that applied, losses from hard surfaces may be as much as 50% (Shepherd and Heather, 1999). Point-source contamination is held to be a major route of entry of herbicides such as isoproturon into streams and river systems in parts of the UK (Table 8.9).

Run-off, especially from pesticide spillages and careless washing of farm sprayers and pesticide containers, may thus result locally in much higher levels of pesticides in natural waters than from leaching through the soil matrix or macropores. These may also occur from application to non-porous surfaces such as roads and pavements as well as gravel and railway permanent way (sleepers bedded in gravel). Run-off from non-agricultural situations is probably the major route of entry by the triazine herbicides atrazine and simazine into natural waters in the UK. These compounds have not been used extensively on crops, particularly maize, in the UK but, from their introduction in the mid 1950s,

Table 8.9 *Contribution of isoproturon contamination from activities observed over 2 months in a farmyard with concrete surface, and draining directly into the River Cherwell, Northamptonshire*

| Activity | Grams of isoproturon produced (mid November-mid January) |
|---|---|
| Tank rinsing | 0.5 |
| Spillage | 1.2 |
| Container washings spill | 0.35 |
| Tractor washing | 0.7 |
| Tractor wheels | 0.4 |

From Higginbotham *et al.* (1999)

they were widely used for total weed control on uncultivated locations such as roads, pavements, gravelled areas and railway tracks, until their withdrawal for this purpose in 1993. Since the late 1970s, atrazine and simazine have been regularly detected at concentrations up to 1 μg/l in natural and drinking waters in the UK. The detection of atrazine in borehole supplies in the UK in the mid 1990s is a localised instance of pesticide contamination of drinking water, and resulted from a change in cropping pattern with farmers switching to maize and using atrazine for weed control – a similar scenario to that in parts of the USA in the 1980s (Hillier and White, 2001). In the UK, the residual compound diuron was recommended as an alternative to atrazine for weed control in urban areas, but this too has been detected in water courses following routine application to road and pavement surfaces (Davies *et al.*, 1995).

The potential for pesticides to enter natural waters forms part of risk assessments associated with approval of compounds in the EU and elsewhere. Since the sampling and measurement of pesticide concentrations in surface and groundwater is time-consuming and expensive, theoretical models to predict the fate of pesticides in soil and water have been developed and refined. For example, the FOCUS (FOrum for Coordination of pesticide fate models and their USe) surface water scenarios working group has considered predicted environmental concentrations (PECs) of pesticides arising from drift, run-off and drainage (FOCUS, 2000). The model involves a tiered approach to risk assessment with stepwise considerations of predicted environmental concentrations of pesticides, which may then be used in aquatic risk assessments during approval processes for registration of pesticides in the EU.

Table 8.10 *Pesticides listed under the EU 'Dangerous Substances' Directive (76/464/EEC)*

| List I substances | Examples of List II substances |
|---|---|
| Hexachlorocyclohexane (HCH, lindane) | Cyfluthrin |
| DDT | Permethrin |
| Pentachlorophenol | Sulcofuron |
| Aldrin | Tributyltin |
| Dieldrin | Triphenyltin |
| Isodrin | |
| Endrin | |
| Hexachlorobenzene | |

## 8.8 Pesticides in natural waters

The quality of water in rivers, reservoirs, wells and boreholes, particularly those from which water is abstracted for drinking purposes, is governed in many countries by stringent regulations. In the past these regulations were primarily designed to prevent the spread of water-borne microbial pathogens of human beings, but more recently have extended to considerations of contamination by chemicals introduced into the environment, and particularly pesticides. In the EU, standards have been specified by directives including the 'Dangerous Substances' Directive (76/464/EEC), the 'Groundwater' Directive (80/68/EEC) and the 'Drinking Water' Directive (80/778/EEC). In the UK, under the Water Resources Act of 1991 it is an offence to allow any 'poisonous, noxious or polluting matter' to enter controlled waters, defined as rivers, lakes, groundwater, estuaries and coastal waters.

EU Directive 76/464/EEC established guidelines for the reduction, or in some cases elimination, of substances from natural waters. Substances falling in List I, the 'black list', of the Directive are considered undesirable in terms of their toxicity and persistence, and the intention is to eliminate these from aquatic environments within the European Community. A subsidiary list has also been drawn up and the levels of substances on this List II, or 'grey list', are required 'to be reduced' in EU waters. Eight pesticides are included in List I, and several more in List II (Table 8.10).

Environmental Quality Standards (EQSs) have been set within the EU's 'Dangerous Substances' Directive. An EQS is the concentration of a substance that must not be exceeded within the aquatic environment in order to protect it

Table 8.11 *Environmental Quality Standards for selected pesticides in the UK*

| Pesticide | Annual average (μg/l) |
|---|---|
| γHCH | 0.1 |
| DDT | 0.025 |
| Total cyclodienes | 0.02 |
| Total triazines | 2 |
| Malathion | 0.01 |
| Trifluralin | 0.1 |
| Permethrin | 0.01 |
| Dimethoate | 1 |

From UK Environment Agency at http://www.environment-agency.gov.uk

for its recognised uses. There is a statutory requirement on EU member states to monitor levels of substances in List I within natural waters and to act if concentrations exceed EQSs set by the EU. Some countries have gone further and established additional non-statutory EQSs. For example, the Environment Agency in the UK has set EQSs for about 50 pesticides in addition to those of List I of the EU 'Dangerous Substances' Directive (Table 8.11). EQSs are specific to individual pesticides and are derived from data on persistence, toxicity and other ecotoxicological properties.

Monitoring of pesticides in UK waters in the early 1990s initially showed a high compliance with EQSs, but further sampling of targetted sites thought to carry a high risk of pesticide contamination has resulted in a higher percentage of samples exceeding EQSs (Chave, 1995; Table 8.12). Pesticides that exceeded EQSs in 1997 were primarily insecticides such as cypermethrin associated with sheep-dipping operations. Few breaches of EQSs for List I persistent compounds, such as the organochlorine insecticides, occurred in the 1990s, and indeed the presence of organochlorines in natural waters within the UK has been steadily declining since the 1970s (Figure 8.8), and in 2000–2002, no sites sampled by the Environment Agency had γHCH or DDT concentrations in excess of the EQS. In 2002, 122 of 1773 (about 7%) freshwater sites in England and Wales monitored by the Environment Agency failed EQSs at least once. Overall percentage failures for 1999, 2000 and 2001 were 10%, 9% and 7%, respectively, indicating a downward trend in non-compliance.

Table 8.12 *Pesticides exceeding environmental quality standards in surface freshwaters of England and Wales during 1995 and 1997*

| | Percentage of sites failing EQS | |
|---|---|---|
| Pesticide | 1995 | 1997 |
| Total HCH | 0.8 | 1.0 |
| Dieldrin | 0.4 | 0 |
| DDT | 0.06 | 0 |
| Permethrin | 1.6 | 42.6 |
| Cyfluthrin | 1.4 | 42.4 |
| Diazinon | 2 | 13.1 |
| Cypermethrin | 1.5 | 45.0 |
| PCSD/Eulan | 0.8 | 15.6 |
| Atrazine | 0.1 | 0 |
| Isoproturon | 0.6 | 0.3 |
| Diuron | 0.3 | 0.5 |
| **Number of sites** | 3.000 | 1419 |
| **Percentage of sites failing an EQS** | 7.9 | 14 |
| **Targetted** | No | Yes |

Data from UK Department of the Environment, Transport and the Regions (1999) and from Environment Agency (2000).

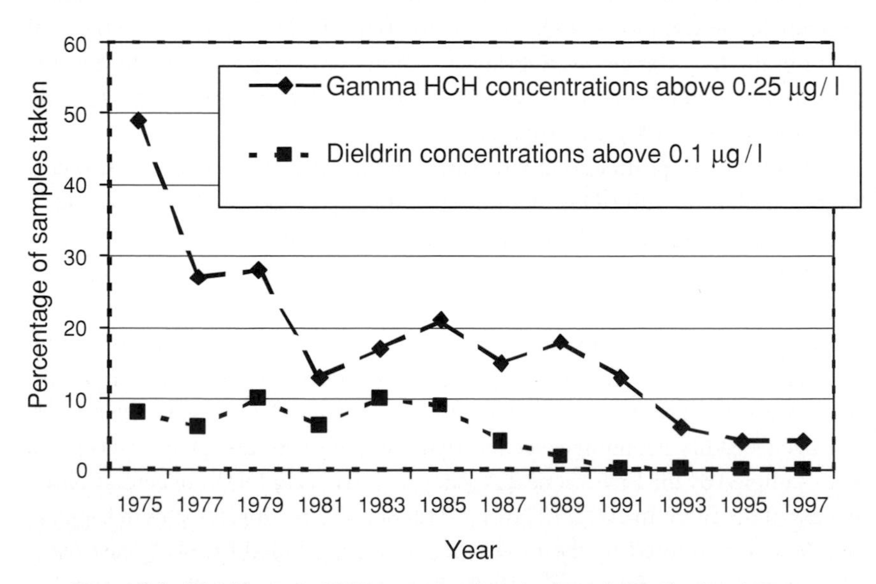


Figure 8.8 Dieldrin and γHCH concentrations in rivers in England and Wales 1975–1997. Data from the UK Environment Agency at http://www.environment-agency.gov.uk.

**Guidance to farmers, managers and advisors**

- Store pesticide containers in a safe, secure place
- Get the dosage right
- Plough and cultivate clay soils thoroughly to prevent rapid flow through cracks
- Implement strategies such as drilling across sloping fields to reduce run off

**Guidance to spray operators**

- Apply only when cereal crops have emerged
- Apply only to moist soils which have few cracks
- Do not apply if heavy rain is forecast
- Don not apply to very wet or waterlogged soils
- Take particular care on sloping fields if the soil is very wet

**Guidance during spray operations**

- Take care not to spill concentrate
- Rinse concentrate from packs/containers into spray tank
- Check continually for drift
- Dispose of tank residues appropriately e.g. on unsprayed crop
- Wash down sprayer only in field
- Store empty containers under cover

Figure 8.9 Recommendations of the UK Isoproturon Task Force intended to minimise the entry of IPU into watercourses.

Pesticides, particularly those used as antifouling agents on ship hulls, have also been monitored in estuarine and coastal waters by the Environment Agency in the UK. The most common compound detected in 1997 was the herbicide diuron, used to prevent algal colonisation of ship hulls, but concentrations of diuron in that year did not exceed the EQS at any sampling point in the UK. However, tributyltin and triphenyltin compounds, also used as antifouling agents for aquatic molluscs (4.3), exceeded the EQS for these compounds at over 60% of sampling sites in 1997 in the UK, only declining to 20–21% in 2001–2002, despite limitations placed on their use.

Measures to reduce pesticide movement into water-courses have been implemented (Bewick, 1994). In addition to the local environment risk assessment for pesticides schemes (LERAPs) noted earlier, specific initiatives have been set up to deal with problems associated with individual pesticides. One such is the isoproturon task force established in the UK in the 1990s and their recommendations, which are applicable to most pesticides, are summarised in Figure 8.9 (White *et al.*, 1997).

# 9
# Pesticides and non-target species

## 9.1 Introduction

Even though dissipation may occur from photochemical, chemical and microbial degradation, the introduction of pesticides into the environment, particularly as sprays, has led to adverse effects on organisms other than the intended target. The duration and intensity of exposure to pesticides of non-target species are affected by many factors related to their habitat and behaviour. These include the population density of non-target species at the time of application, the physical nature of the area treated as well as the type of crop or vegetation (including refuges for non-target species), the time of year when pesticides are applied and the age and sex of fauna. The effects of many pesticides on non-target species are often restricted to the locality in which they were applied. For non-target fauna and flora, knowledge of local environmental concentrations may be useful in predicting the likelihood of acute toxic effects on sensitive species. However, concentrations within areas of application may not be the sole determinants of risk to non-target species. The risk posed by a pesticide depends on the sensitivity of individual organisms and the degree of exposure to the chemical.

For example, soil-applied herbicides may be present in relatively high concentrations and have little direct effect on soil fauna, or accumulate through food webs/chains. Of course such herbicides may pose problems for local flora and even following crops. Conversely, chemicals present in low concentrations or having low persistence could be classified as safe, whereas in practice they may present risks due to their intrinsic high toxicity, bioconcentration or bioaccumulation in certain species. For example, pyrethroid insecticides may be highly toxic at low doses to non-target aquatic arthropod species. Other compounds may present risks to some species at higher trophic levels since they are persistent, leading to long periods of exposure and consequent bioconcentration from initially very low levels in soil and water, or bioaccumulation through food webs and chains, and this may occur in areas remote from the site of application. Thus, the effects of pesticides even if not manifest in individual species may have consequences for the ecosystem as a whole. Constituents of the microflora and microfauna may be food sources for larger animals as part of food webs and chains, and, apart from effects due to reductions in populations at the base of food webs/chains, adverse effects may follow the bioaccumulation and biomagnification of pesticide residues, leading to problems at higher trophic levels.

Another important factor to consider in addition to concentrations of pesticides is the biomass of non-target organisms. Larger organisms may be less affected or even unaffected by concentrations of chemicals that may have adverse effects on smaller species. For example, with respect to mammalian toxicity, ingestion of a small amount of pesticide, perhaps 1 mg, may prove lethal to a small mammal such as a vole weighing 5 g. Here the dosage is 200 mg/kg in terms of body weight. Ingestion of the same amount by a cow weighing 500 kg would give a dosage of 0.002 mg/kg body weight, and here adverse effects may be less pronounced or even absent. Of course, such assertions depend to some extent upon interspecies variation in rates of absorption, metabolism and excretion of compounds.

## 9.2 Early studies and identification of the long-term effects of organochlorines

Although some early studies were undertaken on, for example, the effect of arsenical compounds on wildlife in forest plantations in the USA, concerns about effects on non-target organisms heightened following the increased use of pesticides after 1945. Reports of the detrimental effects of pesticides began to appear almost immediately: studies in the USA in 1946/7 showed dramatic reductions in populations of songbirds and invertebrates in forests treated with DDT. Subsequent studies worldwide have identified among all pesticides the organochlorine insecticides as the group most linked with effects on non-target fauna, and as noted earlier this has led to the prohibition of agricultural use of compounds such as DDT, the cyclodienes and $\gamma$HCH in many countries. Studies in the USA during the late 1940s had indicated that deaths of fish and invertebrates occurred in streams some distance from forested areas treated with insecticides, and this was naturally linked to the water in the streams moving from the areas of application. However, it became clear in the late 1950s and 1960s that deaths of birds, and some other non-target vertebrates were occurring in areas

Table 9.1 *Amounts of DDT and DDE present in soil, leaves and earthworms after the application of DDT to elm trees in summer*

| | DDT | DDE |
|---|---|---|
| Application rate | 0.9 kg per tree | |
| Concentration of residues: | | |
| Soil surface | 1.46 kg/ha | |
| Top 5 cm of soil | 5.9–9.5 ppm | 0.7–5.4 ppm |
| Leaves after spraying | 174–273 ppm | 9–20 ppm |
| Leaves after falling from tree | 20–28 ppm | 1–4 ppm |
| Earthworms, after spraying | 33–164 ppm | 3–73 ppm |
| Earthworms several months later | 86 ppm | 33 ppm |

From Rudd (1964) and quoted in Cooke, Greig-Smith and Jones (1992)

remote from the application of pesticides, and that the tissues of many were contaminated with organochlorine insecticides. Subsequent studies showed the very wide dispersal of residues of organochlorine insecticides in the global biosphere and furthermore that organochlorines (and other persistent pollutants such as polychlorinated biphenyl compounds – PCBs) could move from sites of application in water, air, rain, snow, fog and in migratory species such as birds. Indeed, organochlorines and PCBs have been found in most parts of the globe, including the Antarctic and Arctic (8.4). Exposure of animals at sites of pesticide application may be due to dermal absorption, ingestion or inhalation, but in areas remote from application, the principal source of contamination is food, supplemented in the case of fish by absorption from contaminated rivers, seas and oceans during passage of water across the gills.

In this respect, the phenomenon of biomagnification of residues through higher trophic levels within ecosystems is now well documented. Classical studies carried out in the USA of earthworms in areas treated with DDT in attempts to control the vectors of dutch elm disease illustrate well the bioaccumulation of organochlorine insecticides within food chains. In the 1950s DDT was repeatedly applied in high doses to elms in an effort to control the bark beetle vectors of dutch elm disease. American robins (*Turdus migratorius*) were killed or their breeding success impaired to such an extent that catastrophic declines in local populations occurred. For example, annual treatment of elms on the campus of Michigan State University led to the almost total elimination of robins in the area by 1958 (McEwan and Stephenson, 1979).

Earthworms are not as sensitive to organochlorine insecticides when compared to many other species and can thus accumulate significant residues of DDT and its metabolite DDE from the soil via leaf litter (Table 9.1). However, an American robin eating 100 earthworms could ingest more than 3 mg of DDT, which constitutes a lethal dose. Indeed, post mortem studies of American robins in the late 1950s and early 1960s showed these to have as much as 70 mg of DDT in their brain tissues (McEwan and Stephenson, 1979).

Following the progressive withdrawal of persistent organochlorine insecticides for agricultural use, problems of bioaccumulation of pesticides in wildlife have declined (Peakall, 1996). However, organochlorines are still routinely detected in animals at higher trophic levels such as predatory birds and sea mammals, and occasionally at concentrations that might be expected to cause acute adverse health effects in such species. Furthermore, although the concentrations of organochlorines detected may not produce acute ill-effects, the long-term presence of these compounds in tissues may lead to chronic conditions. However, sublethal influences of pesticides on biochemical and behavioural dysfunction in non-target animal species have often proved difficult to identify. Many environmental influences may contribute to behavioural phenomena in whole ecosystems, and interactions may occur. For example, poor water quality in terms of low oxygen and high organic matter concentrations as well as bacterial contamination may limit the development of invertebrate and fish populations in addition to chemical pollutants. Nevertheless, disruption of endocrine function has been firmly linked to the long-term exposure of some non-target species to pesticides, particularly in aquatic situations.

## 9.3 Regulatory requirements and non-target organisms

The increasingly stringent regulations applied worldwide to pesticide approval and use have included a gradual expansion in the number of non-target species required in toxicological studies. Most regulatory authorities require evaluation of toxicity to species representative of particular ecosystems or of major ecological significance, such as fish, and birds (Table 9.2). Mammalian toxicity is of course covered by the toxicology tests described in Chapter 6. Tests are commonly carried out on individual species and are usually conducted under closely controlled conditions. Varying doses of the active ingredient are used to establish an $LD_{50}$ or $LC_{50}$, the concentration in food or water respectively that will kill 50% of test animals. It is important to note that the lower the figure, the more toxic is the compound to that species. These basic laboratory tests may need to be followed by so-called higher tier tests, generally involving exposure of test organisms to compounds in field situations.

Such studies may be related to analysis of tissues for pesticide residues. Early studies on pesticidal effects on wildlife were often carried out without

Table 9.2 *Ecotoxicological studies required by the European Union for pesticides in 2002*

| Species | Tests required |
|---|---|
| Soil microbial activity – nitrogen transformation and carbon mineralisation | Carried out when the active ingredient is applied to or may contaminate soil |
| Algae | Growth rates: two species must be used for evaluation of herbicides |
| Earthworms – *Eisenia foetida* | Acute toxicity; sublethal effects |
| Aquatic invertebrates – water flea (*Daphnia* spp. preferably *D. magna*) plus one other representative aquatic insect spp., an aquatic crustacean, and an aquatic gastropod mollusc | Acute toxicity always for *Daphnia*, and where the pesticide is to be used on surface water, the others. Chronic toxicity unless repeated exposure unlikely |
| Sediment-dwelling organisms – emergence of adults of *Chironomus* spp. | Carried out where the pesticide is likely to partition into and persist in sediment |
| Bees | Acute oral and contact toxicity; effect on honeybee larvae for insect growth regulators |
| Predators and parasitoids – *Aphidius rhopalosiphi* and *Typholodromus pyri* and others where relevant | Effects on predators/parasitoids in artificial culture |
| Fish – rainbow trout (*Salmo gairdneri*) and a warm water species | Acute toxicity; chronic toxicity; bioconcentration potential |
| Birds (a quail species which may be either *Coturnix coturnix japonica* – Japanese quail or *Colinus virginianus* – bobwhite quail); or mallard – *Anas platyrhynchos* | Acute oral toxicity; short-term dietary toxicity; subchronic toxicity and reproduction unless exposure is not likely to occur during the breeding season |

Adapted from the UK Pesticides Safety Directorate at http://www.pesticides,gov.uk/applicant/registration_guides/data_reqs_handbook/ecotox.pdf

establishing residue levels in dead or injured species, due mainly to a lack of technical equipment and expertise in monitoring institutions. From the late 1950s, the use of gas chromatography followed by sophisticated gas chromatography–mass spectrometry and then liquid chromatography–mass spectrometry procedures allowed the accurate determination of pesticides and their metabolites in animal tissues (as well as in air, soil and waters). Residue data then could be compared with experimentally derived $LD_{50}$s. Correlations between tissue levels of pesticides in wildlife and mortality were established, particularly for birds and organochlorine pesticides, and these were used to support conclusions that deaths were due to poisoning by these compounds (Blus, 1996; Keith, 1996).

Many studies have reported the effects of pesticides on single species, especially those nominated in protocols for registration purposes. Relatively few studies have been undertaken on the effects of pesticides on whole ecosystems, particularly over large areas. Exceptions in the UK include the experiments conducted at Boxworth Experimental Husbandry Farm in Cambridgeshire from 1981 to 1988. Their main aim was to examine the effects of high-input and low-input pesticide regimes on populations of birds, small mammals, soil fauna, crop invertebrates and plants over several years (Greig-Smith, Frampton and Hardy, 1992). The Boxworth project subsequently led to the establishment of further long-term projects including SCARAB (seeking confirmation about results at Boxworth), set up to determine whether effects observed in cereal crops at Boxworth occur elsewhere in the UK in different arable crops and with pesticide inputs typical of the period 1990–1995.

Effects on non-target organisms may result from pesticides in aerial, terrestrial or aquatic environments. Particular focus has been directed towards effects on vertebrate fauna, but increasing attention has been paid to ecosystem components further down the trophic order, and the indirect effects that pesticides may have in food webs and chains. Most work has been carried out within the agricultural systems of developed countries and there is a paucity of data on the effects of pesticides in non-target organisms in less developed parts of the world.

## 9.4 Non-target effects on plants and animals from pesticides in the atmosphere

### 9.4.1 Non-target plants

Direct contact in the form of sprays may have adverse effects on a range of biota according to the type of pesticide being applied. The principal risks with herbicides may be to neighbouring crops, gardens or vegetation, whereas insecticides may pose risks to pollinating and other beneficial species.

As noted earlier (8.2), drift of pesticides away from their intended site of application may cause damage to nearby plants and crops. Abnormal growth and plant death may result from drift of herbicides. In the UK instances of phytotoxicity have been associated particularly with the hormone herbicides, and on occasions legal action has been taken against those held to be responsible for the damage caused by spray drift. Other problems with herbicides may arise from persistence in soil. Residual herbicides applied to soil have on occasions caused damage to following crops. For example, the herbicide trifluralin has been used extensively for weed control in brassicas, with repeated applications during

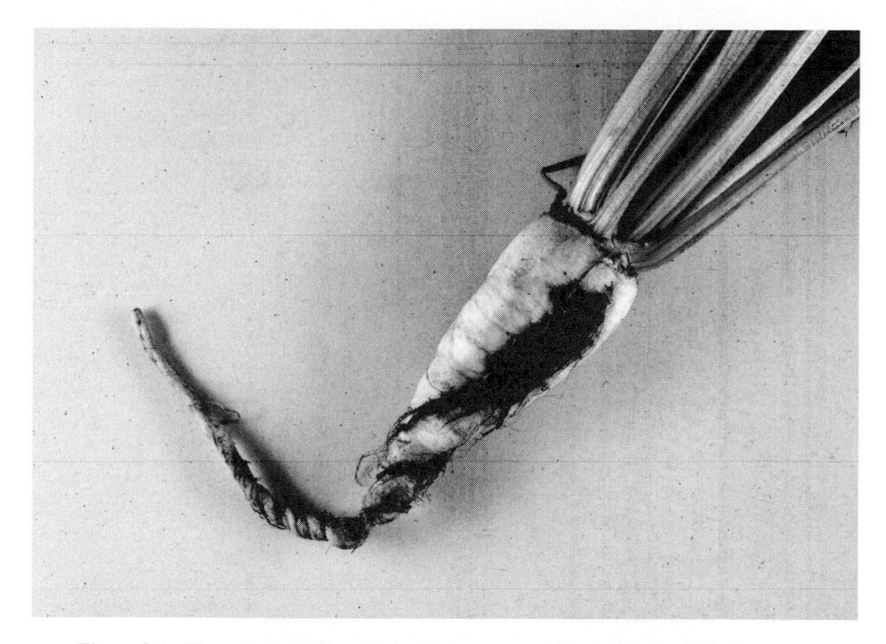

Figure 9.1 The effect of trifluralin residues on sugar beet following brassica crops in which the herbicide had been extensively used.

the growing season. Malformations have occurred in following crops such as sugar beet (Figure 9.1) if residual concentrations are sufficiently high. Similar problems have occurred in broad-leaved crops such as oilseed rape sown after cereals to which herbicides, such as persistent sulphonylureas and carotenoid biosynthesis inhibitors, have been applied.

In addition to these off-target effects, the efficacy of herbicides in achieving weed control has led in the UK to plants formerly regarded as weeds in crops such as cereals becoming locally rare. Indeed, some such as cornflower (*Centaurea cyanus*), shepherd's needle (*Scandix pecten-veneris*) and red hemp nettle (*Galeopsis angustifolia*) have become so rare as to be listed as priority species in the UK Biodiversity Action Plan (Marshall, 2001).

### 9.4.2 Bees

Pesticides in the atmosphere may affect flying invertebrates, and prominent among these are pollinating insects, especially bees. Evaluation of toxicity to bees forms part of the approval and registration procedures for pesticides in all EU and many other countries. Laboratory tests usually involve feeding caged bees with sugar solutions containing a series of concentrations of

Table 9.3 *Acute toxicity of pesticides to honey bees*

| Pesticide | $LD_{50}$ (μg per bee by contact application) |
|---|---|
| **Herbicides** | |
| MCPA | >100 |
| Isoproturon | >100 |
| Glyphosate | >100 |
| Metsulfuron-methyl | >25 |
| Carfentrazone-ethyl | >100 |
| **Fungicides** | |
| Triadimefon | >100 |
| Prochloraz | 50 |
| Epoxiconazole | >100 |
| Kresoxim-methyl | 28 |
| **Insecticides** | |
| Dimethoate | 0.1–0.2 |
| Chlorpyrifos | 0.06 |
| Malathion | 0.71 |
| Cypermethrin | 0.02 |
| Deltamethrin | 0.051 |
| Thiacloprid | 24 |
| Diflubenzuron | >1000 |

the active pesticidal ingredient, and $LD_{50}$ values may be calculated from such studies. If the compound proves toxic in laboratory tests then field studies will normally be required to assess the degree of damage which may occur to foraging bees, as well as to those present in hives. In the UK, these studies are commonly carried out in crops of flowering oilseed rape, field beans or in orchards during the flowering period of apples. The $LD_{50}$ values for bees of some widely used foliar-applied pesticides in the UK are given in Table 9.3.

Few herbicides and fungicides have proved toxic to bees. Predictably many insecticides, especially organophosphates and carbamates, are highly toxic and instances of bee deaths have been reported in the UK where these compounds have been applied to crops in flower. Most organophosphate insecticides are highly toxic to bees in both cage and field tests. Pyrethroids are highly toxic to bees in cage tests, but they may not be as damaging as organophosphates in field tests. For example, the pyrethroid insecticide lambda-cyhalothrin has an acute contact $LD_{50}$ of about 0.05 μg/bee, and is thus classified as highly toxic on the basis of this test. However, field studies have shown that

Table 9.4 *Numbers of incidents involving honey bees and bumble bees reported to the Environmental Panel of the Advisory Committee on Pesticides in the UK in 1991 and 2001 (in parentheses)*

| Pesticides involved | Number of incidents | Colonies affected |
|---|---|---|
| Carbamates | | |
| Bendiocarb | 1(1) | 3 (1) |
| Pirimicarb | 1 | 1 |
| Organochlorines | | |
| Dieldrin* | 1 | 1 |
| Gamma-HCH* | 4 | 8 |
| Organophosphates | | |
| Chlorpyrifos | 1 | 7 |
| Chlorpyrifos-methyl | 1 | 2 |
| Demeton-s-methyl* | 1 | 10 |
| Dichlorvos* | (2) | (5) |
| Dimethoate | 6 (1) | 63 (2) |
| Fenitrothion* | 2 | 55 |
| Pirimiphos-methyl | 1 | 1 |
| Triazophos* | 8 | 103 |
| Pyrethroids | | |
| Alphacypermethrin | 2 | 6 |
| Deltamethrin | 1 | 1 |
| Permethrin* | (1) | (1) |

Data from Pesticide Poisoning of Animals 1991: Investigations of Suspected incidents in Great Britain. Figures in parentheses from Barnett *et al.* (2001); * no longer approved for use on crops in the UK.

lambda-cyhalothrin presents little risk to honey bees at rates of up to 15 g of active ingredient per hectare (three times the normal field rate) when applied to flowering oilseed rape. The foraging activity of bees is influenced by this and other pyrethroids and it appears that these insecticides have a deterrent effect on bees which may then seek pollen elsewhere. Instances of adverse effects on bees are reported to the Wildlife Incident Investigation Scheme in the UK and these show that during the 1990s a marked decline in incidents of bee poisoning occurred (Table 9.4).

### 9.4.3 Predators and parasitoids

Predators and parasitoids of invertebrate pests are now widely used in systems of integrated crop and plant management. Practitioners of integrated pest management (IPM), which may use both pesticides and biocontrol measures, must be aware of the potential harmful effects of insecticides against arthropod predators, and of some fungicides against fungal species used in biocontrol. In this respect, use of some pesticides has resulted in an upsurge of species formerly of little importance. In the UK, the application of organochlorine insecticides in orchards during the 1950s and 1960s for control of coding moth (*Cydia pomonella*) led to major outbreaks of the red spider mite (*Panonychus ulmi*) on apples and pears, on which it has remained a serious pest. Its rise to prominence was associated with elimination of predatory invertebrates which were very susceptible to organochlorine insecticides.

It is thus important to evaluate the effects of pesticides on beneficial organisms, and most regulatory authorities now require studies on potential adverse effects on representative species. Commonly used arthropod species in tests include the aphid parasitoid *Aphidius rhopalosiphi* and *Typhlodromus pyri*, a predator of mite pests in orchard fruit. Many modern insecticides have far fewer adverse effects on such beneficial arthropods compared to the broad-spectrum organophosphate and carbamate compounds used for pest control in orchards and vineyards in the 1970s and 1980s.

## 9.5 Non-target organisms in terrestrial environments

### 9.5.1 Effects of pesticides on soil microorganisms

Most regulatory authorities require effects on non-target soil microorganisms to be evaluated if the pesticide is to be applied to soil, or may contaminate soils under practical conditions of use, such as in seed dressings or indeed foliar sprays. The activity of soil microorganisms is commonly monitored by measuring respiration as output of $CO_2$. Tests are usually carried out by monitoring the effects of the pesticide at recommended rates and ten times this rate on soil respiratory activity, as well as on processes of nitrification and ammonification. Generally, if no effects are discerned at ten times the field application rate, it is presumed that no adverse effects will occur to the soil microbial population during use. In practice, extensive testing has shown that soil microbial activity is little affected by pesticides with the exception of those used as soil sterilants. In fact, as noted earlier (8.6), the population and activity of microorganisms may be enhanced in soil following continued applications of the same pesticide, leading to the unwelcome phenomenon, with respect to weed or pest control, of accelerated degradation.

### 9.5.2 Effects of pesticides on earthworms

Earthworms, in the light of their importance in maintaining soil structure and fertility, are considered to be one of the most important indicator species in

Table 9.5 *Acute toxicity of pesticides to* Eisenia foetida

| Pesticide | $LC_{50}$ mg/kg dry substrate after 14 days at 20 °C | Pesticide | $LC_{50}$ mg/kg dry substrate after 14 days at 20 °C |
|---|---|---|---|
| **Herbicides** | | **Insecticides** | |
| Atrazine | 78 | Tefluthrin | 2 |
| Isoproturon | >1000 | Methiocarb | >200 |
| Metamitron | >1000 | Imidacloprid | 10 |
| Metsulfuron-methyl | >1000 | Deltamethrin | 1 290 |
| Carfentrazone-ethyl | >820 | Thiacloprid | 105 |
| | | Diflubenzuron | 780 |
| **Fungicides** | | **Others** | |
| Benomyl | 0.4 | Metaldehyde | >50 000 |
| Epoxiconazole | >1000 | | |
| Azoxystrobin | 283 | | |

soil. They incorporate decaying organic matter into the soil and improve soil aeration, drainage and water-holding capacity. The use of pesticides that may reduce the numbers or activity of earthworms can result in adverse effects on soil fertility, which in turn may influence crop yields. Secondly, reductions in populations of earthworms may lead to a decrease in food supply for their predators. Thirdly, the presence of pesticide residues in earthworms may lead to the poisoning of predators after consumption. Bioaccumulation may occur at secondary levels with earthworm predators such as birds (Table 9.1), or at tertiary levels, for example where a fox preys upon a bird that has consumed earthworms containing pesticide residues.

Risk assessments of toxicity to earthworms are required for registration of pesticides in many countries and considerable efforts have been made to devise appropriate procedures for evaluating the toxicity of pesticides to this key group of soil-inhabiting invertebrates. A standard method involves mixing a range of concentrations of a chemical into a loam soil containing a population of the earthworm species *Eisenia foetida*, which is easily cultured in the laboratory. The lethal concentration to kill 50% of earthworms ($LC_{50}$) and no observed effect levels (NOELs) may then be measured. If the NOEL is greater than the expected environmental concentration, then it is unlikely that the pesticide will have appreciable effects under field conditions. The $LC_{50}$ values for some pesticides, primarily soil-applied, to *Eisenia foetida* are given in Table 9.5: as previously, the lower the value the more toxic is the compound.

The most important consequences of pesticide contamination of earthworms were seen in the 1960s and 1970s upon species at higher trophic levels in food chains, when adverse effects on predators of earthworms were extensively reported. In addition to the example of local declines in the population of American robins noted earlier (Table 9.1), poisoning of birds after eating DDT-contaminated earthworms has been reported from several countries. In the UK several studies carried out in orchards in the early 1970s reported deaths of blackbirds (*Turdus merula*), song and mistle thrushes (*Turdus philomelos* and *T. viscivorus*), grey partridge (*Perdix perdix*), red-legged partridge (*Alectoris rufa*), pheasant (*Phasianus colchicus*) and tawny owl (*Strix aluco*). Work carried out in a Norfolk orchard during the 1960s showed that thrushes could accumulate DDT in their fat deposits from eating worms and other invertebrates. Deaths of birds were associated with release of DDT into the bloodstream following the mobilisation of fat reserves during periods of stress (Edwards, 1973; McEwan and Stephenson, 1979).

DDT passed to them by consumption of contaminated earthworms may also affect vertebrates other than birds. Frogs of the species *Rana temporaria* may be affected by DDT arising from consumption of earthworms and slugs containing residues of this compound. Moles (*Talpa europaea*) also eat earthworms. Although moles trapped in a Norfolk orchard where DDT had been used contained high levels of this compound, they appeared to be unaffected (McEwan and Stephenson, 1979).

Organochlorines other than DDT may be transmitted through earthworms to predators. Aldrin and dieldrin have been implicated in several instances of bird deaths in the UK from consumption of contaminated earthworms, but these reports were overshadowed by the principal effects on birds of these compounds when used as seed dressings (9.9).

Apart from effects of contamination by the organochlorine insecticides, few instances of harm to earthworms or their predators from pesticide applications have been reported. A number of deaths of black headed gulls (*Larus ridibundus*), common gulls (*Larus canus*) and lapwings (*Vanellus vanellus*) resulted from consumption of granules (which had not been incorporated as recommended into soil) and possibly of earthworms containing the carbamate insecticide aldicarb shortly after the introduction of this compound in the UK in the mid 1970s (McEwan and Stephenson, 1979). Experiments indicate that vertebrates do not accumulate residues of pyrethroid insecticides that may be taken up by earthworms. The first soil-active pyrethroid insecticide, tefluthrin, proved to have a relatively high toxicity to earthworms (at 2 mg/kg soil) in laboratory studies. However, no significant reductions in the populations of *Allolobophora* and *Lumbricus*, the two main genera of earthworms in UK soils, were evident at the approved field rate of the compound in field trials.

One group of fungicides – the benzimidazoles – is highly toxic to earthworms, and studies have shown reductions of the activity in soils to which compounds such as carbendazim have been applied. These fungicides, which inhibit microtubule assembly in annelid and some other invertebrate groups as well as sensitive fungi, have been used to control earthworms on ornamental lawns and golf greens.

Monitoring populations of earthworms formed part of the SCARAB project in the UK, and was carried out in fields where traditionally high-input systems of then current farm practice (CFP) were compared with field sites where a reduced-input approach (RIA) was adopted (Tarrant *et al.*, 1994). Earthworm densities varied considerably between the three farms in different parts of the UK chosen for the investigations, but no significant differences were observed at any site in numbers in CFP and RIA fields. Searches of the surfaces of soils treated with individual pesticides including the organophosphates omethoate and dimethoate found no dead earthworms. No residues of omethoate were found in earthworms from treated soil, but residues of dimethoate up to 0.15 mg/kg were found in some worms (Greig-Smith, Frampton and Hardy, 1992).

### 9.5.3 Effects of pesticides on soil-inhabiting arthropods

Soil arthropods play important roles in breaking down plant remains, controlling crop pests by parasitism and predation, and are food sources for larger animals. For sprays, the concentration of pesticides will be greatest at the soil surface immediately after application, and the species most vulnerable to potential adverse effects are those which live on the soil surface. Examples of such species include spiders, as well as ground and rove beetles belonging to the Carabidae and Staphylinidae families that are predators of aphids and which may contribute to the control of these insect pests. Many regulatory authorities require an assessment of the effects of pesticides on non-target terrestrial arthropods, particularly if the compound is intended for application to soil (Oomen, 1998).

Extensive field studies have been carried out by the Game Conservancy and associated trusts in the UK on the effects of insecticides in particular on ground dwelling beetles, and evaluation of the effects of pesticides on arthropod fauna formed an integral part of both the Boxworth and SCARAB projects (Cilgi and Frampton, 1994). Both these studies and similar investigations over a 7-year period in the Netherlands have found that application of insecticides, particularly in the autumn, has led to reductions in the numbers of ground dwelling beetles in cereals and other crops. Reductions in populations of up to 90% were observed in some trials. A decrease in carabid beetle populations was

Table 9.6 *Proportion of marked insects surviving insecticide applications in fields of winter wheat*

| Species of ground beetle | Year | Proportion surviving: Dimethoate | Lambda-cyhalothrin | Untreated |
|---|---|---|---|---|
| *Nebria brevicollis* | 1986 | – | 1.00 | 1.00 |
| *Nebria brevicollis* | 1987 | 0.18 | 0.93 | 1.00 |
| *Trechus quadristriatus* | 1987 | 0.42 | 1.00 | 1.00 |
| *Bembidion obtusum* | 1987 | 0.30 | 0.48 | 1.00 |

From Brown, White and Everett (1988)

consistently evident in the Boxworth study where full spray programmes were employed in comparison to supervised or integrated systems of pest control. In some cases this decrease in predatory beetles was associated with the increased incidence of aphids, one of their major food sources.

Reductions in populations of ground dwelling beetles have been particularly linked to the use of organophosphorus compounds such as dimethoate applied to cereals in autumn. With some species such as *Bembidion obtusum*, populations had not recovered 1 year after application of organophosphorus compounds. Autumn-applied pyrethroids, which do not persist as long as the organophosphorus compounds, have markedly less effect on most ground dwelling beetles (Table 9.6) and it is thus now common practice in the UK to spray pyrethroid insecticides in the autumn for control of vectors of barley yellow dwarf virus in cereals. However, pyrethroid insecticides such as lambda-cyhalothrin, cypermethrin and especially deltamethrin were shown in the Boxworth and SCARAB studies to markedly affect ground dwelling spiders such as the lycosid species *Trochosa ruricola* and *T. terricola*, although populations of these generally recovered by the summer following autumn applications (Cilgi and Frampton, 1994).

## 9.6 Effects of pesticides on non-target invertebrates in water

As with terrestrial non-target organisms, direct reductions in populations of non-target aquatic invertebrate species may occur following the application of pesticides. Non-target species may also accumulate pesticides at sublethal concentrations and predators of such species may then accumulate residues by feeding to levels that may be damaging or even lethal. Assessment of the effects of new and existing compounds on non-target species representative of aquatic systems forms part of modern regulatory schemes including that of the EU (European Commission, 2002).

The lowest trophic levels or base of the food chain in aquatic systems are phytoplankton and zooplankton. Toxicity tests are not generally required by regulatory authorities for the effects of pesticides on phytoplankton and zooplankton, but assessment of the potential of compounds to inhibit growth of algae is a requirement for product registration in many countries. The tests are usually carried out under sterile conditions using the green alga *Selenastrum capricornutum* as test species. In the EU further tests with macroscopic species such as the pond weed, *Lemna* spp., may be required for herbicides as part of registration processes.

Herbicides, particularly those that act by inhibiting photosynthesis, entering aquatic systems may have direct toxic effects on phytoplankton. The increasing incidence of triazine and urea herbicides in natural waters in for example the UK and USA may have adverse effects in this respect. Paradoxically, in reservoirs for drinking water these effects may be beneficial since algal blooms have been associated with the production of toxins to human beings as well as farm and domestic animals. However, few instances of damage to phytoplanktonic and similar communities have been reported, with concentrations of herbicides such as atrazine and simazine being well below those required to cause growth retardation in microscopic and macroscopic algae (Hedgecott, 1996).

Assessment of effects on non-target aquatic molluscs is not required within most registration procedures. Many studies have been carried out with molluscs, notably filter feeding mussels and other bivalves. In laboratory studies, DDT and other organochlorine insecticides have been shown to reduce shell growth in oysters at less than 1 mg/l. Such effects were not seen with other insecticides including the organophosphates diazinon and malathion, the carbamate carbaryl as well as the herbicide 2,4-D.

In contrast, adverse effects on molluscs have been firmly linked to the tributyltin (TBT) compounds in paints that have been used to prevent colonisation of ship hulls by barnacles, mussels as well as seaweeds. TBT has a very high toxicity to molluscs and crustaceans, and adverse effects occur to their larval stages at very low concentrations. The NOEL for young larvae (spat) of the most sensitive oyster species is very low at 0.002 μg/l (UK Marine SAC, 2003). TBT is highly lipophilic and thus tends to bioaccumulate in molluscs, crustaceans and also fish. Endocrine disruption has been attributed to the effects of TBT, and imposex, the development of male characteristics in females, has been demonstrated in several species exposed to low concentrations, at around 0.05–3 μg/l TBT. This androgenesis is probably due to a build-up of testosterone in affected animals (Matthiessen and Gibbs, 1998). Such endocrine effects are widespread and globally over 100 species of prosobranch molluscs have been found to suffer from imposex in their natural habitat. Many populations of species such as the dogwhelk have been severely depleted in some areas and such effects have led to the establishment of a very low environmental quality standard for TBT of 0.002 ng/l of seawater in the UK. The EU prohibited the use of TBT in 2003 and the Marine Environment Protection Committee of the International Maritime Organisation has drafted mandatory regulations to phase out and eventually prohibit the use of TBT by 2008 as a constituent of antifouling paints for use on ships. Recovery of benthic species has occurred where prohibition of use of TBT as an antifouling agent has been implemented. For example, in the River Crouch in Essex, numbers and diversity of aquatic fauna increased markedly from 1987 to 1992 (UK Marine SAC, 2003).

Tests are required by most regulatory authorities to establish the toxicity of pesticides to aquatic arthropods. The usual species employed is *Daphnia* spp., the water flea, which is especially sensitive to pesticides, particularly insecticides, and has been used as a standard invertebrate indicator for pesticides in water. Table 9.7 shows the acute toxicity of pesticides to *Daphnia* spp.

Table 9.7 *Acute toxicity of selected pesticides to Daphnia* spp

| Pesticide | *Daphnia* $LC_{50}$ | Pesticide | *Daphnia* $LC_{50}$ |
|---|---|---|---|
| **Insecticides** | | **Herbicides** | |
| DDT | 1.1 | Isoproturon | >500 000 |
| Dieldrin | 250 | Atrazine | 87 000 |
| Diazinon | 0.9 | Paraquat | 3 700 |
| Malathion | 18 | Isoproturon | >500 000 |
| Chlorpyrifos | 1.7 | Diuron | 12 000 |
| Carbofuran | 38 | Trifluralin | 245 |
| Tefluthrin | 70 | Amidosulfuron | 85 000 |
| Deltamethrin | 3.5 | Diflufenican | >10 000 |
| Cypermethrin | 0.15 | **Fungicides** | |
| Diflubenzuron | 7.1 | Benomyl | 640 |
| | | Fenpropimorph | 2 400 |
| | | Pyraclostrobin | 5 330 |

$LC_{50}$ values are in μg/l water after 24 or 48 h exposure

The $LC_{50}$ values of pesticides given in Table 9.7 may be compared to the environmental quality standards given in Table 8.11 and examples quoted (8.7). In general the concentrations normally encountered in natural waters are well below the concentrations that might be expected to cause acute effects on the non-target aquatic invertebrates present in these waters. Deaths of non-target species have been associated with the direct spraying of natural waters to kill pest insects such as gnats and mosquito larvae, or with point-source pollution arising from spillages of concentrates or careless disposal of pesticides through sprayer tank rinsing (8.7) or of sheep dip as described below.

Insecticides are toxic to many aquatic invertebrates in clean water but some such as the pyrethroids may be considerably less toxic in waters containing sediment, to which they may be adsorbed (Table 9.8). The latter situation may prevail in most natural aquatic systems.

Studies funded by the Environment Agency and its predecessors in the UK have monitored concentrations of pesticides and their impact on mortality and feeding rates of aquatic invertebrates (Environment Agency, 2000). The change during the 1990s from organophosphate to pyrethroid insecticides for use in sheep dipping in the UK resulted in a dramatic increase in serious incidents of water pollution, resulting in severe reductions and in some instances total elimination of invertebrate populations in streams near sheep-dipping sites (Environment Agency, 2000; Table 8.12; Figure 9.2). Legal action and the introduction of precautionary measures to minimise sheep dip wastes entering watercourses has led to fewer instances of pollution incidents in England and Wales since 1998. In most cases recolonisation of streams by the affected species was rapid, and no long-term declines in populations of these non-target aquatic invertebrates was evident. Nevertheless, although these population crashes are temporary, they may reduce the amount of food available to fish, and killed and dying insects may contain pesticide levels that may be toxic to their predators. In addition to these direct toxic effects, bioaccumulation and transfer of pesticides through

Table 9.8 *Effects of sediment on the toxicity of cypermethrin to aquatic organisms*

| Organism | Clean water | Water/soil mixture |
|---|---|---|
| *Daphnia magna* | 1.4 | 310 |
| *Cloeon diptera* | 0.01 | 2.9 |
| *Asellus aquaticus* | 0.01 | 2.8 |

Figures are the $EC_{50}$ in μg of active ingredient per litre after 72 h. Data from Riley (1990)

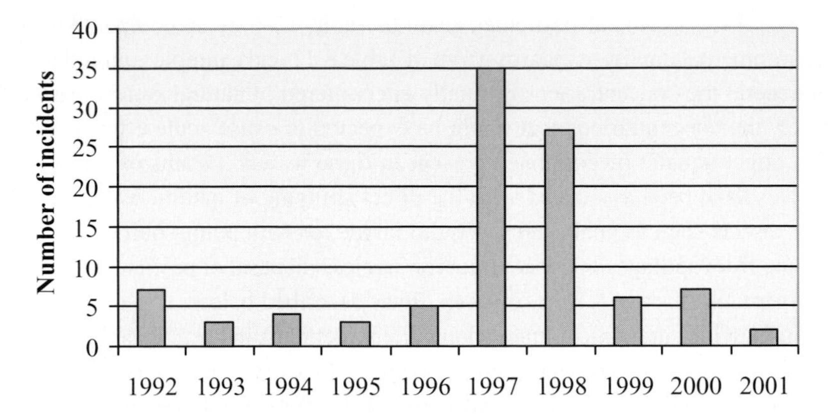


Figure 9.2 Substantiated pesticide-related incidents involving sheep-dip reported by the Environment Agency for England and Wales 1992–2001. (From Annual Pesticide Reports of the Environment Agency 1992–2002 at http://www.environment-agency.gov.uk).

aquatic food chains has led to serious consequences for some non-target organisms at higher trophic levels, and in particular fish and fish-eating mammals. These effects have been associated almost exclusively with the organochlorine insecticides.

## 9.7 Pesticides and fish

Manufacturers of pesticides are required by most regulatory authorities to test new compounds for their direct toxicity to species of fish and for this purpose the most common species employed is the rainbow trout (*Salmo gairdneri*) but the common carp (*Cyprinus carpio*) and bluegill (*Lepomis macrochirus*) are also used. Toxicity is commonly expressed as the concentration present in water (as mg/l) which will kill 50% of test animals – the $LC_{50}$ – over a period of 96 h. The $LC_{50}$ values of selected pesticides to rainbow trout are given in Table 9.9.

Apart from the dinitroaniline compound trifluralin, herbicides are not very toxic to fish when compared to some insecticides and other chemical and biological pollutants of water. The acute toxicity of most herbicides to fish is low, especially those soil-applied types such as atrazine and isoproturon which have been increasingly found in natural waters of Europe and the USA in recent years (8.7 and 8.8). The concentrations of these – on rare occasions as high

Table 9.9 *Acute toxicity of selected pesticides to rainbow trout (Salmo gairdneri)*

| Pesticide | $LC_{50}$ mg/l 96 h | Pesticide | $LC_{50}$ mg/l 96 h |
|---|---|---|---|
| **Insecticides** | | **Herbicides** | |
| DDT | 0.007 | Mecoprop | 150 |
| Dieldrin | 0.019 | Atrazine | 4.5 |
| Malathion | 0.17 | Glyphosate | 52 |
| Dimethoate | 6.2 | Isoproturon | 3.7 |
| Carbofuran | 25 | Amidosulfuron | >320 |
| Cypermethrin | 0.69 | Trifluralin | 0.09 |
| Deltamethrin | 0.91 | **Fungicides** | |
| Imidacloprid | 211 | Benomyl | 0.48 |
| Rotenone | 0.022 | Azoxystrobin | 0.47 |

Figures are the $LC_{50}$ in mg/l after 96 h exposure

as 0.1 mg/l – are at least two orders of magnitude below the level at which they may be acutely toxic to fish. Furthermore, herbicides do not appear to accumulate in fish. Fungicides are rarely detected in natural waters, and most are not very toxic to fish. In experimental systems, fish have been found to metabolise and excrete fungicides and no evidence of accumulation has been found.

As with aquatic invertebrate fauna, most problems to fish result from the presence of insecticides in natural waters. The data in Table 9.9 indicate the high toxicity of organochlorine and pyrethroid insecticides to fish. The only other pesticide of high acute toxicity to fish is the naturally occurring insecticide rotenone, used commercially to clear lakes and ponds of fish prior to restocking (4.3). As with other non-target aquatic species, fish may be affected in an acute sense by point-source spillages. These vary from local pollution that may cause deaths of fish in a few hundred metres of streams and brooks, to massive pollution events such as the virtual elimination of fish from the river Rhine in 1990 after accidental release of large quantities of pesticides from a manufacturing plant in Basel, Switzerland during a fire.

DDT was shown to be acutely toxic at low dose to fish shortly after its introduction. Despite the availability of this information in the early 1940s, the benefits of DDT were considered to outweigh any adverse effects it may have had on fish. However, massive levels of fish mortality were seen in the 1950s in the USA and Canada in streams and rivers flowing through forests that had been aerially sprayed with organochlorine insecticides to control spruce budworm and other forest pests. Salmon (*Salmo trutta*) in cages placed in oversprayed rivers died within 3 weeks, and huge numbers of young salmon (parr) were found dead or dying. In addition to acute toxic effects, deaths were caused by biomagnification of DDT in fish tissues (McEwan and Stephenson, 1979).

The biomagnification of DDT and other organochlorines by fish is largely associated with the passage of water across their gills, and the partitioning between DDT in water and the blood in the gills followed by a further partitioning between the blood and lipids internally. Pesticides carried in the blood may then be transported internally, and many fish seem to concentrate organochlorine pesticides in the liver. About 700 l of water has been estimated to pass across the gills of a 1-kg fish per day. Even if DDT is present at 0.01 mg/l, fish may absorb most of this giving a tissue concentration of 7 mg/kg in a single day. Fish in general metabolise and excrete organochlorines only slowly, and ultimately the concentration of the pesticide in tissues may be lethal. Indeed, water containing DDT at 0.001 mg/l has proved fatal to salmonid fish (McEwan and Stephenson, 1979).

The extent of bioaccumulation varies according to the level of exposure and the ability of the species to detoxify or excrete the compound (Huckle and Millburn, 1990). Very high concentrations of DDT were found in many freshwater fish from Canadian and American rivers, lakes and streams in areas treated with this insecticide in the 1950s. Concentrations of 100 mg/kg and more were found in lake trout from Clear Lake in California and some specimens contained over 1000 mg/kg. Following the withdrawal of DDT and other organochlorines for treatment of natural waters in the USA, levels of organochlorine insecticides in fish began to decline in the 1960s. In a comprehensive survey carried out by the United States Bureau of Sport Fisheries and Wildlife during the mid 1970s fish from 50 sites were examined for several organochlorine insecticides. DDT was found in all fish but the highest concentration determined was of 57 mg/kg and the mean of all samples was around 1 mg/kg (McEwan and Stephenson, 1979). In the UK, by the mid 1990s organochlorine residues in fish liver were detected at microgram levels with median values for DDE between 15 and 91 μg/kg tissue; and DDT between 13 and 28 μg/kg tissue (UK Marine SAC, 2003).

As well as exerting toxic effects on fish organochlorine compounds may also be further bioaccumulated by predators of fish such as birds. Furthermore, sublethal effects of pesticides on fish have been well documented. When exposed to sublethal concentrations of DDT salmon and trout may change their behavioural responses to cold, by avoiding colder waters. This can result in altered migratory patterns in these species.

Other insecticides do not accumulate in fish to the extent of the organochlorines. Fish exposed to organophosphorus insecticides can temporarily accumulate them, but usually metabolise and excrete these compounds. Indeed, some organophosphates such as dichlorvos have been used to treat farmed fish for ectoparasitic lice. The pyrethroid insecticides are highly toxic to fish in laboratory tests (Table 9.9) and brain tissue in fish is much more sensitive to pyrethroids than tissue from birds and mammals. Furthermore, fish appear to detoxify pyrethoid insecticides slowly compared to mammals and birds. However, pyrethroids may not present as great a risk to fish as organochlorines under natural conditions since they are rapidly adsorbed to particulate matter in natural waters. In rice paddies under natural conditions deaths of fish (common carp – *Cyprinus carpio*) in areas treated with the pyrethroid alphacypermethrin applied at a recommended rate of 15 g of active ingredient per hectare were not significantly different from those in untreated paddies (Stephenson, 1988). The chitin synthesis inhibitor benzoylphenylurea insecticides, widely used in some parts of the world for mosquito control within natural waters, are rapidly metabolised in several species of fish, and do not appear to bioaccumulate.

Sublethal effects on fish include potential endocrine disruption, leading to imposex and intersex conditions. Studies in the UK and elsewhere have shown that endocrine disruption frequently occurs in fish (as well as aquatic invertebrates) downstream from sewage works. Attention has been focused on certain industrial by-products such as nonylphenol and bisphenol but also some pesticides, notably organochlorine and latterly pyrethroid insecticides. Concern has also been expressed that such effects may be widespread in marine life. However, a comprehensive study of endocrine disruption in the coastal and marine environment around the UK has concluded that although some feminisation of species such as blenny (*Lipophrys pholis*) and flounder (*Pleuronectes flesus*) was observed in estuaries, feminisation of migratory fish such as salmon (*Salmo trutta*) does not seem a problem. The report further concluded that endocrine-disrupting substances other than pesticides were most likely to be responsible for the observed effects. (EDMAR, 2003).

## 9.8 Amphibians, reptiles and pesticides

When compared to the amount of research carried out with fish, birds and mammals, rather less work has been undertaken on the effects of pesticides on amphibians and particularly reptiles (Portelli and Bishop, 2000; Sparling, 2000). Tadpoles of both frogs and toads seem to be more tolerant of pesticides such as DDT than fish, having an $LD_{50}$ of around 0.5 mg/l, with adults at 1–5 mg/l. Sublethal effects DDT including morphological abnormalities have been observed in tadpoles, and bioconcentration of DDT and other organochlorines occurs in frogs and some other amphibians (Sparling, 2000).

Observations of decreases in populations of amphibians in some parts of the world, and in particular increases in limb deformities in some amphibian species during the 1990s, led to investigations into potential causes. A trematode parasite, *Ribeiroia*, has been linked with these deformities, but exposure to pesticides and in particular the herbicide atrazine have been implicated in changes in the physiology of amphibian species, possibly increasing susceptibility to infection by the parasite. Furthermore, increases in hermaphroditism have been detected in amphibians at concentrations of atrazine as low as 0.1 μg/l (Hayes *et al.*, 2002), although earlier studies did not detect such effects (Ouellet, 2000). As noted earlier (8.7), atrazine concentrations can be much higher than 0.1 μg/l in natural waters.

Some reptiles such as snakes, turtles and crocodiles may accumulate organochlorine insecticides to higher levels than amphibians. Consumption of prey from lower trophic levels may result in bioaccumulation and bioconcentration of organochlorine insecticides. Dead snakes have been found in some cases to have over 500 μg/g body weight of DDT (Portelli and Bishop, 2000).

Endocrine disruption was held to be responsible for the demasculinisation and 'super-feminisation' of alligators in Lake Apopka, Florida following a pesticide spillage there in 1980 involving dicofol, DDE and DDT (Portelli and Bishop, 2000). Both turtles and alligators have been shown to accumulate DDT, but the effects of this bioaccumulation on populations of these species are uncertain.

## 9.9 Pesticides and birds

The effect of pesticides on birds has been the subject of much debate. The general public interest in birds in both the UK and USA is high and has been promoted and supported by the scientific studies of the British Trust for Ornithology, and Royal Society for the Protection of Birds in the UK as well as the Audubon Society in the USA. These groups have extensively reported the effects of pesticides on avian species.

As with other fauna, the concentration and level of exposure to pesticides as well as the ability of individual species to metabolise or excrete the compound determine the degree of toxicity to birds. A few insecticides do pose acute toxic problems to birds but the majority of pesticides including most herbicides and fungicides as well as many insecticides have a low avian acute toxicity (Table 9.10), and are metabolised and excreted rapidly by many species of birds.

Regulatory authorities require studies of avian toxicity as part of pesticide registration, and the species most commonly used for this purpose are the

Table 9.10 *Variation in acute toxicity of pesticides to birds*

| | Bobwhite quail (*Colinus virginianus*) | | Mallard (*Anas platyrhynchos*) | |
|---|---|---|---|---|
| Pesticide | Acute oral $LD_{50}$ | Dietary $LC_{50}$ | Acute oral $LD_{50}$ | Dietary $LC_{50}$ |
| **Insecticides** | | | | |
| Dieldrin | | 39 | 381 | 520 |
| DDT | | | 2240 | 1025 |
| Dimethoate | 84 | | 40 | 1000 |
| Chlorpyrifos | | 423 | 75 | |
| Malathion | | 3500 | 1485 | >5000 |
| Fipronil | 11 | 49 | >2000 | >5000 |
| Carbofuran | 2.5–5 | | 0.4 | |
| Cypermethrin | | | 10 000 | >2000 |
| Deltamethrin | | >5620 | >4640 | >8039 |
| Imidacloprid | 152 | 2225 | | >5000 |
| Diflubenzuron | | >4640 | | >4640 |
| **Herbicides** | | | | |
| Atrazine | | >2000 | | >2000 |
| 2,4-D | | | 1000 | >5000 |
| Trifluralin | >2000 | | >2000 | |
| Diflufenican | 2150 | | >4000 | |
| Amidosulfuron | | >2000 | | >2000 |
| **Fungicides** | | | | |
| Captan | | 3000 | | >5000 |
| Azoxystrobin | >2000 | >5200 | >2000 | >5200 |

Figures are mg/kg body weight

bobwhite quail (*Colinus virginianus*) and mallard (*Anas platyrhynchos*). Both short-term and long-term dietary studies are usually necessary. As Table 9.10 shows, considerable variation exists in the response of these two species to some pesticides and to take this into account regulatory authorities impose considerable safety margins from studies with birds. For example, in the EU, 5-day acute toxic studies are required, and where the acute toxicity is less than 500 mg/kg body weight, the test must be performed on a second unrelated species. Studies of the potential chronic effects of pesticides are focused particularly on reproductive effects on birds.

The organochlorine insecticides have presented the highest risk to avian populations, but their effects vary greatly between species. One of the commonest of test species used in early studies, chickens (*Gallus domesticus*) are adversely affected by these compounds at much higher doses than many other avian species, and this may have led to erroneous conclusions on the safety of organochlorines to birds. Some other species are also relatively resistant to organochlorines: in field studies even populations of species at the apex of food chains, such as herons (*Ardea cinerea*), remained largely unaffected by concentrations of these insecticides that severely affected other birds (Mellanby, 1992).

However, organochlorines are unquestionably toxic in both an acute and chronic sense to many species of birds. Shortly after its introduction, DDT was associated with the deaths of songbirds in some forested areas in the USA where the insecticide was aerially applied at high rates for control of insect pests of hardwood trees. In the USA, DDT was the principal compound associated with bird mortality, whereas in some other parts of the world the cyclodiene insecticides dieldrin and aldrin were linked to deaths.

Until the UK joined the European Economic Community (now the EU) in 1972, cereal seed was primarily sown in spring and from the mid 1950s seed was routinely treated with organochlorine insecticides, principally dieldrin and aldrin, to control soil-borne insect pests. During sowing spilt and uncovered seed was consumed by grain-eating birds and a high mortality of wood pigeons (*Columba livia*), stock doves (*Columba oenas*), pheasants (*Phasianus colchicus*) and partridges (*Perdix perdix*) was seen. A particularly high mortality was evident in 1960–1961 due to delays in sowing owing to bad weather. In one area of 600 ha of woodland near Tumby in Lincolnshire where corpses were collected, 5668 wood pigeons, 118 stock doves, 59 rooks (*Corvus frugilegus*) and 89 pheasants as well as 10–15 owls and sparrowhawks (*Accipiter nisus*) had been killed (Sheail, 1985). Workers calculated that if a wood pigeon ate 50 g of cereal seeds treated with aldrin at the recommended rate of 2 oz of 40% insecticide per bushel it would ingest 40 mg or about 80 mg/kg body weight, which would exceed the $LD_{50}$ for this species. As a result of these findings, from 1961 only autumn-sown wheat was treated with organochlorines in the UK. Farmers agreed not to use these compounds on spring cereals, and from 1962 the number of dead birds was drastically reduced. Further to this in 1974 the use of dieldrin and aldrin as seed treatments in the UK was discontinued, eliminating the problem permanently.

Many examples of such food chain effects have been documented with the organochlorines. Some of the most striking examples of the effects of organochlorine insecticides on birds result from contamination of prey animals. One of the most cited is that associated with populations of the western grebe (*Aechmophorus occidentalis*) in Clear Lake, California in the 1950s (Edwards, 1973). At this tourist resort, a small midge (*Chaoborus astictopus*) breeds in the mud at the bottom of the lake and its adults reach such high numbers that they become a nuisance to visitors. To control the gnat, the lake was sprayed with the organochlorine TDE in 1949, 1954 and 1957. Before the spraying programme began over a thousand pairs of the western grebe regularly bred on the lake. During the breeding seasons from 1958 to 1961 only a few pairs were seen and no young were fledged. Grebes continued to visit the lake in winter but many of these died and analyses of dead grebes showed up to 1600 mg/kg TDE in their fat. Further analyses indicated TDE levels in sunfish, the major food of the grebes, between 40 and 2500 mg/kg body weight. The sunfish may have bioaccumulated TDE, from very low concentrations in the lake water during movement of water across their gills, but may also have acquired residues from feeding on plankton, where TDE concentrations were recorded at up to 5 mg/l.

Further examples, more serious in terms of reductions in population, involved birds of prey. Populations of golden eagles (*Aquila chrysaetos*) in the Cairngorms and other inland highland areas of Scotland began to decline in the early 1950s and dead birds were found to have high residues of dieldrin and DDT. In these areas the main food source of the eagles is sheep carrion. DDT and later dieldrin were used as sheep dips from 1946 onwards, and it appears that the eagles picked up these compounds during feeding. After prohibition of organochlorines for sheep dipping in the UK from 1966, a marked improvement in the breeding success of golden eagles occurred (McEwan and Stephenson, 1979). Deaths of eagles were also seen in the USA and Canada, notably of the bald eagle (*Haliaeetus leucocephalus*). Here, even as late as 1971 when the dangers of organochlorines to predatory birds were well recognised, several deaths of bald eagles were associated with dieldrin poisoning.

One of the best-documented examples of the effects of pesticides on birds comes from studies on the peregrine falcon (*Falco peregrinus*) in the USA, Canada and the UK (Ratcliffe, 1993). In the 1930s the USA and Canada had a peregrine falcon population of between 5000 and 9000 nesting pairs. The population began to decline in the late 1940s and by 1975 only 325 nesting pairs were left on the North American continent. In Britain the decline began in the early 1950s and spread northwards. By 1961 the once common peregrine was extinct in southern Britain and only the inland populations in the Scottish Highlands remained unaffected. Both deaths of adult birds and a lack of breeding success contributed to the decline of the peregrine falcon. Although peregrine falcons were persecuted during 1939–1945 because of their habit of preying upon pigeons, some of which were used to carry messages, the deaths of adult birds in the UK post 1945 were linked to the acute toxic effects of dieldrin and aldrin. It is likely that these lethal doses were obtained from prey animals such as voles, field mice and possibly wood pigeons that had consumed dieldrin through eating dressed cereal grains. Wood mice (*Apodemus sylvaticus*), for example, have been shown to accumulate over 20 mg/kg body weight of dieldrin in their

Table 9.11 *Organochlorine residues and changes in eggshell thickness in British birds over two decades from 1947*

| Species | Region (UK if not stated) | Change in thickness index | Organochlorine residue in eggs |
|---|---|---|---|
| Peregrine falcon | East & Central Highlands | −4 | 4.0 |
| Peregrine falcon | Other regions | −19 | 15.2 |
| Sparrowhawk | South East England | −21 | 37.2 |
| Sparrowhawk | North and West England | −14 | 14.0 |
| Golden eagle | Western Scotland | −10 | 1.8 |
| Golden eagle | Eastern Scotland | −1 | 0.2 |
| Merlin | | −13 | 16.4 |
| Kestrel | | −5 | 4.1 |
| Rook | | −5 | 0.4 |
| Guillemot | | 0 | 2.5 |
| Kittiwake | | −1 | 0.7 |
| Shag | | −12 | 3.7 |
| Golden plover | | −1 | 1.8 |

Organochlorine residues are total mg/kg wet weight BHC, heptachlor, dieldrin, DDE, TDE, DDT and DME. Data from Ratcliffe (1970)

livers within a week of sowing of dieldrin-dressed winter wheat, and predatory birds need to consume only a few mice carrying such residue levels to acquire a lethal dose.

The bioaccumulation of organochlorine insecticides resulted in further, more subtle adverse effects for many birds. The reproduction of birds is affected by organochlorines, particularly DDE, the principal metabolite of DDT. Breeding failures evident in the peregrine falcon, and also other raptors such as sparrowhawks (*Accipiter nisus*) and kestrels (*Falco tinnunculus*), were linked with thinning of eggshells. Ratcliffe at Oxford University observed large numbers of broken eggs in the nests of peregrine falcons during his studies in the 1950s. He then measured the eggshell thickness of peregrine falcons in museum collections and compared these with measurements made during the 1940s and 1950s. Ratcliffe found that the mean eggshell thickness remained constant until 1947 and thereafter declined (Table 9.11). This decline in eggshell thickness began at about the time of introduction of the organochlorines in the UK. Similar studies in the USA also demonstrated eggshell thinning, in peregrine falcons as well as other raptors, associated with the introduction and use of organochlorine insecticides. Some species of birds, including barn owls (*Tyto alba*) and American kestrels (*Falco sparverius*), fed DDE in experimental studies were shown to lay eggs with significantly thinner eggshells than those receiving DDE-free diets. The brown pelican (*Pelecanus occidentalis*) and cormorants (*Phalacrocorax auritus*) were particularly affected in some parts of the USA and Canada where bioaccumulative effects were linked to their diet of fish, especially anchovies, contaminated with organochlorine insecticides.

The mechanism of eggshell thinning is not fully understood. Birds of prey are very difficult to breed in captivity and thus most studies have been carried out with quails and ducks. However, the eggs of some gallinaceous species such as quails, pheasants and chickens do not experience shell thinning in the presence of organochlorine insecticides. Indeed the use of test species such as hens was the reason why shell-thinning effects were not picked up during the development and testing of the organochlorines in the early 1940s. Chickens fed up to 300 μg/g DDT showed no difference in eggshell thickness compared to those receiving a DDT-free diet. In those species that experience eggshell thinning, interference in the mobilisation of calcium from the bones of birds at the time of shell formation seems to occur. Alternatively, or in addition, inhibition of carbonic anhydrase, which regulates the supply of carbonate to the shell at the sites of synthesis, may contribute to eggshell thinning. Adverse effects on the thyroid glands of affected species may also influence shell thickness (McEwan and Stephenson, 1979).

Although the cyclodienes and DDT have been withdrawn as agricultural products in the UK and North America, residues of these pesticides may still be found in predatory birds, albeit at much reduced levels. Following the withdrawal of these insecticides, populations of the peregrine falcon and other raptors have increased. In 1961, 385 pairs of peregrine falcons were recorded in Britain; in 1971, 489; 1981, 728; and in 1991, 1283 pairs were recorded, which is an increase over the 1930–1939 average of 820 (British Trust for Ornithology, 2002). Although numbers have increased, the recolonisation of some areas, notably south-east England, by the peregrine falcon has been rather slow. Also, recovery has been slower on the North American continent, with 1150 breeding pairs recorded in 1985. The bald eagle, however, has made a very quick recovery and by 1982 the breeding success of this species in North America was back to that seen prior to the introduction of organochlorine insecticides.

Fewer studies have been carried out in tropical and subtropical regions where organochlorines such as DDT are still employed, for example to control the malarial mosquito and tsetse fly. Studies in countries as far apart as Zimbabwe (Douthwaite, 1995) and Vietnam (Minh *et al.*, 2002) have shown relatively high levels of DDT and other organochlorines in resident birds, with local reductions in populations evident in heavily sprayed areas.

Insecticides other than the organochlorines have also been shown to have adverse effects on birds. After the withdrawal of the organochlorines, organophosphate insecticides were increasingly used in seed dressings, with carbophenothion and chlorfenvinphos being employed extensively in the UK during the 1970s and 1980s. However, many reports of deaths of Canada, pink-footed and greylag geese were reported in the UK after feeding on wheat seed treated with this insecticide (Stanley and Bunyan, 1979). Acetylcholinesterase activity in brains of these species was severely impaired indicating organophosphate toxicity, and carbophenothion proved toxic in feeding tests with Canada geese. Following these and similar studies the use of carbophenothion as a seed dressing was discontinued.

Sporadic problems of acute toxic effects of insecticides on birds have continued even up to the present century. In the late 1990s aerial spraying of the organophosphate fenthion for mosquito control in the USA was associated with the deaths of wading birds in Florida and California. In the same decade, the introduction of the pyrrole insecticide chlorfenapyr for cotton pest control caused much concern in ornithological circles, since this compound is toxic in low dose to birds, with an $LD_{50}$ of around 10 mg/kg to mallard duck, and furthermore it has been shown to affect the reproductive success of this species (Albers and Melanoon, 2002).

The so-called second-generation rodenticides (5.3) including bromidalione, brodifacoum, difenacoum and flocoumafen may present risks to owls (*Strigiformes*). When second-generation rodenticides were introduced in oil palm plantations in Malaysia, some areas experienced drastic reductions in barn owl populations, and deaths were accompanied by haemorrhaging – a typical symptom of the anticoagulant rodenticides.

Second-generation rodenticides may be toxic to owls at cumulative doses of as little as 2–5 mg/kg body weight and consequently in the UK the compounds are formulated in wax blocks, which are concealed and unlikely to come into direct contact with owls. The major danger to owls is therefore likely to be consumption of treated prey. Researchers at the Institute of Terrestrial Ecology (now the Centre for Ecology and Hydrology) in the UK found that the proportion of barn owls found with anticoagulant rodenticides present in liver tissue increased from 5% in 1983–1984 to around 40% in the mid 1990s. In 1998, 52% of owl carcasses examined were found to have residues of rodenticides and the concentrations of rodenticides in liver tissue of about 10% of carcasses were considered to be in the lethal range (Shore *et al.*, 2002). The owls may have acquired these residues through consumption of dead rats and

Table 9.12 *Grey partridge brood sizes on cereal fields with sprayed and selectively sprayed headlands in Eastern England 1984–1987*

| Year | Sprayed headlands | Selectively sprayed headlands | Significant difference |
|---|---|---|---|
| 1984 | 4.7 | 7.8 | $p < 0.001$ |
| 1985 | 2.7 | 4.0 | $p < 0.05$ |
| 1986 | 4.8 | 8.7 | $p < 0.001$ |
| 1987 | 4.0 | 7.1 | $p < 0.01$ |

Figures are mean brood sizes. Selectively sprayed headlands were sprayed with graminicides and chemicals to control troublesome weeds such as cleavers, *Galium aparine*. From Sotherton (1988)

mice whose corpses had not been disposed of, as well as from individuals resistant to compounds such as difenacoum. Although concentrations of these compounds appeared high enough to be responsible for deaths of only 5% of owls containing residues, mortalities may become more widespread as the incidence of rodenticide resistance increases. The threat to raptors is of particular concern and in this respect some corpses of red kites (*Milvus milvus*), a rare species reintroduced into England and parts of Scotland in the late 1990s, have been found to contain residues of brodifacoum, bromidalione and other rodenticides.

Indirect effects of pesticides may have also contributed to declines in the population of some avian species in arable habitats. During the 1980s and 1990s drastic reductions in the numbers and range of some farmland bird species were observed in the UK. Loss of food sources such as weed seeds and insects after application of herbicides and insecticides has been proposed as a major factor in the decrease in species abundance. Reductions in weed seeds and insects following applications of herbicides and insecticides in arable crops have been linked to reductions in the number of surviving progeny in species such as the grey partridge (*Perdix perdix*), yellowhammer (*Emberiza citrinella*) and skylark (*Alauda arvensis*). In the UK the Game Conservancy has monitored populations of the grey partridge in Sussex and elsewhere (Potts, 1986): population declines have been clearly linked to the lack of invertebrate food during the early life of offspring (Table 9.12). In complementary studies by the Royal Society for the Protection of Birds, sprays of pyrethroid insecticides were strongly correlated with reduction in chick food supply for both skylark and yellowhammer. Chick starvation in skylark has been associated with summer applications of insecticides (Morris, 2002). Removal of food sources for birds by herbicides and insecticides has also been implicated in the decline of avian populations in other parts of the world. (O'Connor, 1992).

## 9.10 Pesticides and mammals

In general, fewer adverse effects have been reported with pesticides and mammalian species when compared to the problems that have arisen with other non-target species such as insects, fish and birds. Direct effects of pesticides may occur on mammals at lower trophic levels. Woodmice and voles that consume grain treated with insecticides may be killed in this way, and, as noted earlier in the case of organochlorine compounds, may also pass on residues to their predators.

With a few exceptions, DDT and the organochlorines appear to have had little effect in either the short or long term on domestic and wild mammals, although some aquatic mammals have been shown to bioaccumulate these compounds. Comprehensive studies in the USA and the UK have found little changes in populations of species from a wide range of mammalian orders exposed to DDT. However, bats (Chiroptera) are sensitive to DDE, and effects may be linked to the seasonal variation in fat content of these mammals. Residues after hibernation have been shown in East Anglia to increase in some species to almost 50 mg/kg, close to the $LD_{50}$ for bats. Bats may acquire organochlorines from their insect prey, but a much more important source is treatment of roof timbers in or near roosts with lindane, dieldrin as well as other compounds such as dichlofluanid and pentachlorophenol to control wood-boring insects and wood-rotting fungi (Shore *et al.*, 1990). These compounds and possibly also pesticides such as the pyrethroids – used increasingly in timber treatment – are likely to persist for far longer than in soils, where they may be degraded by soil microorganisms. In the generally drier conditions of for example roof spaces, degradative processes may be much slower or even absent.

Problems to wild and domestic mammals have resulted from the use of cyclodiene insecticides, and deaths of cats, dogs and wild mammals, particularly foxes, were reported in the UK in the early 1960s. Most of these were attributed to the eating of dead and dying pigeons that had consumed wheat seed treated with organochlorine insecticides. Deaths of predatory mammals were observed in the UK at about the same time as those of grain eating birds referred to above (9.9). Autopsies of foxes and badgers revealed high levels of dieldrin, probably originating from their prey – small mammals such as voles and woodmice and birds that had consumed grain treated with organochlorine insecticides (Sheail, 1985).

Bioaccumulation of organochlorine compounds by marine and freshwater mammals has been extensively studied. Cetaceans such as whales, porpoises and dolphins have been found to contain DDT and other organochlorines in their tissues, and these have most probably come from consumption of contaminated prey – fish, marine invertebrates and even plankton. The long life span of many cetaceans, more than 50 years for some whales, may enable bioaccumulation of organochlorines and other persistent compounds to relatively high concentrations. Although concentrations of organochlorines in cetaceans in some parts of the world, including the UK, are relatively low at less than 10 μg/g (DEFRA, 2002), in other parts concentrations have been detected at much higher levels, up to 380 μg/g lipid weight in dolphins near Hong Kong (Parsons, 2002). Even in areas considered to be less polluted, occasional high levels of organochlorines have been recorded in the carcasses of whales and other cetaceans.

The insulating blubber layer of cetaceans is rich in lipids into which organochlorines readily pass, and here organochlorines may not exert adverse effects. However, when lipids in the blubber are broken down and utilised by cetaceans during times of food shortage or lactation after birth, organochlorines may be released into the blood and other tissues and adverse effects may result. In particular, organochlorine and other chlorinated compounds have been associated with immunosuppressant activity, leading to increased risks of infection in cetacean species (Parsons, 2002).

The decline in otter (*Lutra lutra*) populations in the 1950s and 1960s in the UK has also been partly attributed to the increasing use of organochlorines, particularly aldrin and dieldrin at that time, and associated with residues in fish, the principal diet of the otter. After withdrawal of aldrin and dieldrin in the UK, recovery of otter populations has been slow compared to that of predatory birds, and it is possible that organochlorine compounds were not the sole reason for its decline in population. Human disturbance and destruction of habitats may have influenced recovery. However, the National Otter Survey for England reported a sustained recovery in otter populations from 1977–1979, when only 6% of sites surveyed indicated the presence of otters, to 2000–2002, when 35% of sites surveyed showed evidence of otter activity (Environment Agency, 2003).

Granular insecticides and molluscicides may present a localised threat to some species. Several instances of cats and dogs suffering acute effects after consumption of slug pellets have been reported to the Wildlife Incident Inspection Scheme in the UK. Hedgehogs (*Erinaceus europaeus*) may consume and possibly be killed by slug pellets. Granular pesticides might also be consumed by grain-eating small mammals, and the compound methiocarb has been shown to cause severe, but transient, decreases in populations of wood mice near sites of application.

## 9.11 Monitoring of the effects of pesticides on non-target species in the UK

Since the problems associated with organochlorine insecticides in the 1950s, procedures for investigation of wildlife incidents following pesticide poisoning have been established in many countries. In some cases, monitoring programmes have continued over several decades. The Wildlife and Pollution contract of the Joint Nature Conservation Committee (JNCC), undertaken by staff at the Centre of Ecology and Hydrology at Monks Wood in Cambridgeshire, has monitored the levels of pollutants such as organochlorines in selected wildlife of the UK since the late 1960s, and is the longest running scheme of its type in the world. Their annual reports focus particularly on the effects of persistent compounds such as organochlorines and rodenticides on populations and breeding success of raptor birds such as hawks, kites and eagles.

The Wildlife Incident Investigation Scheme (WIIS) operated by the Wildlife Incident Units in Scotland, England, Wales and Northern Ireland was established in the 1960s by the former Ministry of Agriculture, and has been maintained by this organisation and its successors. In England the Unit is based at the Central Science Laboratory near York (Anon, 2002b). Following reports to the Unit, a field enquiry is established; post mortems are usually carried out, and chemical analyses of tissues from dead animals performed to identify and quantify any pesticide residues. Information regarding the circumstances and extent of exposure may also be obtained. Results are then evaluated to establish the likely cause of the incident, and these passed to the Environmental Panel of the Advisory Committee for Pesticides for consideration of further action. Data gathered may be used to alter the conditions of use or approval status of a compound.

In the WIIS, reports in which pesticides are involved are assigned to one of four categories. These are:

1. **Abuse** of a product where deliberate and illegal attempts are made to poison animals
2. **Misuse** of a product by accidental, careless or wilful failure to adhere to correct practice
3. During **approved use** of the pesticide
4. **Unspecified use**, where the cause cannot be assigned to one of the above categories.

Table 9.13 *Number and nature of Wildlife Incidents investigated in England 1988–1990 and 1998–2003*

| | 1988 | 1989 | 1990 | 1998 | 1999 | 2000 | 2001 | 2002 | 2003 |
|---|---|---|---|---|---|---|---|---|---|
| Vertebrate wildlife | 277 | 343 | 373 | 256<br>(49) | 204<br>(57) | 231<br>(68) | 176<br>(55) | 232<br>(63) | 208<br>(63) |
| Livestock | 47 | 56 | 67 | 10<br>(4) | 4<br>(2) | 9<br>(0) | 4<br>(0) | 1<br>(1) | 5<br>(0) |
| Companion animals | 206 | 200 | 222 | 235<br>(90) | 149<br>(48) | 160<br>(58) | 109<br>(34) | 150<br>(45) | 130<br>(42) |
| Beneficial insects | 108 | 109 | 100 | (43)<br>(12) | (28)<br>(1) | (48)<br>(13) | (23)<br>(5) | (25)<br>(5) | (24)<br>(8) |
| Suspicious baits and suspicious substances | 24 | 38 | 27 | 62<br>(29) | 67<br>(22) | 64<br>(28) | 35<br>(16) | 47<br>(20) | 31<br>(14) |

Figures in parentheses from 1998 onwards indicate incidents where pesticides were held to be the primary cause. From British Agrochemical Association Annual Reviews for the above years and Pesticide Poisoning of Animals at the website of the UK Pesticide Safety Directorate (http://www.pesticides.gov.uk/citizen/wiis.htm)

The number of pesticide poisoning incidents reported to the Wildlife Incident Units in the UK for 1988 to 2001 is shown in Table 9.13. The table shows that a large number of incidents reported to the Wildlife Incident Units are not attributed to pesticide poisoning. Some are often found to be due to disease, stress or starvation. For example, in England between 1994 and 1996 about 1000 incidents were investigated, and pesticide poisoning confirmed in 33% of these. Misuse of compounds occurred in 19% of confirmed incidents, and unspecified use in 23%. Over half of the confirmed incidents were from deliberate misuse of pesticides, and such abuse was still prevalent in 2001, being responsible for 74% of pesticide incidents reported to the WIIS in that year, and this abuse also occurs in other parts of the UK (Barnett and Fletcher, 1998). In Scotland during the 1990s over 70% of confirmed incidents were due to deliberate abuse of pesticides in efforts to poison animals. Birds of prey were the most common targets of misuse, including hawks, buzzards (*Buteo buteo*) and golden eagles. The pesticides involved include the general poison strychnine, used to control moles (*Talpa europaea*), the organophosphate alphachloralose, also used in vertebrate pest control (5.2), and the carbamate carbofuran.

If deliberate abuse is excluded, then most vertebrate poisoning incidents in the UK have arisen from exposure to anticoagulent rodenticides, the molluscicide metaldehyde, and anti-acetylcholinesterase insecticides. Bromadiolone and difenacoum are the principal rodenticides involved and the species most commonly affected are cats, dogs and badgers. As with birds, toxic effects on these animals may be due to consumption of the carcasses of rodents in which pesticide residues may remain after death. Dogs were also the principal species affected by metaldehyde in the form of bran-based pellets for snail and slug control: the $LD_{50}$ of metaldehyde for most breeds of dog is about 100 mg/kg body weight, and easily obtained from consumption of a few pellets. Snail/slug pellets are attractive to many animals, and incidents have been often associated with spillages of pellets at or near the site of application, or with poor storage conditions.

It is considered that, despite promotion of the UK Government's campaign against illegal poisoning of animals, underreporting of incidents undoubtedly occurs. However, within the reported figures, incidents arising from the use of pesticides as approved by the manufacturer have consistently remained at less than 1% of total incidents reported – a negligible number according to the Environmental Panel of the Advisory Committee for Pesticides in the UK.

# 10

# Public perceptions, comparative risk assessment, and future prospects for pesticides

## 10.1 Introduction

In Europe after the conflict of 1939–1945 food was scarce and the achievements of farmers in raising the yield of staple products, most notably cereals, was closely associated with high levels of fertilisation, better yielding cultivars and the development of pesticides that allowed the selective control of weeds. The intensification and specialisation in farming that followed during the second half of the 1900s resulted from a desire for self-sufficiency in basic foodstuffs, and this approach was widely welcomed. In Western Europe following the establishment of the European Economic Community, or European Union as it is now, this drive was aided by generous subsidies raised by taxation and given to farmers in order to encourage output. An impetus was thus provided to maximise yields and consequently, as noted above, fertiliser use greatly increased, higher yielding cultivars of crops were developed, and pesticides came to be increasingly and widely used to protect crops from weed competition as well as attack by pests and diseases. Yields rose in many countries to the point where production exceeded national requirements, and surpluses began to accumulate. For example, wheat yields in the UK increased from an average of 2.5 tonnes/ha in 1947 to 7.5 tonnes/ha in the mid 1990s. The so-called grain mountains that developed in the EU during the 1980s were a direct result of the subsidised drive to increase crop yields.

Similar increases have been achieved with crops in other parts of the world. Yields of rice and wheat increased markedly in parts of Asia during the 1970s and 1980s. Introduction of better cultivars, assisted by high inputs of plant nutrients and routine use of pesticides helped to create the so-called green revolution reflected by the high yields obtained within such intensive growing systems. In 1965, wheat yields were 12.3 million tonnes in India; by 1970, after the introduction of highly productive dwarf wheats, yields had risen to

20 million tonnes; by 2002 India was producing 73.5 million tonnes and is not only self-sufficient in this staple food but also exports wheat. In some parts of Asia, such as Japan, the introduction of subsidies and trade barriers (a similar situation to that of the EU) promoted the production of high yields, again with attendant inputs of fertilisers and pesticides.

Globally, the public health impacts of pesticide use have been overshadowed by those of vaccine development for viral diseases such as polio and small-pox, as well as antibiotics for control of microbial diseases, but the control of vectors of diseases such as malaria, typhus, yellow fever and other tropical ailments was welcomed not least by people in poorer countries suffering from these maladies. The staggering success in many countries in reducing and in some cases eradicating the malarial mosquito, largely through the use of DDT, gave pesticides a very positive public profile in the 1950s. Vaccine and antibiotic use has served also to improve the health of farm and domesticated animals, but pesticide use in veterinary and animal health has reduced the incidence of some long-standing pests such as sheep scab, and relieved many households of the irritating problem of fleas associated with pet animals.

Changes in the public perception of pesticides began in the late 1950s, and the principal reason was the excessive and sometimes unnecessary use of compounds. Pesticides, particularly insecticides, came to be regarded in some countries, notably the USA, as a panacea for all pest problems. Consequently pesticides were used very widely in both rural and urban settings, and often applied by aerial spraying. The blanket aerial spraying of some urban areas in the USA for control of species as diverse as the bark beetles that transmit dutch elm disease to the common house fly led to public disquiet. These worries were crystallised in the publication by Rachel Carson of '*Silent Spring*' in 1962 (Carson, 1962). This landmark text may be identified as the single most significant contribution to the change and indeed reversal of public attitudes to pesticide use. The tone of Carson's text has been criticised for its emotive nature, and the text itself, produced many years before analyses of hazard and risk became common considerations, makes little reference to the relative risks of pesticides to either the environment or public health. However, it was one of the first attempts to collate the adverse effects that indiscriminate use of pesticides may have on non-target organisms, public health and the environment in general, and its populist and often dramatic tone reached a large audience.

Thus, from the mid 1960s, pesticides began to be regarded with suspicion and eventually by some as a major threat to public health and the environment, and this attitude is prevalent today (Beaumont, 1993; Petty, 2001). Several factors explain this negative image. The adverse effects of organochlorine insecticides (outlined in Chapters 7 and 9) have attracted particular attention over the last

50 years, and have been prominently aired in the media, leading to a general mistrust of all pesticides among the public. The negative image of pesticides has been fostered and maintained by pressure groups such as Friends of the Earth, Greenpeace and a group devoted entirely to pesticides, the Pesticides Action Network. In the USA the National Resources Defense Committee is a body that devotes much of its time in anti-pesticide activities. Whereas Rachel Carson at least partially acknowledged that pesticides might have a role to play in public health and agriculture if used responsibly, modern pressure groups rarely indicate the benefits that may accrue from pesticide use.

To the general public, organisations such as Greenpeace and Friends of the Earth, supported mainly by their membership and charitable donations, have laudable aims and no vested interests, and their information is often accepted without demur. In the late twentieth century, pressure groups became highly adept at exploiting the principal media outlets to promote their views, and with respect to many issues, including pesticides, this still continues. Frequently, through appropriately timed press releases, information from pressure groups will be received and prominently aired or published, often uncritically, in both local and national media. For example, in the UK, press releases from pressure groups to the general media are often timed to coincide with the publication of the Annual Reports of monitoring bodies such as the UK Pesticides Residue Committee (PRC). Attention is almost exclusively drawn to examples of breaches of maximum residue limits detected in the surveys commissioned by the PRC, but without reference to either the level by which limits have been breached or the numbers of breaches as a percentage of the total number of analyses (7.4) and reference is rarely made to the actual risks posed by doses likely to be encountered. Such exaggerated risks may be emphasised by media coverage. In the general media, little attention may be paid to information that a new pesticide may present a low risk to operators and the environment: safety is regarded as bland, not particularly newsworthy, and unlikely to attract the attention of editors in the mass media. However, information that a pesticide is unsafe is viewed as threatening, with potential elements of danger, and thus judged more likely to capture the attention of readers.

The debate over dithiocarbamate fungicide use in the early 1990s provides a good example of selective media reaction to an almost fraudulent claim made by pressure groups (Berry, 1990). In 1990 the pressure groups Friends of the Earth and Parents for Safe Food announced in a UK press release under the heading 'Dangerous Chemicals in Food' the results of analyses of pesticide residues in certain foodstuffs. The groups claimed to have found ethylene thiourea, a major metabolite of ethylenebisdithiocarbamate fungicides (3.3), in bread. Ethylene thiourea has been implicated as a carcinogen. This initial announcement was extensively reported in the popular media, and the then chairman of the UK Advisory Committee on Pesticides received several calls from journalists. However, when tests were repeated on the samples no residues could be found. Pesticide residues had been overestimated by 150-fold in some foods and the groups appeared to have spiked one sample (Berry 1990). The subsequent rebuttal of these, at best, misleading results was not reported in the popular media.

This example illustrates the difficulty in establishing positive features of pesticides. Information about the benefit of pesticides or refutations of claims made by pressure groups usually come from manufacturers or farmers who are seen to have a vested interest in selling or using pesticides and are thus regarded less favourably by the general public. The profit motive may be perceived as the overriding aim of manufacturers and users of pesticides, above that of human health and effects on the environment. Consequently the overall public perception of pesticides in the early twenty-first century is more that of a threat rather than in any beneficial sense.

As noted in Chapter 1, public perceptions are affected by value judgements and the willingness to take personal action in order to reduce risk (Gray, 2001). For example, foodstuffs high in saturated fats and salt may contribute to increased rates of heart disease and cancer, and those high in sugar associated with diabetes in later life. In many cases consumer preferences for the taste of products high in these parameters may outweigh considerations of the risk of long-term harm. Both pesticides and diets high in fat, sugar or salt may be perceived as harmful to human health, but the desire to consume such foodstuffs may override such considerations. However, eating foods that may contain pesticide residues gives no additional benefit such as improved taste.

For some potential hazards to health, individuals may be prepared to bear the costs themselves of changing their behaviour; for example, changing to a low-fat diet. In the case of pesticides, however, others such as farmers and manufacturers may be judged to bear the cost of corrective action, since consumers themselves may have little direct control over the occurrence of pesticides in food and water. Indeed, many perceive that the only way to avoid pesticides is to have recourse to organic food, specially treated or bottled drinking water and fabrics made from untreated cotton, wool and other materials. However, organic foods and fabrics are invariably more expensive than those produced without recourse to pesticides.

Attempts to counter the claims of pressure groups and others that pesticides present serious risks to human health and the environment have been made. Prominent among those advocating the benefits, and seeking to counter the claims of pressure groups include Dennis Avery, who has produced the controversial text '*Saving the Plant with Pesticides and Plastic*' (Avery, 1995), and Bjorn Lomberg, author of '*The Skeptical Environmentalist*' (Lomberg, 2001). Both dispute the tenet that pesticides markedly affect public health, and firmly extol their benefits, in agriculture as well as public health. Avery also advocates that intensive farming, including optimal pesticide use to attain the highest possible yields on the best quality agricultural land, will prevent the need to cultivate poorer land which can then be left for nature conservation and recreation. He does not, however, consider to any great extent the potential for off-target effects resulting from intensive pesticide application to land of good agricultural potential, or the overall long-term effects that may result from such intensification. Equally Lomberg, whilst considering at length human health in relation to long-term exposure to pesticide residues in food and water, does not consider the problems associated with pesticides in their concentrated form on human health or the environment.

The differences between those who advocate or see few problems with pesticide use and dissenting groups may be considered in the light of comparative risk analysis: evaluating pesticidal effects alongside other causes of health and environmental problems. The principal concerns identified by the Pesticides Action Network are outlined in Table 10.1, and these may be considered in terms of related risks.

## 10.2 Obsolete and unusable stocks/unwanted imports

As identified in earlier chapters, the greatest risks to both human health and the environment are likely to arise from exposure to pesticides in their concentrated form. Thus, the presence in many poorer countries of large quantities of pesticide concentrates, stored for many years in containers that may deteriorate in warm, moist climates, may pose a high risk to both human health and the environment. The problem is further exacerbated by the nature of the pesticides that are stored: many are obsolete with revoked permissions for use; some are of high mammalian toxicity and poorer countries often do not have the facilities or technology to dispose of these unwanted concentrates (Figure 10.1). Most of these compounds were donated in good faith by aid agencies but many are acute and cumulative environmental toxicants such as organochlorine insecticides. The Food and Agriculture Organisation of the United Nations (FAO) estimated in 1999 that over 100 000 tonnes of obsolete pesticide concentrates were held by developing countries, and has called for action from the international community to assist in disposing of these compounds that clearly present a considerable risk to health.

Table 10.1 *Problems with pesticides*

| | |
|---|---|
| Human poisonings and health risks | Environmental hazards |
| Loss of biodiversity | Wildlife deaths |
| Animal and livestock deaths | Interference with natural pest control |
| Resistance among pests | Unwanted imports |
| Obsolete and unusable stocks | Residues in food |
| Water pollution | |

From the website of the Pesticides Action Network at http://www.pan-uk,org/Reviews/2000/ar00p2-3.htm

Figure 10.1 An obsolete pesticide store showing corroded metal drums (photograph by kind permission of FAO & A. Wodeageneh).

An associated problem has resulted from the increase in global trade in pesticides. In some cases industrial chemicals such as pesticides that have been withdrawn or have severely restricted uses in some, usually developed, countries have been exported to poorer countries. The establishment of pesticide-manufacturing facilities in countries such as China, and other Asian countries has led to more aggressive marketing of products to developing nations, again leading to accumulation of concentrates there.

Table 10.2 *Pesticides subject to the Prior Informed Consent procedure in 2002*

| Organochlorine insecticides | Organophosphate insecticides |
|---|---|
| Aldrin | Monocrotophos (formulations above 60% active ingredient) |
| Chlordane | Methamidophos (formulations above 60% active ingredient) |
| DDT | Phosphadimidon |
| Dieldrin | Methyl parathion |
| HCH (mixed isomers) & lindane | Parathion |
| Heptachlor | |
| Hexachlorobenzene | |
| **Fungicides** | **Herbicides** |
| Captafol | 2,4,5-T |
| Mercury compounds | Chlordimeform |
| Pentachlorophenol | Dinoseb |
| | Fluoroacetamide |

From http://www.epa.gov/oppfead1/international/pic.htm

To assist in avoiding the importation and use of pesticide concentrates with high acute toxicity, stricter regulations relating to international trade in chemicals have been introduced. Both pesticide suppliers and many exporting/importing countries have welcomed the implementation of prior informed consent (PIC) procedures developed by the FAO and Environment Programme of the United Nations (UNEP). These procedures enable importing countries to be aware of the characteristics of potentially toxic chemicals that may be shipped to them, and consequently allow the countries to decide whether they wish to allow importation of these substances. Acutely toxic pesticides such as those in the World Health Organization Class I (extremely hazardous) are included in PIC (Table 10.2). Once a chemical has been identified for inclusion in the PIC procedure, a decision guidance document is sent to participating countries. The document summarises the toxicological and environmental characteristics of the compound, its known usage, exposure routes and any special regulations pertinent to its use.

Such measures may help to reduce problems of human health and the environment in poorer countries. Although better storage and security measures would assist in reducing suicide attempts from ingestion of pesticide concentrates, a move towards the use of compounds of lower acute toxicity, or indeed other measures of control (10.9), might eliminate this problem altogether. However, newer less toxic compounds are nearly always more expensive than older, often off-patent materials and thus less likely to be used in poorer countries.

## 10.3 Comparative analysis of risks to human health: pesticides and other hazards

Hazards to human health include pesticides, but in global terms pesticide use or abuse is linked to a very small proportion of premature mortality and illness. However, the effects on health from spraying of pesticides in the third world are probably vastly underreported, and these may constitute one of the principal risks to health in rural communities there. In 1990, the World Health Organisation (WHO) estimated that there might be as many as 50 cases unreported for every case reported (Harris, 2000). In order to place the problems that pesticides present to public health in perspective, it is appropriate to consider the principal causes of premature death and long-term ill health.

The World Health Report of 2002 identified ten major factors that provide the greatest risks to human health (Table 10.3). According to data produced by WHO in 1997, of a global total of 52 million premature deaths, 17 million were due to infectious and parasitic diseases, 15 million to circulatory diseases, 6 million to cancer, 3 million to respiratory diseases and 2.4 million to perinatal conditions. Rising causes of death included HIV/AIDS with 2.7 million deaths and road traffic accidents with 1.3 million fatalities in 1997.

The WHO considers that, worldwide, HIV/AIDS and smoking are the two principal avoidable causes of premature death. As noted in Chapter 7, WHO estimated in 1990 that acute deaths worldwide from pesticide poisoning were about 100 000: few more recent estimates are available.

In addition to the acute effects of pesticides, the potential long-term effects for human health that may arise following exposure to pesticides have attracted much interest. One of the most contentious claims made by pressure groups is

Table 10.3 *Principal risks to human health*

| | |
|---|---|
| Underweight | Unsafe water, sanitation and hygiene |
| Unsafe sex | Iron deficiency |
| High blood pressure | Indoor smoke from solid fuels |
| Tobacco consumption | High cholesterol |
| Alcohol consumption | Obesity |

From the World Health Report 2002 published by the World Health Organization at http://www.whi.int/whr/2002/en/Overview_E.pdf

Table 10.4 *Principal cancer types and their associated risk factors*

| Cancer | Deaths in 2000 | Principal risk factors |
|---|---|---|
| Lung | 1 190 000 | Smoking; exposure to industrial chemicals and asbestos |
| Colon/rectal | 509 000 | Family history; diet |
| Liver | 589 000 | Smoking |
| Breast | 467 000 | Family history; ageing |
| Prostate | 255 000 | Ageing |
| Stomach | 801 000 | Smoking |
| Oesophagus | 381 000 | Smoking; behavioural |
| Leukaemia | 298 000 | Largely unknown; possibly viral or exposure to some chemicals, e.g. benzene |
| Lymphoma | 295 000 | Exposure to HIV; possible exposure to industrial solvents and herbicides |

From Cancer Facts and Figures 1994, American Cancer Society and the WHO website at http://www.whi.int/whr/2002/en/Overview_E.pdf

that exposure to pesticides from consumption of food and water containing very small doses of pesticide residues may lead or predispose to long-term adverse conditions such as cancer.

Epidemiological studies have indicated that the principal risk factors associated with cancer in the human population are smoking, poor diet, ageing and a genetic predisposition. The principal cancers of the human population in the early twenty-first century are listed in Table 10.4 along with epidemiological assessments of the principal factors that lead to an increased risk of developing the disease.

Overall cancer death rates are decreasing in many countries but with some exceptions, notably lung, breast, prostate and skin cancers and non-Hodgkins lymphoma (NHL). As seen in Table 10.4, major contributing factors to the development of cancers include a genetic predisposition, lifestyle and chronic infections. For example, lung cancers are primarily linked to tobacco smoke, and skin cancers with exposure to the sun. Poor diet, particularly lack of fruit and vegetables, and obesity are associated with increased risks of cancers of the intestinal and rectal tracts. The incidence of some cancers such as prostate is associated with ageing, and since many people are now living longer, an increase in such cancers has occurred.

However, some cancers are firmly linked with occupational exposure to chemicals, such as mesothelioma resulting from exposure to certain types of asbestos. Soft tissue cancers and lymphomas are major cancers that have been associated with occupational exposure to pesticides (Hoar *et al.*, 1986). There is a considerable body of evidence suggesting that phenoxy hormone herbicides may pose an increased risk of developing NHL. For example, several studies in the USA and Sweden produced strong evidence for an increased risk of NHL in farmers and forestry workers who had been exposed to phenoxy herbicides. However, similar investigations in New Zealand did not reveal any greater risk to NHL from exposure to these compounds. The Environmental Protection Agency in the USA, having considered the above studies, concluded that epidemiological studies relating to 2,4-D have shown no increased risk of cancer, indicating that some studies on NHL imply an association with farming rather than pesticides alone (USEPA, 1994).

Although some studies appear to have shown a link between NHL and exposure to phenoxy herbicides, the role of herbicides in predisposition to soft tissue sarcomas is much less clear. A review of North American, Swedish and New Zealand studies has concluded that, at most, very small risks of soft tissue sarcoma may result from exposure to herbicides (Morrison *et al.*, 1992).

Kogevinas *et al.* (1997) carried out an extensive study of cancer mortality in workers exposed to hormone herbicides. They examined the medical records of 21 863 male and female workers from 12 countries who had been exposed to phenoxyacetic acid hormone herbicides, as part of a study that also included assessments of chlorophenol and dioxin exposure. In workers exposed to phenoxy herbicides with minimal or no contact with dioxins, mortality from cancer was similar to expected values, with a very slight rise in soft tissue sarcomas: two deaths were recorded in the studies as opposed to an expected value of slightly less than one. However, exposure to herbicides in the presence of dioxins, which have often appeared in the past as contaminants during the manufacturing process, resulted in a small but significant increase in cancer risk.

A similar situation exists with insecticides, particularly the organochlorine group and the aetiology of breast cancer. Increases of breast cancer are occurring in many countries, but there are considerable geographical differences, with a much lower incidence in Asian countries. Early studies with small numbers of individuals appeared to indicate a link between polychlorinated biphenyls, DDT and its metabolites, particularly DDE and breast cancer. However, the comprehensive study of Krieger *et al.* (1994) found no link between serum DDT levels, polychlorinated biphenyl concentrations and the risk of human breast carcinoma.

Some epidemiological studies have shown a link between occupational exposure to pesticides and prostate cancer (Van Maele-Fabry and Willems, 2003).

The Department of Health Committee on Carcinogenicity in the UK (Department of Health, 2004) has concluded that a small increase in risk of prostate cancer exists among farmers and farmworkers who regularly use pesticides, although the evidence did not identify any single pesticide or groups of pesticides that may be responsible.

In 2004 the Ontario College of Family Practitioners published a systematic review of the epidemiological literature on the possible chronic health effects of pesticides (Sanborn *et al.*, 2004) and their review was widely circulated and uncritically accepted by many pressure groups. Whilst reiterating links between pesticides and cancer of the prostate, soft tissue sarcomas and NHL, the authors also stated that there are positive associations between solid tissue tumours and exposure to pesticides. This last assertion has been severely criticised by, among others, medical toxicologists and epidemiologists of the UK Advisory Committee on Pesticides (ACP) (Anon, 2004b). Furthermore, the ACP and others have pointed to a strong bias in the selection of literature and its interpretation by the Ontario College of Family Practitioners. In particular, a failure to include many studies that showed no links between occupational pesticide use and long-term illness was noted, and some major epidemiological studies, including those of Kogevinas *et al.* (1997) and Krieger *et al.* (1994) referred to above, were overlooked or ignored.

In addition to problems associated with occupational exposure to relatively high concentrations of pesticides, there is much public concern about the effects of long-term exposure to residues of pesticides in food and water. The potential for exposure to low doses of pesticides in food and drinking water to cause long-term problems to human health may be reviewed in perspective alongside potential risks to health from other substances ingested or drunk by human beings. The chemicals tested for potential carcinogenic effects have been primarily synthetic; far fewer naturally occurring compounds have been examined using standard rodent carcinogen tests. The significance and risk to human health of naturally occurring substances have not been studied to anywhere near the same level as the potential long-term effects of synthetic pesticides. Many of these naturally occurring compounds in plants are involved in resisting consumption by herbivores as well as attack by pests and diseases and in this sense may be construed as naturally occurring pesticides. Indeed, some of these natural toxins with pesticidal activity have caused adverse health effects.

The beneficial effects to health of consuming fruit and vegetables are unquestionable. In addition to their mineral and vitamin provision, the protective effect of vegetables such as crucifers in reducing the risk of cancers of the intestinal tract is well established. However, when subjected to the same toxicological

Table 10.5 *Chronic effects of some naturally occurring substances extracted from plants*

| Food plant | Substance | Chronic effects |
|---|---|---|
| Cabbage, oilseed rape, other crucifers | Isothiocyanates formed from glucosinolates | Goitrogenic; mutagenic in *Salmonella typhimurium* tests |
| Lettuce, cranberry | Quercetin | Mutagenic in *S. typhimurium* tests |
| Comfrey | Pyrrolizidine alkaloids | Hepatocarcinogenic |
| Potato | Glycoalkaloids | Inhibition of cholinesterase |
| Parsley, celery, parsnips | Furanocoumarins | Dermatitis, carcinomas |
| Sweet potato | Ipomeamarone | Hepatocarcinogenic; lung oedema |

From Beier (1990)

tests required for registration of pesticides, many chemicals from foods are seen to have adverse toxicological properties. In studies of chronic toxicity, many chemicals extracted from common food plants including herbs and vegetables have been shown to produce adverse health reactions in test animals, sometimes at low dose, ranging from increases in cancer to mutation as well as interference with the immune and endocrine systems (Table 10.5).

Indeed, some of these natural toxins with pesticidal activity have caused adverse health effects in human beings. Psoralens are antifungal compounds that occur in celery and parsnips, but are also potent skin photosensitisers as well as mutagens (for which purpose they are sold commercially). The concentrations of psoralens in vegetables rise markedly on infection. In acute terms, occasional deaths have occurred among people who have handled celery, particularly after visiting sun tan parlours, when severe burning of the skin has occurred (Beier, 1990).

The herb comfrey is used in the preparation of some medicinal teas, and comfrey-pepsin tablets are sold without regulation through health food outlets (as are many other natural products). Comfrey contains hepatotoxic pyrollizidine alkaloids and acute incidents of ill health and deaths have been linked to consumption of herbal teas prepared from comfrey.

Many other naturally occurring substances exhibit adverse effects in tests used in the toxicological evaluation of pesticides. As the national beverage of several countries, coffee and its constituents have been toxicologically investigated, and of the 30 or so naturally occurring substances so far

Table 10.6 *Carcinogenicity in rats of substances from roasted coffee*

**Positive in rat tests:**
Acetaldehyde; benzaldehyde; benzene; benzofuran; benzpyrene; caffeic acid; catechol; 1,2,5,6-dibenzanthracene; ethanol; ethylbenzene; formaldehyde; furan; furfural; hydrogen peroxide; hydroquinone; limonene; styrene; toluene; xylene

**Not positive in rat tests:**
Acrolein; biphenyl; choline; eugenol; nicotinamide; nicotinic acid; phenol; piperidine

**Uncertain** – caffeine

**But approximately 1000 others yet to be tested**

Adapted from Ames (1998)

identified and studied, two-thirds have proven carcinogenic in rodent tests (Table 10.6)

Cooking foods, particularly by roasting, frying or barbecuing, produces burnt material that contains many substances such as benzpyrenes and acrylamide that have adverse effects such as mutagenicity and carcinogenicity when subjected to the same toxicological scrutiny as pesticides (Ames, 1998). It has been estimated that, on average, people in Western societies consume on average about 2 g per day of burnt material produced by cooking, and again this can be compared to the quantities of synthetic pesticides ingested at the submilligram, or even submicrogram level.

Attempts have been made to rank both natural and carcinogenic hazards based on exposure levels that human beings might encounter in order to evaluate the risk posed by different chemical compounds. Analyses have been extrapolated from the human exposure/rodent potency (HERP) index. This is derived from studies of the ability of chemicals to cause cancer in rats in terms of $TD_{50}$ values: the concentration of a chemical that will produce tumours in 50% of a population of rodents. The HERP index is based on what percentage of the $TD_{50}$ (in milligrams per kilogram of body weight per day) a human being is likely to receive from a daily lifetime exposure. Table 10.7 shows the ranking by HERP for rodent carcinogens including a range of natural and synthetic compounds including pesticides. From this information, it seems likely that some naturally occurring substances and again those associated with certain lifestyles (e.g. alcohol) may present a far greater risk to health in chronic terms than pesticides. Indeed, for those pesticides listed in the US carcinogenicity potency database, the risks would appear to be very small indeed.

Table 10.7 *Carcinogenic risk estimates expressed as HERP for selected natural and synthetic substances*

| Substance | Origin and average daily exposure for USA | $TD_{50}$ (rat) ($mg \cdot kg^{-1} \cdot day^{-1}$) | Risk as indicated by HERP (%) |
|---|---|---|---|
| Pyrrolizidine alkaloids | (Comfrey-pepsin tablets – up to 9 may be taken daily) | 626 | 6.2 |
| Ethyl alcohol | Beer 257 ml/day | 9110 | 2.1 |
| Caffeic acid | Coffee – 500 ml/day; lettuce – 14.9 g/day | 297 | 0.1<br>0.04 |
| Safrole | Saffron & other spices – 1.2 mg/day | 440 | 0.03 |
| d-Limonene | Orange juice – 138 g/day | 204 | 0.03 |
| Coumarin | Cinnamon – 65 μg/day | 14 | 0.007 |
| Furfural | White bread – 67.6 g/day | 683 | 0.004 |
| Allyl thiocyanate | Brown mustard – 68.4 mg/day | 96 | 0.0009 |
| Chloroform | Tap water – 1 litre/day | 262 | 0.000 3 |
| Carbaryl | Carbaryl – 2.6 μg/day | 14 | 0.000 3 |
| Celery | Methoxypsoralen – 4.9 μg/day | 33 | 0.000 2 |
| DDE/DDT | DDE/DDT at 659 ng/day | 12.5* | 0.000 08 |
| Lindane | Lindane at 115 ng/day | 31* | 0.000 001 |
| Captan | Captan at 115 ng/day | 2080 | 0.000 000 08 |

* Figures for mice; DDT and lindane have not been shown as carcinogenic in rats
From Gold and Slone (1999)

Only a very small number of naturally occurring chemicals in plants have been evaluated for their potential to cause cancer, but of those that have been studied, about 50% prove to be carcinogenic. Natural substances capable of causing cancer in rodent carcinogen tests are likely to be present in all fruits and vegetables, but the risks from these to human health are, as noted earlier, far outweighed by the beneficial effects of consuming fruit and vegetables.

Furthermore, in addition to the immune responses that may counteract microbial attack, human beings possess natural defences against exposure to environmental toxins. The proposition has been made that mechanisms which protect against naturally occurring chemicals have evolved throughout human history, and that these may be insufficient to cope with modern synthetic compounds. However, many defence mechanisms are by necessity general in nature to counter the huge range of substances and situations encountered in everyday existence. For example, skin and the external layers of the lungs and digestive tract are shed continuously. Enzymes exist that repair DNA,

Table 10.8 *Acute toxicity of natural compounds expressed as $LD_{50}$s*

| Compound | Origin | $LD_{50}$ (mg/kg body weight – oral rat) |
|---|---|---|
| Aconitine | Wolfsbane (*Aconitum napellus*) | 0.05 |
| Colchicine | Autumn crocus (*Colchicum autumnale*) | 0.3 |
| Veratrine | False hellebore (*Veratrum* spp.) | 0.3 |
| Atropine | Nightshade (*Atropa belladonna*) | 1 |
| Scopolamine | Nightshade (*Atropa belladonna*) | 1 |
| Strychnine | *Strychnos nux-vomica* | 2 |
| Coniine | Hemlock (*Conium maculatum*) | 5 |

From Strong (1973) and summarised in *The Agrochemical Industry and the Environment* published by Rhone-Poulenc Agrochemie, 1990

irrespective of the source of damage. Detoxification enzymes such as cytochrome mono-oxidases have a low substrate specificity and can thus deal with a range of toxins. Such defence mechanisms no doubt evolved to deal with a whole range of environmental toxins, and these detoxification mechanisms that deal with plant metabolites are also involved in metabolism of pesticides in human beings (6.3).

Notwithstanding the long time-span of human and animal evolution, there are still many naturally occurring compounds for which detoxification mechanisms have not evolved. Acute poisons such as atropine (belladonna), coniin (hemlock) and strychnine may be lethal at low dose, and have been used throughout recorded history to deliberately kill (Table 10.8). Interestingly, the first of these is used widely as a medication at very low dose rates. Several leguminous plants contain globulin proteins that if ingested may cause agglutination of red blood cells. Deaths have been associated with compounds such as ricin from the castor oil bean (*Ricinus communis*). Agglutinins in soya and kidney beans have caused severe toxic effects and occasionally death; and cyanogens present in uncooked cassava have often caused deaths among people in poorer countries. Lathyrogens from species of peavine (*Lathyrus sativus*) cause the condition known as neurolathyrism, a degenerative disorder of the spinal chord still prevalent in the Indian subcontinent (Beier, 1995).

There are some naturally occurring compounds that act as chronic lung poisons in an analogous manner to the herbicide paraquat (Beier, 1995). Ipomeamarone and related compounds are produced by yams (*Ipomoea batata*) in response to attack by fungi, and these have been identified as the principal causes of the oedema of the lungs that has been long associated with excessive consumption of sweet potatoes, particularly in Pacific island communities.

Bioaccumulative effects have also been seen with some naturally occurring plant metabolites, and these have led to deaths. The white snakeroot (*Eupatorium rugosum*) causes the disease known as milk sickness in human beings and was a major cause of death among early settlers in the USA. It results from drinking milk from animals that have fed on the white snakeroot. The compound tremetone was initially considered to be responsible for the disease, but in fact is metabolised in human beings by cytochrome mono-oxygenases to a more active compound that has proved highly toxic to mammalian cells. This activation is analogous to that of the organophosphate parathion to the more toxic paraoxon (6.3).

Furthermore, despite the detoxification mechanisms that exist in human beings and other vertebrate species, there are some natural toxins that cause cancer at low doses. The most notorious of these are mycotoxins: chemicals produced by fungi that may cause cancer and other long-term illnesses (EMAN, 2003).

Species of the fungal genera *Aspergillus*, *Penicillium* and *Fusarium* colonise a range of foodstuffs in store and under some circumstances can produce secondary metabolites that are highly potent carcinogens. The aflatoxins produced by *Aspergillus flavus* are some of the most potent of known carcinogenic substances, and are held to be a major cause of liver cancer, particularly in poorer countries. Ochratoxins are genotoxic metabolites produced by *Aspergillus ochraceus* and *Penicillium verrucosum* that colonise the ears of cereals. *Penicillium expansum* infects apples and the presence of patulin, which causes oedema in rodents at low dose, has caused concern to producers of apple juices and alcoholic drinks made from the crop. Other moulds of cereal ears such as *Fusarium culmorum* may produce zearalenone which has pronounced oestrogenic activity (10.4) and fumonisins, implicated in oesophageal cancer. The concentrations of some of these fungal metabolites permitted by law are given in Table 10.9 and these may be compared to the ADIs for pesticides given in Table 6.5, and especially for substances in drinks to the EU limit for pesticides in water of 0.1 μg/l.

Potential problems to human health from consumption of mixtures of pesticides present in foodstuffs have been highlighted by pressure groups. Exhaustive studies in the UK (Anon, 2000a, 2002a) undertaken on the potential for mixtures of pesticides and other chemicals to cause adverse effects to health concluded that the probability of health hazards due to additive or potential interactions between pesticides in the human diet at the low doses experienced is likely to be small, since the dose of pesticides to which humans are exposed is generally much lower than the NOAEL (6.2.3). Aside from the low doses involved, it is pertinent to consider interactions that may occur between natural

Table 10.9 *Recommended limits in the European Union for mycotoxins in food and drink*

| Mycotoxin | Source | Foodstuff and maximum permitted concentration |
|---|---|---|
| Aflatoxin $B_1$ | *Aspergillus flavus* | 2 μg/kg in groundnuts, dried nuts and nut products |
| Total aflatoxins | *Aspergillus flavus* | 4 μg/kg in groundnuts, dried nuts and nut products |
| Ochratoxin A | *Aspergillus ochraceus; Penicillium verrucosum* | 5 μg/kg in cereals; 3 μg/kg in cereal products; 10 μg/kg in vine fruits |
| Aflatoxin $M_1$ | *Aspergillus flavus* | 0.05 μg/kg in milk |
| Patulin | *Penicillium expansum* | Apple juice- proposed level 50 μg/kg |

From the European Mycotoxin Awareness Network at http://www.lfra.co.uk/eman2/fsheet6_1.asp

compounds in diet, and perhaps between these and pesticides. Again, the much greater doses of natural compounds experienced in the human diet may present greater risks to health, but few problems have been identified.

The concentration of pesticides in water can also be compared to other substances that may pose risks to human health. Chlorine is added to drinking water for the very good reason that it eliminates many water-borne diseases varying from coliforms and *Escherichia coli* in temperate countries, to typhoid and cholera in tropical countries. However, one of the consequences of chlorination is that small quantities of chloroform and other trihalomethane compounds may form in drinking water (UK Drinking Water Inspectorate, 2001). These are carcinogenic at low dose in rat tests of the type used in toxicological evaluation of pesticides, and there is concern over the risks that this may present to the general public. Most authorities have concluded that the benefits in terms of controlling water-borne diseases vastly outweigh the small risks to health that may ensue from possible carcinogenic effects. Nevertheless, the EU recommended limits of 100 μg/l for these chlorine by-products might be compared to the EU limit of 0.1 μg/l for pesticides in drinking water.

Algal toxins are another naturally occurring hazard, present particularly in reservoirs and rivers used as sources for drinking water (Minting, 1998). They include molecules with neurotoxic, hepatotoxic and carcinogenic properties and instances of poisoning and death of human beings have occurred. Saxitoxin, an alkaloid produced by species of the blue-green algae *Anabaena* and

*Aphanizomenon*, is a potent shellfish poison, acting as a sodium channel blocker in the same way as the organochlorine and pyrethroid insecticides. Human fatalities have occurred following consumption of shellfish contaminated with this highly toxic substance. Microcystins are liver toxins produced by *Anabaena*, *Microcystis* and *Oscillatoria* and have been implicated in human fatalities such as that in Carueru, Brazil in 1996 where 60 people died from the use of contaminated water in dialysis (Carmichael *et al.*, 2001). In the UK in 1989, a canoeing group experienced severe adverse health effects after exercising on Rudyard Lake in Staffordshire during a phase of high algal incidence (Anon, 2002c). In rat assays, the $LD_{50}$s of saxitonin and microcystin are 9 μg/kg and 50 mg/kg body weight respectively, and these figures can be compared to those in Table 6.2 for pesticides: the WHO's limit for microcystin in drinking water is set at 1 μg/l.

## 10.4 Pesticides and endocrine disruption

Endocrine systems release hormones that act as chemical messengers, and these in turn interact with specific receptors in cells to influence growth, development and reproduction in human beings and other animals. The precise regulation of endocrine function is not fully understood, but external stimuli including physical and chemical stresses may alter endocrine activity. Chemicals such as alcohol, even in moderate amounts, and common salt can influence endocrine function. These and other substances may interfere with endocrine function by mimicking the behaviour of hormones, affecting hormone receptors at the cellular level or by influencing the synthesis and/or degradation of hormones. Effects that have been attributed to endocrine disruption in human beings include lowered sperm counts, increases in testicular cancer and birth defects. Interest in the disruption of endocrine function by chemicals present in the environment and diet increased markedly in the 1990s. Colborn and her co-authors (Colborn, Dumanoski and Myers, 1996) have drawn much attention to this subject following the publication in '*Our Stolen Future*' and parallels have been drawn between this publication and '*Silent Spring*' by Rachel Carson.

Feminisation of fish has been reported in many rivers around the world and has been associated with the presence of synthetic chemicals such as pesticides leaching into natural waters. In some cases, such as that of the tributyltin compounds used to discourage barnacles and mussels from colonising the hulls of boats (9.6), a clear association with endocrine disruption has been shown. However, endocrine disruption may be equally linked to the presence of natural and synthetic female hormones present in sewage outfalls (UK Department of the Environment, Transport and the Regions, 2001). The widespread use of oral contraceptive pills from the mid 1960s is the principal source of oestrogenic activity in sewage outfalls (Desbrow *et al.*, 1998). However, locally high concentrations of endocrine-disrupting substances, such as alkylphenols discharged from wool-cleaning plants, may enter watercourses.

The current highly stringent testing procedures for pesticides (6.2) including long-term carcinogenic, reproductive and teratological studies would appear sufficient to reveal any endocrine-disrupting ability in test compounds. Attention has thus turned to older compounds that may not have been tested to modern standards and in particular the organochlorine and pyrethroid insecticides, and dicarboximide fungicides (Mattheissen, 1998). Some pyrethroid insecticides, notably permethrin, are potentially endocrine disruptors, albeit at high dose, and local contamination of rivers in the UK has resulted following discharge of these compounds from, for example, the premises of carpet manufacturers who treat their products with pyrethroids to prevent cat and dog flea infestations.

The potential endocrine-disrupting effects of compounds such as DDT have been linked with breast cancer. In this respect, a comparative risk analysis has been carried out by the UK Department of Health Committee on Carcinogenicity (Department of Health, 1999) of potential endocrine-disrupting effects arising from various sources. They considered that DDT, its metabolites and other organochlorine compounds present a negligible or, in the case of HCH, no risk of disruption of endocrine function, particularly when compared to the risks arising from other sources (Table 10.10).

Similar comparisons to those made between naturally occurring carcinogens and pesticides can be made for oestrogenic substances in the human diet (Tables 10.10 and 10.11). The endocrine-disrupting effects, for both wildlife and human beings, of natural and synthetic oestrogens may be far more important than the effects of very small doses of pesticides present in most natural

Table 10.10 *Estimated mass balance of human exposures to environmental and dietary oestrogens*

| Source of oestrogen | Oestrogen equivalents (μg/day) |
|---|---|
| Morning after pill | 333 500 |
| Birth control pill | 16 675 |
| Flavonoids in foods | 102 |
| Environmental organochlorine oestrogens | 0.0000025 |

From Safe (1995)

Table 10.11 *Concentrations of natural oestrogens in foodstuffs, and residues of (alleged) endocrine-disrupting pesticides in water*

| Compound/source + intake | Concentration in diet |
| --- | --- |
| Isoflavones in soya-based infant formulae* | 18–41 mg/l made up infant food |
| Atrazine, at highest level detected in UK | 0.005 mg/l in drinking water |
| Lindane at highest level detected in UK | 0.005 mg/l in drinking water |

* From *Plant Oestrogens in Soya-based Infant Formulae*. MAFF Food Surveillance Information Sheet 167, November 1998

and drinking waters. Certain foods such as soya beans and cabbage contain far more potent oestrogens than pesticidal molecules, and the concentrations of these in the diet may vastly exceed the concentration of pesticide residues in drinking water. Indeed because of concerns about soya-based oestrogens in infant foods, several studies have been commissioned to evaluate the risks to infant health from such naturally occurring, potentially endocrine-disrupting, substances. All have concluded that there are no significant risks to health from these compounds, to which infants may be exposed in much greater amounts than pesticides alleged to have endocrine-disrupting capacity.

## 10.5 Environmental hazards, wildlife deaths and loss of biodiversity

Evolution of life over millions of years has led to a range of ecosystems across the globe, characterised by a diverse range of species. Human activity over the last thousand years, and especially during the last century has resulted in the alteration of habitats and in some cases their destruction with consequent risks to the viability of some species. Extinction and reduction in the numbers of species have resulted for various reasons. Introduction of non-indigenous species has resulted in competition and sometimes predation of indigenous species, often leading to extinction. The demise of most of the flightless birds of New Zealand following the introduction of rats, foxes and opossums is a good example of such human interference.

Such direct effects may occur within a short period of time, as do the deliberate acts of poisoning of species that are occasionally carried out with pesticides. This misuse of pesticides, for example to kill raptor birds perceived to pose a

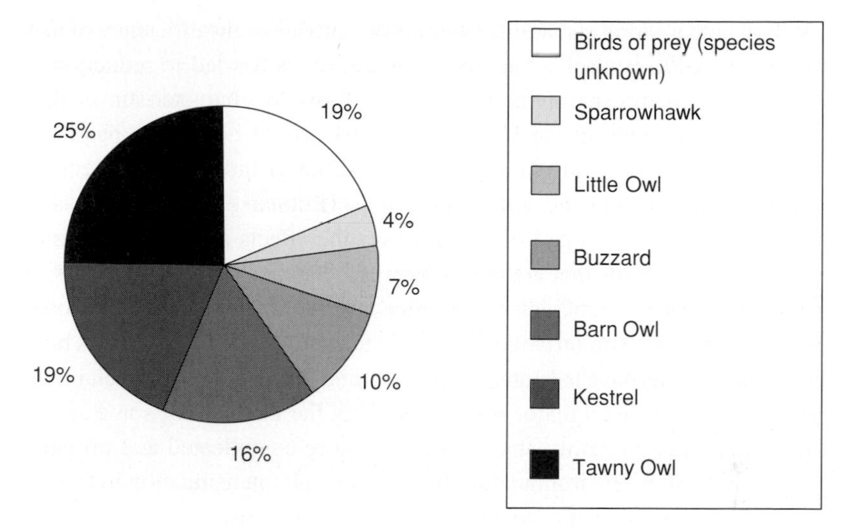


Figure 10.2 Birds of prey recorded as road casualties, 2001–2002. Total number of deaths was recorded as 142. From The Mammal Society (2002) *National Survey of Road Deaths* at http://www.abdn.ac.uk/mammal/road_deaths.htm.

threat to young livestock on farms in the UK, is still reported, with an annual total of 40–50 deaths of raptors between 1992 and 2002 (Barnett *et al.*, 2002). However, deaths in road accidents appear to be a more important cause of raptor mortality in the UK (Figure 10.2), with 142 recorded in the comprehensive survey carried out by the Mammal Society in 2001–2002.

Before the twentieth century, the principal change to natural ecosystems arose from the development of agriculture. The growth of a single species for food clearly needed a reduction or elimination of competing species: clearance of land for cropping, pastoral use or forest development obviously involves a major change in species diversity. The felling of mature woodlands in Western Europe during the Middle Ages and subsequent use of land for agriculture as well as the more recent inroads into rain forests in tropical countries have resulted in major changes in habitats with attendant consequences for indigenous species of plants and animals. Paradoxically, one of the most sterile of environments resulting from this policy is that of mature coniferous woodland offering niches for very few species. Other habitats have also suffered. The ploughing of marginal, often species-rich grasslands, in the UK for cultivation of more profitable, subsidised cereal crops during the 1970s and 1980s may be seen as one of the less desirable consequences of EU support for crop production.

Within both pastoral and arable systems of cultivation, the efficiency of measures employed to control weeds, pests and diseases has led to reductions in populations of plants and animals that were formerly characteristic of these habitats. For example, in the UK, the effective use of herbicides has led to arable weeds becoming rare in many localities (9.4.1) and has led to demands for control of these only above threshold limits (Lutman *et al.*, 2003). Considerable interest has developed in the UK over the effects of intensive farming on populations of birds that are commonly associated with arable habitats. As noted earlier (9.9), insecticides may reduce invertebrate food for insectivorous birds and herbicides similarly reduce weed seed supplies for granivorous birds. These observations have led some authorities and pressure groups to claim that pesticides have played a major role in reducing the numbers of some species of farmland birds. In reality, the reasons are more complicated and provide a good example of the environmental effects of overall intensification in farming practices driven by subsidies to encourage yields of crops.

The change in populations of farmland birds that occurred in the UK during the last three decades of the twentieth century has been well documented by the British Trust for Ornithology (Table 10.12). Of 28 species classified as farmland birds, 24 (86%) declined in abundance between 1970 and 1990 (Evans, 1997). Five interrelated factors are held to contribute to the decline of farmland birds.

Table 10.12 *Resident farmland bird species of high conservation concern in the UK*

| Species | Decline in numbers 1969–1994 | Decline in range | Eats seeds in winter |
|---|---|---|---|
| Grey partridge (*Perdix perdix*) | 82% | 19% | + |
| Skylark (*Alauda arvensis*) | 58% | 2% | + |
| Songthrush (*Turdus philomelos*) | 73% | 2% | |
| Tree sparrow (*Passer montanus*) | 89% | 20% | + |
| Linnet (*Acanthis cannabina*) | 52% | 5% | + |
| Bullfinch (*Pyrrhula pyrrhula*) | 73% | 7% | + |
| Cirl bunting (*Emberiza cirlus*) | No data | 83% | + |
| Reed bunting (*Emberiza schoeniclus*) | 62% | 12% | + |
| Corn bunting (*Miliaria calandra*) | c. 80% | 32% | + |

Data collected from British Trust for Ornithology's Common Bird Census (1969–1994), plus surveys done for the Atlas of British Breeding Birds, and given in Evans (1997)

1. The decrease in availability and abundance of weed seeds following application of herbicides. This may have contributed to the decline in numbers of seed-eating birds.
2. The decreased availability and abundance of ground-dwelling invertebrates due to insecticide use may reduce brood survival. In the early 1950s, populations of the grey partridge in the UK began to decline and this has been associated with reductions in populations of aphids and sawfly larvae on which the chicks feed.
3. An increase in fertiliser use. The decline in use of farmyard manure on mixed farms may also contribute to the reduction in populations of farmland birds. Earthworms are far more prevalent in soils to which manure is regularly applied. Fertilisers may also encourage a swift growing taller crop – disadvantageous to birds such as skylarks and lapwings which prefer open, sparse vegetation as nesting sites.
4. A decrease in boundary habitats. Removal of hedgerows during the latter part of the twentieth century in the UK has reduced nesting habitats and shelter for birds such as finches, as well as depriving these of a food supply.
5. Changes in cropping patterns. These now appear to be the major factor contributing to the decline of many farmland species in the UK. The move from spring- to autumn-sown cereals and the dense cover of these, exacerbated by increasingly warmer springs which promote crop growth, has led to a decrease in the habitat to which many farmland species became adapted viz. open ground in winter, frequently covered in bare stubble.

After the UK joined the EU in 1972, many farmers, bolstered by subsidies to increase output, switched from spring- to autumn-sown cereals that offered higher yield potential. The consequent disappearance in spring of bare stubble, often containing spilt seed from the combine harvester, and providing weeds and their seeds in early spring as well as being a source of ground-dwelling invertebrates, is held to be one of the most important changes in farming practice that have led to the decline in numbers and abundance of farmland birds. Studies on the distribution of the corn bunting in the UK reinforce these views. In 1992–3 volunteers in the UK visited 1313 tetrads (2 km × 2 km) throughout the range of the corn bunting (*Miliaria calandra*) in the UK. The species was located in 160 tetrads and a total of 2909 individuals were counted. Stubble fields held approximately twice as many birds as 'clean' fields. Further studies carried out in 1993–1995 by two groups, one working in Oxfordshire and the other in both Devon and East Anglia, are reported in Table 10.13.

Data such as that above have been used to advocate changes in farm management practices. Cooperation between the Royal Society for Protection of

Table 10.13 *Results of farmland bird distribution studies*

| Species | Stubble | | Winter cereal | | Grazed grass | | Set aside | |
|---|---|---|---|---|---|---|---|---|
| | 1 | 2 | 1 | 2 | 1 | 2 | 1 | 2 |
| Grey partridge | + | + | | − | − | | nd | + |
| Skylark | + | + | − | − | − | − | nd | + |
| Song thrush | | | − | | | − | nd | |
| Linnet | + | + | − | − | − | − | nd | + |
| Reed bunting | + | nd | − | nd | − | nd | nd | nd |
| Cirl bunting | nd | + | nd | − | nd | − | nd | + |

From data quoted by Evans (1997)
Columns 1 and 2 refer to separate studies cited by Evans (1997)
nd is not determined

Birds and just 50 farmers in the UK Countryside Stewardship Scheme included a provision for winter stubble. Even with this very small increase in availability of stubble fields, the cirl bunting (*Emberiza cirlus*) population in the UK rose from 120 pairs in 1988 to 373 in 1995, and in 2003 the population was 700 pairs.

Further problems to farmland birds arise from the failure to find nesting sites within autumn-sown cereals due to the height and thickness of the crop in spring. Creation of unsown patches in autumn-sown wheat has been shown to enhance the breeding prospects of skylarks by providing both nesting and feeding sites (Morris, Bradbury and Evans, 2003).

An additional factor that has contributed to the decline of some farmland species is a reduction of nesting sites due to removal of hedgerows to accommodate larger farm sprayers on crop land. Re-creation of hedgerow habitats and their associated field margins is one of the major objectives of the UK Department of Food, Environment and Rural Affairs through its Countryside Stewardship Schemes.

The examples above serve to illustrate the interrelated factors that may be involved in causing a decline of species, as well as the procedures that can be adopted to enhance the survival of the species in question. Pesticides and fertilisers are components of high-input systems that have unquestionably contributed to the decline of some species in countries such as those of the EU, but the principal reason for reductions in biodiversity has been the overall high-input, high-output farming systems adopted.

Another major factor of global concern influencing the environment is climate change. Pesticides have few direct effects on climate change; their manufacture may include energy consumption and output of $CO_2$, but these contributions are negligible compared to other sources. Ozone depletion in the stratosphere may contribute to climate change and bromide ions from, for example, the general biocide methyl bromide has been implicated in depletion of the ozone layer above the Earth and so, as noted in Chapter 5, this chemical will be prohibited worldwide by 2015. The principal synthetic chemical agents linked to depletion of the ozone layer are held to be large quantities of chlorofluorocarbons resulting from the disposal of obsolete cooling systems and aerosols.

## 10.6 Animal and livestock deaths

Although the Wildlife Incidence and Investigation Scheme assiduously investigates reports of pesticide-related deaths, the principal causes of death and disease in livestock and companion animals in the UK are parasitic in nature, and include microbial diseases such as mastitis and brucellosis in cattle, enzootic infections in sheep and distemper and toxoplasmosis in cats and dogs (Table 10.14). In addition, there are invertebrate pests such as blowfly and scab

Table 10.14 *Prevalence of some major diseases of cattle and sheep in the UK compared to reports of adverse effects of pesticides on livestock and companion animals reported to the UK Wildlife Incident Reporting Scheme*

| Host and disease | Numbers of cases annually in UK |
|---|---|
| Cattle – enteric disease caused by *Escherichia coli* and viruses | 2000–3000 |
| Cattle – bovine diarrhoea and mucosal disease complex | 400–700 |
| Cattle – Fascioliasis (liver fluke) | 3000–7000 |
| Sheep – enzootic abortion of ewes (EWE) | 3000–12000 |
| Sheep – toxoplasmosis | 1500–3000 |
| Sheep – blow fly strike (myiasis) | 3000–6000 |
| Total incidents where pesticides were identified as a likely cause of livestock poisoning | 1993, 2; 1995, 2; 1997, 1; 1999, 2 |
| Total incidents where pesticides were identified as a likely cause of poisoning of companion animals | 1993, 85; 1995, 91; 1997, 86; 1999, 48 |

Partly derived from Bennett, Christiansen and Clifton – Hadley (2003)

Figure 10.3 Ragwort (*Senecio jacobaea*).

in sheep, trematode flukes in both sheep and cattle, and fleas, ticks and toxocara worms of domestic animals, all of which may be treated with pesticides.

In addition, livestock and wild animals suffer widely from the toxic effects of certain plants. In the UK the ubiquitous weed Oxford ragwort, *Senecio jacobaea* (Figure 10.3), is the commonest cause of premature death in horses that may graze this weed, particularly if it has been cut. The pyrrolizidines present in *S. jacobaea* may be acutely fatal to horses, and according to the British Horse Society resulted in over 6500 deaths of horses and ponies in the UK in 2001.

The most striking examples of livestock poisoning by plants come from the Australian continent, where non-native mammals have not developed immunity to many of the metabolites within native plants. The indigenous marsupial fauna have evolved along with the native flora and most species are immune to their metabolites. Cattle and sheep introduced by European settlers in the eighteenth century were often killed after consuming native plants, and this is still the commonest cause of death today. In Australia, herbicides are widely used to clear pastures of plants toxic to livestock.

Even allowing for underreporting of pesticidal effects, risks to wildlife may be small compared to other hazards. As with birds (Figure 10.2), road casualties are a major cause of death of animals in the UK. A survey conducted by the Mammal Society in the UK during 2000–2001 estimated that annual road casualties included 100 000 foxes, 100 000 hedgehogs (*Erinaceus europaeus*) and 50 000 badgers (*Meles meles*) annually (Mammal Society, 2002). These figures may be compared to the 20–30 incidents of adverse effects on wildlife reported annually to the Wildlife Incident Investigation Scheme in the UK where pesticides are implicated (Barnett *et al.*, 2002).

## 10.7 Problems of resistance and interference with natural pest control

Pressure groups, farmers, growers and pesticide producers all consider resistance as a major problem. Pressure groups allude to the sometimes ineffective deployment of pesticides where resistance has occurred with a knock-on effect of unnecessarily releasing chemicals into the environment. Pesticide manufacturers equate loss of efficacy with loss of income. To the farmer or grower resistance can mean a loss in both crop yield and/or quality, along with wasted expenditure on a product that does not work.

The development of resistance in target organisms is one of the major problems associated with the specific action of pesticides. Resistance is indicated by the failure of a compound to control a weed, pest or disease at the dose initially used for this purpose. In some cases, resistance may develop slowly, but in many instances has developed suddenly and spread quickly. Resistance by certain bacteria to antibiotics presents a considerable risk to health from a failure to control human pathogens. Problems of resistance to pesticides in microbial, weed and pest populations are analogous to those of antibiotic resistance. The specific sites at which pesticides and antibiotics act (and that confer selectivity) may be circumvented by mutation arising through variation within populations of target organisms.

Single base changes in the structure of DNA may lead to slight changes in protein structure at the site of action of pesticidal compounds. Pesticides may not bind to their biochemical targets, and thus a resistant strain emerges. Most pests, diseases and weeds often have highly efficient means of reproduction and variants are likely to arise as in any normal population of organisms. However, the huge numbers of offspring, spores and seeds produced by many pests, diseases and weeds respectively increase the chances of mutants arising that are resistant to pesticides. Overuse and continual exposure to pesticides with a single mode of action will select out resistant biotypes and their high rates of reproduction may allow the rapid establishment and spread of resistant strains.

Table 10.15 *Some examples of resistance to pesticides*

| Pesticide & date of introduction | Principal type of resistance | Resistance detected |
|---|---|---|
| **Herbicides** | | |
| Atrazine/Simazine (1950s) | Mutation at single site | 1970s |
| Aryloxyphenoxypropionates (1970s) | Combination – mutation and detoxification by glutathione conjugation | 1990s |
| **Fungicides** | | |
| Benzimidazoles (1970s) | Mutation at single site | 1970s |
| Phenylamides (1980s) | | 1980s |
| Strobilurins (1996) | | 1999 |
| **Insecticides** | | |
| Organophosphates (1950s) | Enzyme detoxification | 1960s |
| Pyrethroids (1960s) | | 1970s |
| **Rodenticides** | | |
| Warfarin | Mutation at a single site | 1958 |
| Brodifacoum and other second-generation rodenticides | Behavioural changes – bait avoidance | 1990s |

Indeed, as noted in the chapters on herbicides, fungicides and insecticides, resistance has occurred to many specifically acting pesticides (Table 10.15), and some have been rendered virtually useless for the problems they were introduced to combat. The aminopyrimidine fungicide dimethirimol was withdrawn within 10 years of its introduction for control of powdery mildews in cucurbits due to resistance problems. The benzimidazole fungicides, which gave excellent control of the grey mould pathogen *Botrytis cinerea* upon their introduction in the late 1960s, are rarely used for control of this pathogen due to widespread resistance. Resistance has also developed in populations of other pathogens formerly controlled by the benzimidazole fungicides. Worldwide, resistance has arisen in many weed populations to the triazine herbicides atrazine and simazine. Where pesticides have a common site of action, resistance to one compound often means that resistance extends to others that act in the same manner – the phenomenon of cross resistance.

Other mechanisms of resistance include the detoxification of pesticides by target organisms. As seen in Chapter 4, metabolism is the basis of selectivity of several groups of insecticides. Unfortunately, elevated levels of enzymes capable of detoxifying insecticides such as organophosphates and pyrethroids have developed in some target insect species, again due to selection pressure followed by the rapid increase in resistant biotypes. Resistance is thus due to rapid metabolism of these compounds, and in many cases is linked to persistent use of compounds with the same mode of action.

A diversity of pesticidal compounds assists in combating resistance since, where target-site resistance occurs, a compound with a different mode of action may be used. However, multiple target-site resistance has developed to some insecticides and multiple resistance through detoxification mechanisms to both herbicides and insecticides has developed in some species. Extreme examples include the diamond back moth (*Plutella xylostella*), which in parts of Eastern Asia has developed resistance to all groups of insecticides, as well as *Bacillus thuringiensis*; in Europe, *Alopecurus myosuroides* (blackgrass) has become resistant to fops, dims and other herbicides and this weed has become a major problem to cereal farmers.

Strategies to prevent the establishment of resistant biotypes have focused on preserving the selective activity of pesticides, and recommendations have been made by industry-coordinated resistance action groups including HRAC (Herbicide Resistance Action Committee), IRAC (Insecticides) FRAC (Fungicides) and RRAG (Rodenticide Resistance Action Group). All groups advocate similar policies and these are summarised in Table 10.16.

Biological control agents are now routinely used in integrated pest-management strategies, particularly for glasshouse pests. Some insecticides as noted in Chapter 4 (methoprene, some acaricides) have little effect on many beneficial and non-target species and can be used in integrated pest-management programmes. The fungicide fenhexamid (3.7.4), widely employed for control of the grey mould pathogen *Botrytis cinerea*, does not affect the growth of many non-target fungi such as entomopathogenic species. However, many older compounds are detrimental to organisms used in biocontrol programmes.

Table 10.16 *Strategies to minimise the occurrence of resistance*

- Restricting the numbers of applications of pesticides
- Effective timing of pesticide application, linked to assessment of the weed, pest or disease problem
- Changing spray programmes where an 'at-risk' molecule is alternated with a compound having a different mode of action
- Using mixtures of an 'at-risk' molecule with one having a different mode of action
- Using appropriate dose rates

Adapted from Russell (2003)

Companies that research and develop new pesticidal molecules now routinely test these for their activity against a range of predatory insects and mites. In some cases, the level of selectivity achieved is such that biocontrol agents may be related to the target organism and yet remain largely unaffected. For example, the insecticide pyridalyl developed in the early twenty-first century for control for lepidopteran pests is of low toxicity to many beneficial and non-target insect species (De Maeyer *et al.*, 2002); newer acaricides such as spirodiclofen are able to give good control of pest mites with minimal effects on their predators, which also belong to the acarine group (Saito *et al.*, 2002).

## 10.8 Risks arising from not using pesticides

Although this text is primarily concerned with evaluating risks of health and the environment arising from pesticides, it is pertinent to consider problems that may occur from not using these compounds. Risks to human and animal health have arisen with the restriction or withdrawal of some pesticides, and widespread adoption of pesticide-free systems of growing would lead to a requirement of more land for agricultural production.

Most authorities agree that between 2000 and 2030, world populations will increase by 1.5 to 2 billion. Recent pressures on the environment result from the need to accommodate increasing populations. The use of land for housing, commercial and industrial activity as well as the associated development of transport links, predominantly roads, in the late twentieth century has led to much urban encroachment and loss of countryside habitats. In line with increases in population, urbanisation is rapidly increasing in developing countries and the FAO estimates that across the world arable land per head of population has decreased from 0.38 ha in 1970 to 0.23 ha in 2000 and is predicted to decrease further to 0.15 ha by 2050 (FAO, 2001).

To achieve the same or increased levels of crop production needed to feed an increasing global population, Avery (1995), Goklany and Trewavas (2003), and others have proposed that the best agricultural land should be used effectively, deploying fertilisers, pesticides and transgenic crop plants to ensure maximum yields. If the tripling of crop yields brought about by technological advances during 1960–1992 had not occurred, some 10–12 million square miles of additional land, over and above the current six million currently cultivated, would have been needed to achieve the same output. From 1961 to 1963, the yield of grain in India averaged 0.91 tonnes per hectare of land; by 1991–1993 this had risen to 1.96 tonnes per hectare. If yields had remained close to levels in the early 1960s, some 218 million extra hectares of land would have been needed to deliver the quantities of grain produced in the early 1990s. Even so, in many countries, the hectarage devoted to agricultural land is still expanding, often at the expense of species-rich rainforests and other areas rich in biodiversity (Potts, 2003).

Organic systems of growing are generally less reliant on pesticides, and synthetic products are generally not permitted in such systems (although naturally occurring compounds such as sulphur, pyrethrins and rotenone are allowed by some organic authorities). The yield of many crops, and especially cereals, in organic systems is much lower than that in pesticide-aided production (Table 10.17).

Table 10.17 *Yields of crops from organic and pesticide-aided production systems in the UK. Yields are tonnes/hectare with the conventional yields being median values*

| Crop | Organically produced | Conventional practice with synthetic pesticides and fertilisers |
|---|---|---|
| Autumn-sown wheat | 4.0 | 8.0 |
| Autumn-sown barley | 3.7 | 6.35 |
| Potatoes | 25 | 42.5 |

Yields are tonnes/ha with the conventional yields being median values
From Lampkin, Measures and Padel (2003) and Nix (2002)

Thus, widespread adoption of organic systems with their current lower yields would increase the need for extra land, possibly leading to cultivation of marginal areas of conservation value. Organic systems of growing are generally more labour-intensive than pesticide-aided production systems, and in affluent countries such labour, given the poor salaries of agricultural workers and the often tedious nature of duties such as hand-weeding, is often difficult to find. Organic systems often involve extra tillage and mechanical weeding operations: accidents involving tractors and other farm machinery are responsible for most deaths and injuries sustained on farms in the UK. The risks of such accidents may be greater in organic production systems involving increased tillage.

In the area of public health, limitations placed on the use of pesticides has had considerable impact in some parts of the world; and the most notable example is the restricted use of DDT as an aid to control the malarial mosquito. In the early twenty-first century, the United Nations Environmental Programme (UNEP), as part of a move to eliminate so-called persistent organic pollutants (POPs),

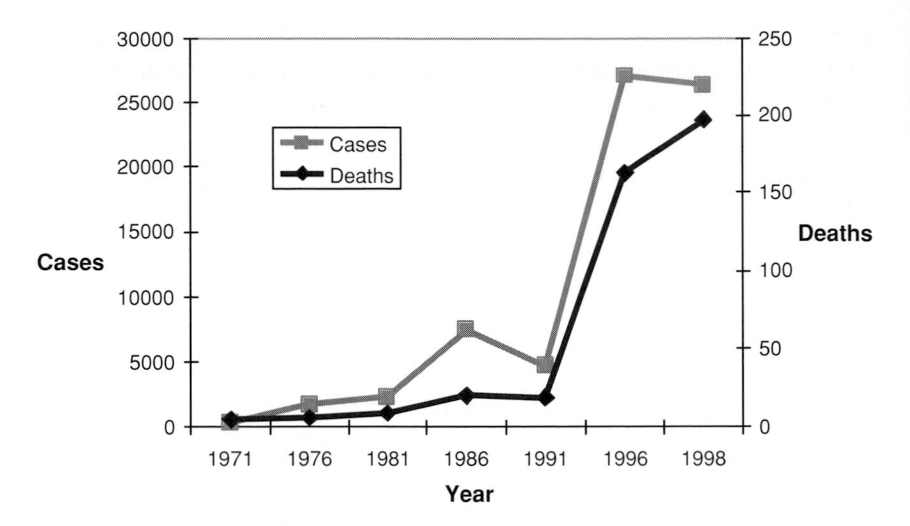


Figure 10.4 Notified cases and deaths from malaria in South Africa 1971–1998. From: Tren, R. *The Economic Costs of Malaria in South Africa*, at http://www.malaria.org/tren.html.

proposed to prohibit the use of DDT worldwide. Pressure from environmental groups has resulted in many countries abandoning the use of DDT in public health, and in particular its use as a house wall spray (Roberts *et al.*, 2000). Alternatives to DDT are more expensive, and with discontinuation of its use has come a major increase in the incidence of malaria in many countries, and deaths resulting from the disease (Figure 10.4).

Prohibition of certain pesticides has been associated with resurgence of some animal health problems. Organochlorine pesticides, notably $\gamma$HCH, used against sheep scab (*Psoroptes ovis*) eliminated this disease on the UK mainland between 1952 and 1972. In the light of environmental concerns, organochlorine use was reduced: sheep scab returned (probably on imported stock) and has been present in the UK since 1972. From the early 1970s, control strategies switched first to organophosphorus insecticides, which were not as effective as $\gamma$HCH and resulted in health problems to sheep dippers (7.3), and then to pyrethroids with their associated adverse effects on non-target aquatic fauna (9.6).

From 2000 onwards in the EU and elsewhere, many older pesticides have been subjected to reviews of their toxicological and environmental impact with respect to modern protocols (1.4). If financial support from manufacturers did not materialise, and product approval lapsed, the pesticide was thus withdrawn from the market. In most cases, older pesticides whose patent life had expired have not been supported for review. Toxicological and environmental studies might have revealed adverse effects of some of these pesticides, but many compounds may have been lost to farmers and growers, some of which may have been useful in anti-resistance strategies (10.7). For example, the withdrawal of many protectant fungicides, most of which acted in an unspecific manner after uptake by pathogens, has left farmers and growers heavily reliant on highly selective, specifically acting compounds to which resistance is more likely to arise. Many older compounds were relatively cheap, and where alternatives exist, these may be more expensive.

## 10.9 Risk assessment for pesticides – a summary

The effects of pesticidal molecules on both target and non-target species is dependent on the sensitivity of the organism, dose experienced and duration of exposure. Ideally, target organisms should be killed at doses and exposures well below those that produce adverse effects in non-target species, including human beings. In fact most pesticides are effective against their targets at low concentrations: the amount of the appropriate compound required to kill for example an insect by topical application or to inhibit photosynthesis in isolated chloroplasts is often very small indeed. However, to get the pesticide to its target in a field situation or other site of application it is applied at a much higher concentration, usually in the form of a spray. Furthermore, for obvious reasons of convenience in transport and storage, pesticides are provided to farmers, growers and pest control operatives in a highly concentrated form.

Pesticides in their manufactured, concentrated forms clearly have the potential to cause harm to the health of human beings, and also other non-target organisms. The risks of exposure to pesticides in their concentrated form is much greater in developing countries, giving rise to many of the problems outlined in Chapters 7 and 9. Regulations aimed at reducing the risk of exposure to pesticides may exist in these countries, but application and enforcement of legislation regarding pesticides are often lax. Even the manufacture of concentrates presents an increased risk in developing countries. The dangers of lax safety procedures in a pesticide-manufacturing facility, around which homes and other urban infrastructure were allowed to develop, were seen in the awful consequences of the explosion at the pesticide production plant at Bhopal, India in 1983 when over 3000 people lost their lives, hundreds suffered acute health problems, and where many thousands living in the vicinity have experienced chronic health conditions since the disaster. Deliberate consumption of pesticide concentrates continues to be a major feature in suicide attempts, and again is especially associated with poorer countries. Such events would clearly be reduced in these countries if compounds that present high risks to the health of human beings were phased out, or at least made much less accessible.

Pesticides in their concentrated form are also environmental hazards, and if released in this form the risks to wildlife, domesticated stock and other animals may be great. The acute effects of pesticides are often dramatically evident after spillages of concentrate, particularly into watercourses. The need to prevent such point-source release of pesticides in their concentrated form is obvious.

Thus, strict precautions to reduce the risk of exposure of human beings and other non-target organisms to pesticide concentrates are clearly warranted, and in some instances greater reinforcement of restrictions would seem appropriate. The risks associated with the production and use of pesticides in their concentrated form thoroughly justify measures such as the training of personnel in handling concentrates, the adoption of safe working practices, use of personal protective equipment and ensuring safe and secure storage. In developing countries, the substitution of compounds of high toxicity by pesticides with better safety and environmental profiles is a desirable aim. Whilst legislation may exist in many countries requiring these procedures to be met, it is the enforcement of regulations that is needed to minimise risks to health and the environment.

At lower concentrations, in dilute form as sprays, dips or dusts, pesticides may present fewer acute risks to health than those of concentrates. However, risks even at these concentrations may still exist with some pesticides, and long-term risks to human health and populations of other non-target organisms need to be considered. In the case of pesticides at spray strength, assessment of exposure may be more pertinent to those in the vicinity of the area being treated rather than to the spray operator, who should be wearing appropriate personal protective clothing. Indeed, the potential chronic risks in the UK to health of bystanders and residents living in rural areas adjacent to sprayed fields have been highlighted (RCEP, 2005). However, in some situations spray operators and users of diluted concentrates may be at considerable risk: sheep dippers are an obvious example along with those applying pesticides in poorer countries where the technology and precautions used by spray operators in wealthier nations may not be available. Adverse effects are of course linked to the selectivity of compounds used: the herbicides and fungicides that dominate the European market, for example, present fewer risks than insecticides used in tropical and subtropical countries.

Precautions to protect the general public as well as the environment in general from the off-target effects of diluted pesticides are therefore justified. In addition to the safety procedures adopted by spray operators, developments in nozzle and spray technology that reduce the risk of drift, strict imposition of recommendations concerning re-entry to sprayed areas and compliance with harvest intervals and maximum residue limits on foodstuffs are all important considerations in reducing the risks to health from pesticide sprays, dusts and dips.

At a tertiary level, long-term exposure to low concentrations of pesticides in the form of residues has given rise to concern. Whereas only part of the population at large may encounter pesticides in their concentrated form, some in the form of spray drift, dips or dusts, almost all human beings have been exposed to pesticides at low concentrations in for example foodstuffs, tap water, clothing, fabrics as well as treated surfaces in the home and at work.

A retrospective consideration of pesticide concentrations in food, water and the tissues of human beings enables an assessment of the risks to human health, and indeed some other non-target organisms. Initial concerns focused on the organochlorine insecticides and their bioaccumulative properties: the build up in human tissues of pesticides such as DDT attracted particular attention. Quite apart from comparative risks to health resulting from chemicals other than pesticides, and considered earlier in this chapter, it is salient to consider the concentrations of pesticides present in foodstuffs in the 1950s and 1960s, and especially the levels in human tissues. Some of these concentrations reached levels that would be considered extremely high by current standards. Almost extraordinary concentrations of DDT at milligram quantities per kilogram of body weight were not uncommon. Despite such concentrations, it has been very difficult to establish any long-term adverse effects on health directly attributable to ingestion of pesticide residues. As the use of organochlorines has declined, so have residue levels in the tissues of human beings (7.6).

Of course, as noted in Chapter 9, bioaccumulation of residues was linked with population reductions of some, but by no means all, avian species. Again, with the decrease in use of these compounds, problems of bioaccumulation have declined and in most cases populations of species at the apex of food webs and chains have recovered.

The risk to human health and non-target species from residues is much lower today than in the 1950s and 1960s, particularly in developed countries. Within the steady increase in stringency of regulations relating to pesticides, comprehensive monitoring of foodstuffs and water for pesticide residues is common practice in many parts of the world. The global market for food has also resulted in routine testing of produce from around the world by government agencies as well as multiple retailers to ensure maintenance of appropriate safety standards, in particular by not exceeding maximum residue limits. Comparative risk analysis and the current monitoring systems with sophisticated analytical procedures enabling detection of most compounds at microgram or even submicrogram levels per kilogram of produce or litre of water, beverage, wine or beer suggest that the risks to health from pesticide residues must be considered as low. As with sprays and concentrates, a caveat applies regarding exposures in different parts of the world. Action still appears necessary in some developing countries to address problems of consumption of produce with residues above maximum residue limits and contamination of drinking water by pesticides to levels above maximum admissible concentrations.

A common theme evident throughout this text is that risks to both health and the environment from pesticides depend to a great extent on the geographic area of use. Poorer countries located in subtropical and tropical areas experience far more problems with insects and invertebrates in public health and agriculture. In these countries, cheap insecticides and herbicides such as paraquat are widely used, some of which, in terms of selectivity, provide the lowest margin between doses likely to kill the target and those likely to cause health and environmental problems. The herbicides and fungicides that make up the principal pesticides used in more affluent parts of the world such as the EU and USA pose fewer problems, and additionally those using these compounds are likely to be better resourced. Efforts may therefore be directed at reducing problems resulting from pesticide deployment in developing countries, and approaches are outlined in Atkin and Leisinger (2000), as well as the major review of Konradsen *et al.* (2003).

## 10.10 Current perspectives and future prospects

Consistent improvements in the selectivity of pesticides were achieved during the latter part of the twentieth century. During that time, compounds were developed that are effective at lower dose with very favourable environmental profiles, and that present very low risks to human health. However, during the same period, the cost of developing and registering pesticides has risen enormously. In the early twenty-first century the cost of bringing a new compound to market has been estimated at over $180 000 000 (Phillips and McDougall, 2003). Only large corporations are able to support the level of funding required to support the discovery and development of new compounds, and the escalating costs have resulted in many mergers and takeovers among companies involved (1.6). Indeed some have withdrawn from the industry altogether.

Reductions in the amount of pesticides used are likely to continue because of the introduction of more potent compounds, effective at a lower dose than currently available materials. Pressures from environmental lobbies allied to registration costs in what many perceive as a maturing rather than expanding market may reduce the appeal to investors and decision makers in the pesticide industry. Consequently and notwithstanding advances in high throughput screening, genomics and proteomics, the development of novel pesticides may slow down with fewer compounds reaching the market.

Legislative requirements, particularly associated with registration and restrictions on the use of pesticides, may lead to further reductions in the quantities applied to crops, and perhaps also in other areas of use. Part of the EU Directive 91/414/EEC required regulators and the pesticide industry to implement a new data requirement programme not only for new active ingredients but also for the 850 substances currently used in member states by 2008. Reviews are also taking place in other parts of the world, and incentives to place compounds that are safer than existing products on the market have also been instigated. Pesticides may be registered as having 'reduced-risk' status in the USA and manufacturers eagerly seek this label (United States Environmental Protection Agency, 2003). The withdrawal of older compounds not supported for testing to modern standards means that the number of compounds available to the farmer, grower, timber treatment operator, carpet manufacturer, veterinarian and even medical practitioner is likely to decrease.

Pressures for reduced reliance on pesticides, especially in agricultural systems, are likely to continue. National policy initiatives have been taken to reduce pesticide use. In Europe, Sweden, Denmark and the Netherlands programmes to reduce pesticide use by up to 50% have been implemented (Pettersson, 1997). The Meerjarenplan in the Netherlands – an area of highly intensive horticulture and agriculture – led to a reduction in pesticide use of some 50% in the years 1992–1997 (Figure 1.8). In some of these countries, pesticide taxes have been introduced. In Denmark, a tax of 3% was introduced in 1986, and by 1998 this had been raised to 33% for herbicides and fungicides, and 50% for insecticides. From 1986 to 1994 pesticide use, in terms of active ingredient tonnages, fell from 5678 tonnes to 2026 tonnes – a reduction of over 60%. Pesticide taxes have been proposed in the UK, but had not been implemented by 2005. Instead, the UK Crop Protection Association developed a so-called Voluntary Initiative to reduce the impact of pesticides. The agreement proposes to implement crop protection management plans designed to ensure that competent decisions are made with respect to pesticide use on individual farms.

In agriculture, and to some extent in other areas of use, in order to preserve the life of remaining pesticides these compounds should be deployed in a rational manner. The benefits offered by their selectivity may well be extended through coordinated use with complementary control measures. Traditional measures such as crop rotation and intercrop cultivation enable control of unwanted weeds through mechanical means in addition to deployment of selective herbicides within for example monocotyledonous and dicotyledonous crops. Sanitation and rotation procedures may also help to reduce the amount of disease and pests surviving between crops. Resistant cultivars, to plant pathogens or in transgenic form to herbicides or pests, may be used in conjunction with pesticides. In glasshouse and outdoor vegetable growing systems, combinations of biological control and pesticides may prove highly efficient in achieving control of pests and diseases whilst preserving the efficacy of compounds.

Of course, such strategies involving integrated pest management (IPM) have long been advocated. The approach has obvious benefits in reducing pesticide use and also assists in prolonging the effective life of pesticides by delaying the onset of resistance. Such policies offer much promise in less affluent parts of the world. For example, investment in Indonesia into IPM programmes for control of rice pests, notably the brown plant hopper, has led to a reduction in pesticide use of over 65%, whilst at the same time yields have increased by over 10%. Programmes of IPM include that launched by the FAO in 1999 for cotton in China, India, Pakistan, Bangladesh, Vietnam and the Philippines. Other programmes have been funded by the USA and the EU. Cotton growing in Asia and elsewhere has often involved multiple applications of pyrethroids and organophosphate insecticides, the latter posing considerable health risks to workers. Additionally pest resistance to both organophosphates and pyrethroids has arisen. Implementation of the FAO programme is directed through field training to farmers, who are educated in pest identification, economic threshold levels, identification of beneficial insects and, from these, selection of the best strategy to deal with pests (Sagenmuller and Hewson, 2000). A typical set of results from trials conducted to IPM principles compared to conventional practice is shown in Table 10.18, where it can be seen that fewer pesticide applications as well as higher yields were associated with the IPM programme.

Clear benefits to poorer countries in terms of reductions in pesticide use may also result from advances in plant breeding through transgene technology and its implementation. Genetic engineering of plant species through insertion of specific genes can achieve a very high degree of selectivity with respect to control of weeds, pests and diseases. One such example is the insertion of the Bt gene referred to in Chapter 4 (4.5) into crop plants. Here the crop plant may be fully resistant to insect attack if the gene is expressed in all tissues. In poorer countries, use of such insect-resistant plants would reduce the need to use neurotoxic pesticides for pest control.

Incorporation of genes conferring resistance to fungal diseases has formed part of strategies to control plant pathogens for over a century. Successes include the incorporation of genes giving control of potato wart disease (*Synchytrium endobioticum*), black stem rust (*Puccinia graminis f.sp. tritici*) in wheat and

Table 10.18 *Pesticide use, yields, profitability and risk to health in IPM crop management schemes compared with conventional production techniques in India*

| | Punjab | Tamil Nadu | Andhra Pradesh | Maharashtra |
|---|---|---|---|---|
| Reduction in pesticide use % (a.i./ha) | 29 | 42 | 69 | 92 |
| Reduction in plant protection cost % | 21 | 39 | 55 | 88 |
| Yield increase % | 49 | 17 | 31 | 70 |
| Net increase in profitability $/ha | 40 | 93 | 125 | 226 |
| Reduction in health hazard %* | 48 | 77 | 89 | 92 |

* Calculated on a human $LD_{50}$ dose reduction
Adapted from Russell *et al.* (2000)

powdery mildew (*Blumeria graminis*) in barley, the latter through the durable mlo gene. Although fungicides pose fewer problems in terms of acute toxicity than insecticides, the incorporation of resistance genes, either by crossing and selection or through transgene technology, would reduce dependence in poorer countries on imported, often expensive fungicides. The application of GM technology with respect to herbicides (2.10) still requires the use of these compounds; however, compounds used with herbicide-resistant crops, such as glyphosate and the sulphonylureas, are, in toxicological terms, some of the most benign of all pesticides.

Despite the apparent advantages of GM crops, there is resistance to the deployment of these, notably in the EU. During the 1990s and early twenty-first century, a strong lobby has existed in Europe and especially the UK against the deployment of crops produced by in vitro transgene technology. Indeed, even more so than pesticides, the negative image of GM crops has been almost consistently emphasised, and their deployment fanatically opposed. The principal concerns of the media, at least in the UK, are focused upon possible health effects, but other concerns are the potential movement of herbicide- and insect-resistance genes from crops, via cross-pollination, to natural vegetation, with possible consequences for the environment as a whole. Such concerns have not arisen to the same extent in some other parts of the world such as the USA, Canada, China and Argentina, where GM crops occupy large hectarages. In terms of human health, pesticide- and insect-resistant crops would appear to pose very low risks: crops and produce have been exhaustively tested for potential adverse long-term and short-term effects. However, hybridisation of GM crops with related plants in the environment is a current cause of concern. Self-sown oilseed rape has become a pernicious weed in the UK, and herbicide-resistant biotypes may exacerbate problems with this and other species.

Pesticides will continue to form an important part of strategies for control of weeds, pests and diseases in agricultural systems, public health and medicine for the foreseeable future. A principal aim of those who recommend and use pesticides may therefore be to preserve the life of currently available compounds that are acceptable within strict modern regulatory systems, in order to form part of a coordinated therapeutic strategy of pest, weed and disease management. The use of pesticides in for example pre-planned spray programmes for cereal crops is still common in the UK and other parts of the world. Such use without reference to the diagnosis of incidence and threshold levels for treatment of weeds, pests or diseases may come to be regarded as undesirable, in terms of unnecessary environmental inputs as well as increasing the risk of resistance. Decisions on pesticide deployment followed by their timely and accurate application at precise dose rates may become standard practice in the future. The selectivity associated with many compounds may be thus maintained, a low risk of environmental contamination ensured and the benefits of pesticides conserved for weed, pest and disease control in agriculture and public health.

# References


Albers, P. H. and Melancon, M. J. (2002). Effect of chlorfenapyr on adult birds. At http://www.pwrc.usgs.gov/resshow/Albers/albers1.htm.

Albert, A. (1985). *Selective Toxicity*, 7th edn. London: Chapman and Hall.

Ames, B. (1998). *Misconceptions about Environmental Pollution, Pesticides and the Causes of Cancer*. NCPA Policy Report 214. At http://www.ncpa.org/studies/s214/s214b.html.

Anon. (1990). *Pesticides in Drinking Water Wells*. United States Environmental Protection Agency Report 20T-1004. At http://www.pueblo.gsa.gov/cic_text/housing/water-well/waterwel.txt.

Anon. (1999). *Epidemiological Study of the Relationship Between Exposure to Organophosphate Pesticides and Indices of Chronic Peripheral Neuropathy and Neuropsychological Abnormalities in Sheep Farmers and Dippers*. Reports TM/99/02a; TM/99/02b; Tm/99/02c. Edinburgh: Institute of Occupational Medicine.

Anon. (2000a). *Design of a Tax or Charge Scheme for Pesticides*. Report by the Department of Environment, Food and Rural Affairs. At http://www.defra.gov.uk/environment/pesticidestax/19.htm.

Anon. (2000b). *Committee on Toxicology of Chemicals in Food Consumer Products and the Environment: Risk Assessment of Mixtures of Pesticides and Similar Substances*. At http://www.swan.ac.uk/cget/ejget/articleb.htm.

Anon. (2002a). *Annual Report of the Pesticides Residue Committee 2001*. At http://www.pesticides.gov.uk/prc.asp?id=751.

Anon. (2002b). *Annual Reports of the UK Wildlife Incident Investigation Scheme*. At http://www.pesticides.gov.uk/citize/wiis.htm.

Anon. (2002c). *Blue-green Algae (Cyanobacteria) in Inland Waters: Assessment and Control of Risks to Public Health*. Report by Scottish Executive Health Department. At http://www.scotland.gov.uk/library5/environment/bgac.pdf.

Anon. (2003a). *Transcription of the Third Open Meeting of the Advisory Committee on Pesticides*, 10 July 2002. At http://www.pesticides.gov.uk/committees/ACP/Open_ACP_2002/ open2002update.html.

Anon. (2003b). *Illnesses and Injuries in California Associated with Pesticide Residue in Agricultural Fields 1982–2001*. Report by California Department of Pesticide Regulation, Pesticide Illness and Surveillance Programme. At http://www.cdpr.ca.gov/docs/whs/pdf/hs1843.pdf.

Anon. (2004a). *Annual Report of the Pesticides Residue Committee 2003*. At http://www.pesticides.gov.uk/prc.asp?id=796.

Anon. (2004b). *Advisory Committee on Pesticides Statement on the Pesticides Literature Review by the Ontario College of Family Physicians*. At http://www.pesticides.gov.uk/acp.asp?id=1387.

Arlian, L. G. (2002). Arthropod allergens and human health. *Annual Review of Entomology*, **47**, 395–433.

Atkin, J. and Leisinger, K. M. (2000). *Safe and Effective Use of Crop Protection Products in Developing Countries*. Wallingford: CABI.

Atkinson, R., Kwok, E. S. C. and Arey, J. (1992). Photochemical processes affecting the fate of pesticides in the atmosphere. *Proceedings of the Brighton Crop Protection Conference, Pests and Diseases*, 469–476.

Avery, D. T. (1995). *Saving the Planet with Pesticides and Plastic*. Washington, DC: Hudson Institute.

Baldwin, B. C. and Corran, A. J. (1995). Inhibition of sterol biosynthesis: application to the agrochemical industry. In: *Antifungal Agents. Discovery and Mode of Action. New York: Bios Scientific*, pp. 59–68.

Barnett, E. A. and Fletcher, M. R. (1998). The poisoning of animals from the negligent use of pesticides. *Proceedings 1998 Brighton Crop Protection Conference, Pests and Diseases*, 279–284.

Barnett, E. A., Fletcher, M. R., Hunter, K. and Sharp, E. A. (2001). *Pesticide Poisoning of Animals 2001. Investigations of Suspected Incidents in the United Kingdom*. At http://www.pesticides.gov.uk/citizen/wiis 2001.pdf.

Barnett, E. A., Fletcher, M. R., Hunter, K. and Sharp E. A. (2002). *Pesticide Poisoning of Animals 2002: Investigations of Suspected Incidents in the United Kingdom*. Report by DEFRA, UK. At http://www.pesticides.gov.uk/citizen/wiis02.pdf.

Bartlett, D. W., Clough, J. M., Godwin, J., Hall, A. A., Hamer, M. and Parr-Dobrzanski, B. (2002). The strobilurin fungicides. *Pest Management Science*, **58**, 649–662.

Baylis, A. D. (2000). Why glyphosate is a global herbicide: strengths, weaknesses and prospects. *Pest Management Science*, **56**, 299–308.

Beaumont, P. (1993). *Pesticides, Policies and People. A Guide to the Issues*. London: The Pesticides Trust.

Beesley, W. N. (1994). Sheep dipping, with special reference to the UK. *Pesticide Outlook*, February 1994, 16–21.

Beier, R. (1990). Natural pesticides and bioactive components in foods. *Reviews of Environmental Contamination and Toxicology*, **113**, 48–136.

Beier, R. (1995). Role of metabolism and the toxic response of some naturally occurring chemicals in food. In: Hutson, D. H. and Paulson, G. D. (eds.) *The Mammalian Metabolism of Agrochemicals*. New York: John Wiley and Sons, pp. 309–332.

Beitz, H., Schmidt, H. and Herzel, F. (1994). Occurrence, toxicological and ecological significance of pesticides in groundwater and surface water. In: Borner, H. (ed.) *Pesticides in Ground and Surface Water*. Berlin: Springer-Verlag.

Bennett, R. M., Christiansen, K. and Clifton-Hadley, R. S. (2003). *Economics of Livestock Diseases*. University of Reading and Veterinary Laboratories Agency at http://www.rdg.ac.uk/livestockdiseas/index.htm.

Berg, D., Tietjen, K., Wollweber, D. and Hain, R. (1999). From genes to targets: impact of functional genomes on herbicide discovery. *Proceedings of the Brighton Crop Protection Conference, Weeds*, 491–500.

Berry, C. (1990). The hazards of healthy living – the agricultural component. *Proceedings of the Brighton Crop Protection Conference, Pests & Diseases*, 3–13.

Beulke, S., Brown, C. D. and Dubus, I. G. (1998). *Evaluation of the Use of Preferential Flow Models to Predict the Movement of Pesticides to Water Sources under UK Conditions*. MAFF project PLO516 available at http://www.pesticides.gov.uk/applicant/registration_guides/data_reqs_handbook/Supporting/PL0516%20-%20Evaluation%20of%20preferential%20flow%20models.pdf.

Bewick, D. (1994). The mobility of pesticides in soil. Studies to prevent groundwater contamination. In: Borner, H. (ed.) *Pesticides in Ground and Surface Water*. Berlin: Springer-Verlag.

Bloomqvist, J. R. (1996). Ion channels as targets for insecticides. *Annual Review of Entomology*, **41**, 163–190.

Blus, L. J. (1996). DDT, DDD and DDE in birds. In: Nelson Beyer, N., Heinz, G. H., Redmon-Norwood A. W. (eds.) *Environmental Contaminants in Wildlife – Interpreting Tissue Concentrations*. Boca Raton, FL: Lewis Publishers, pp. 49–71.

Boger, P. and Sandmann, G. (1998). Carotenoid biosynthesis inhibitor herbicides – mode of action and resistance mechanisms. *Pesticide Outlook*, **8** (6), 29–35.

Boger, P., Matthes, B. and Schmalfuss, J. (2000). Towards the primary target of chloracetamides-new findings pave the way. *Pest Management Science*, **56**, 497–508.

Bramley, P. M. and Pallett, K. E. (1993). Phytoene desaturase: a biochemical target of many bleaching herbicides. *Proceedings of the Brighton Crop Protection Conference, Weeds*, 713–722.

British Trust for Ornithology (2002) at http://www.bto.org/birdtrends/wcrpereg.htm.

Brown, V. K. (1980). *Acute Toxicity in Theory and Practice*. New York: John Wiley and Sons.

Brown, R. A., White, J. S. and Everett, C. J. (1988). How does an autumn applied pyrethroid affect the terrestrial arthropod community? In: *Field Methods for the Study of the Environmental Effects of Pesticides*. BCPC Monograph 40. Farnham: BCPC Publications, pp. 137–145.

Burges, H. D. (2001). *Bacillus thuringiensis* in pest control. *Pesticide Outlook*, June 2001, 90–97.

Butters, J. A., Kendall, S. K., Wheeler, I. E. and Holloman, D. W. (1995). Tubulins: lessons from existing products that can be applied to target new antifungals. In: *Antifungal Agents. Discovery and Mode of Action*. New York: Bios Scientific, pp. 131–141.

Carmichael, W. W., Azevedo, S. M. F. O., An, J. S., Molica, R. J. R., Jochimsen, E. M., Lau, S., Rinehart, K. L., Shaw, G. R. and Eaglesham, G. K. (2001). Human fatalities from cyanobacteria: chemical and biological evidence for cyanotoxins. *Environmental Health Perspectives*, **109**, 663–668.

Carson, R. (1962). *Silent Spring*. New York: Pelican.

Carter, A. D. (2000). Herbicide movement in soils: principles, pathways and processes. *Weed Research*, **40**, 113–122.

Casida, J. E. and Quistad, G. B. (1998). Golden age of insecticide research: past, present or future. *Annual Review of Entomology*, **43**, 1–16.

Chave, P. A. (1995). Pesticides in freshwaters from arable use. In: Best, G. A., Ruthven, A. D. (eds.) *Pesticides, Developments, Impacts and Controls*. London: Royal Society of Chemistry, pp. 100–111.

Chilton, P. J., Stuart, M. E., Gooddy, D. C., Williams, R. J. and Williamson, A. R. (1995). The occurrence of herbicides in and the modelling of their transport to the chalk aquifer beneath arable land in Southern England. In: *Pesticide Movement to Water*. BCPC Monograph 62. Alton: BCPC Publications, pp. 111–116.

Cilgi, T. and Frampton, G. K. (1994). Arthropod populations under current and reduced input pesticide regimes: results from the first four treatment years of the MAFF 'SCARAB' project. *Proceedings of the Brighton Crop Protection Conference, Pests and Diseases*, 653–660.

Cobb, A. H. (1991). *Herbicides and Plant Physiology*. London: Chapman and Hall.

Coggon, D., Pannett, B., Winter, P. D., Acheson, E. D. and Bonsall, J. (1986). Mortality of workers exposed to 2-methyl-4-chlorophenoxyacetic acid. *Scandinavian Journal of Work and Environmental Health*, **12(5)**, 448–454.

Colborn, T., Dumanoski, D. and Myers, J. P. (1996). *Our Stolen Future*. London: Penguin.

Cole, D. J., Pallett, K. E. and Rodgers, M. (2000). Discovering new modes of action for herbicides and the impact of genomics. *Pesticide Outlook*, **9**, 223–229.

Conway, G. R. and Pretty, J. N. (1991). *Unwelcome Harvest*. London: Earthscan at Pesticides Action Network website http://www.pan-uk.org/articles/pn47p12.htm.

Cooke, A. S., Greig-Smith, P. W. and Jones, S. A. (1992). Consequences for vertebrate wildlife of toxic residues in earthworm prey. In: Greig-Smith, P. W., Becker, H., Edwards, P. J. and Heimbach, F. (eds.) *Ecotoxicology of Earthworms*. Andover: Intercept, pp. 139–155.

Corbett, J. R., Wright, K. and Baillie, A. C. (1984). *The Biochemical Mode of Action of Pesticides*, 2nd edn. London: Academic Press, pp. 128–131.

Crossley, S. J. (1997). Recent international activities in the methodology for dietary risk assessment. Proceedings of the IBC UK Conference, Café Royal, London, 'Pesticide residues and dietary risk assessment'.

Davies, A. B., Joice, R., Banks, J. A. and Jones, R. L. (1995). A stewardship programme on diuron aimed at protection of water quality. In: *Pesticide Movement to Water*. BCPC Monograph 62. Alton: BCPC Publications, pp. 311–316.

Davies, J. (2001). Herbicide safeners – commercial products and tools for agrochemical research. *Pesticide Outlook*, **10**(2), 10–15.

Dayan, F. E. and Duke, S. O. (1996). Porphyrin-generating herbicides. *Pesticide Outlook*, **6(5)**, 22–27.

Dayan, F. E. and Duke, S. O. (1997). Overview of protoporphyrinogen oxidase inhibiting herbicides. *Proceedings of the Brighton Crop Protection Conference, Weeds*, 83–92.

De Maeyer, L., Peeters, D., Wijsmuller, J. M., Cantoni, A., Brueck, E. and Heibjes, S. (2002). Spirodiclofen: a broad-spectrum acaricide with insecticidal properties: efficacy on *Psylla pyri* and scales *Lepidosaphes ulmi* and *Quadraspidiotus perniciosus*. *Proceedings of the Brighton Crop Protection Conference, Pests and Diseases*, 65–72.

DEFRA (2002). *Quality Status Report of the Marine and Coastal Areas of the Irish Sea and Bristol Channel 2000.* At http://www.defra.gov.uk/environment/marine/quality/05.htm.

Dekeyser, M. A. and Downer, R. G. H. (1994). Biochemical and physiological targets for Miticides. *Pesticide Science*, **40**, 85–101.

Dennis, R. (1991). The rat – Part II. Rodenticides and rat control. *Pesticide Outlook*, **2**, 41–42.

Department of Health (1999). *Committee on Carcinogenicity: Breast Cancer Risk and Exposure to Organochlorine Insecticides: Consideration of the Epidemiology Data on Dieldrin, DDT and Certain Hexachlorocyclohexane Isomers.* At http://www.doh.gov.uk/cocbreast.htm.

Department of Health (2004). *Committee on Carcinogenicity: Statement on Prostate Cancer. CoC/04/56.* At http://www.advisorybodies.doh.gov.uk/coc/prostate.htm.

Desbrow, C., Routledge, E. J., Brighty, G. C., Sumpter, J. P. and Waldock, M. (1998). Identification of estrogenic chemicals in STW effluent. 1. Chemical fractionation and *in vitro* biological screening. *Environmental Science and Technology*, **32**, 1549–1558.

Dhadialla, T. S., Carlson, G. R. and Le, D. P. (1998). New insecticides with ecdysteroidal and juvenile hormone activity. *Annual Review of Entomology*, **43**, 545–569.

Dinham, B. (ed.) (1995). *The Pesticide Trail: The Impact of Trade Controls in Reducing Pesticide Hazards in Developing Countries.* London: The Pesticides Trust.

Dinham, B. (2003). Growing vegetables in developing countries for local urban populations and export markets: problems confronting small-scale producers. *Pest Management Science*, **59**, 575–582.

Dogheim, S. M., El-Marsafy, A. M., Salama, E. Y., Gadalla, S. A. and Nabil, Y. M. (2002). Monitoring of pesticides in Egyptian fruits and vegetables during 1997. *Food Additives and Contaminants*, **19**, 1015–1027.

Douthwaite, R. J. (1995). Occurrence and consequences of DDT residues in woodland birds following tsetse fly spraying in NW Zimbabwe. *Journal of Applied Ecology*, **32**, 727–738.

Dubus, I. G., Hollis, J. M., Brown, C. D., Lythgo, C. and Jarvis, J. (1998). Implications of a first-step environmental exposure assessment for the atmospheric deposition of pesticides in the UK. *Proceedings of the Brighton Crop Protection Conference*, pp. 273–278.

Ecobichon, D. J. (1991). Toxic Effects of Pesticides. In: Amdur, M. O., Doull, J. and Klaasen, C. D. (eds.) *Casarett and Doulls Toxicology – The Basic Science of Poisons*, 4th edn. Oxford: Pergamon.

Eddlestone, M. (2000). Patterns and problems of deliberate self-abuse in the developing world. *Quarterly Journal of Medicine*, **93**, 715–731.

Eddlestone, M., Rezvi-Sheriff, M. H. and Hawton, K. (1998). Deliberate self-harm in Sri Lanka: an overlooked tragedy in the developing world. *British Medical Journal*, **317**, 133–135.

EDMAR – Endocrine Disruption in the Marine Environment. (2003). DEFRA, Environment Agency, Sniffer, CEFIC available at http://www.defra.gov.uk/environment/chemicals/hormone/report.htm.

Edwards, C. A. (1973). *Environmental Pollution by Pesticides.* New York: Plenum Press.

EMAN (2003). European Mycotoxin Awareness Network Factsheets at http://www.lfra.co.uk/eman2/factsheet.asp.

Environment Agency (2000). *Monitoring of Pesticides in the Environment*. London: Environment Agency.

Environment Agency (2003). *The Fourth Otter Survey of England (2000–2002)*. At http://www.environment-agency.gov.uk/commondat/105385/otte_introduction.pdf.

European Commission (2002). *Guidance Document on Aquatic Toxicology*. At http://www.pesticides.gov.uk/applicant/registration_guides/data_reqs_handbook/supporting/3268_rev4_final.pdf.

Evans, A. D. (1997). Seed eaters, stubble fields and set aside. *Proceedings of the Brighton Crop Protection Conference, Weeds*, 907–914.

Evans, D. A. (1999). How can technology feed the world safely and sustainably. In: Brooks, G. T. and Roberts, T. R. (eds.) *Pesticide Chemistry and Bioscience: The Food–Environment Challenge*. London: Royal Society of Chemistry, pp. 3–24.

FAO. (1997). *FAO Manual on the Submission and Evaluation of Pesticides Residue Data for the Estimation of Maximum Residue Levels in Food and Feed.* Rome: Food and Agriculture Organisation of the United Nations. At http://www.fao.org/docrep/x5848e/X5848e00.htm#Contents.

FAO. (2001). *Food Security and Environment*. Document AD/I/Y1 303 E1/7.01/36000 for the World Food Summit.

FAO/WHO. (1999). *Understanding the Codex Alimentarius*. Report by Food and Agriculture Organisation of the United Nations/World Health Organisation. At http://www.fao.org/docrep/w9114e00/htm#TopOfPage.

Ferrer, A. and Cabral, J. (1989). Epidemics due to pesticide contamination of food. *Food Additives and Contaminants*, **6**, Supplement 1, 895–898.

Flynn, D. J. (1999). Pesticide registration in Europe – current status and future developments. *Proceedings of the Brighton Crop Protection Conference, Weeds*, 905–912.

FOCUS (2000). *FOCUS Groundwater Scenarios in the EU Plant Protection Product Review Process*. Report of the FOCUS groundwater Scenarios Workgroup. EC document Sanco/321/2000.

Goklany, I. M. and Trewavas, A. J. (2003). How technology can reduce our impact on earth. *Nature*, **423**, 115.

Gold, L. and Slone, H. S. (1999). Publications from the carcinogenic potency report. *Ranking Possible Toxic Hazards of Dietary Supplements Compared to Other Natural and Synthetic Substances*. At http://potency.berkeley.edu/text/fdatestimony.html.

Graham-Bryce, I. (1977). Crop protection: a consideration of the effectiveness and disadvantages of current methods and the scope for improvement. *Philosophical Transactions of the Royal Society, London*, **B281**, 163–179.

Gray, G. M. (2001). *Risk Characterisation and Public Perceptions of Risk*. At http://www.ag.ohio-state.edu/~plantdoc/pubpest/6feature.html.

Gregor, D. J. (1990). Deposition and accumulation of selected agricultural pesticides in Canadian arctic snow. In: Kurtz, D. A. (ed.) *Long Range Transport of Pesticides*. Boca Raton, FL: Lewis Publishers, pp. 373–386.

Greig-Smith, P. W., Frampton, G. and Hardy, T. (1992). *Pesticides, Cereal Farming and the Environment*. London: HMSO.

Grossman, K. (1998). Quinclorac belongs to a new class of highly selective auxin herbicides. *Weed Science*, **46**, 707–716.

Gullan, P. J. and Cranston, P. S. (2000). *The Insects. An Outline of Entomology*, 2nd edn. Oxford: Blackwell.

Harris, C. A. and Hill, A. R. C. (2004). Variability of residues in unprocessed food items and its impact on consumer risk assessment. In: Marrs, T. C. and Ballantyne, B. (eds.) *Pesticide Toxicology and International Regulation.* New York: Wiley, pp. 413–428.

Harris, G. L., Turnbull, A. B., Gilbert, A. J., Christian, D. G. and Mason, D. J. (1992). Pesticide application and deposition – their importance to pesticide leaching to surface water. *Proceedings of the Brighton Crop Protection Conference, Pests and Diseases*, 477–486.

Harris, J. (2000). *Chemical Pesticide Markets, Health Risks and Residues.* Wallingford: CABI Publishing.

Harwood, J. L. (1999). Graminicides which inhibit lipid biosynthesis. *Pesticide Outlook*, **10(4)**, 154–158.

Hatzios, K. K. (2000). Herbicide safeners and synergists. In: Roberts, T. (ed.) *Metabolism of Agrochemicals in Plants.* New York: Wiley, pp. 259–294.

Hawkes, T. R. (1993). Acetolactate synthase; the perfect herbicide target? *Proceedings of the 1993 Brighton Crop Protection Conference, Weeds*, 723–730.

Hay, J. V. (1999). Herbicide discovery in the 21st century – a look into the crystal ball. In: Brooks, G. T. and Roberts, T. (eds.) *Pesticide Chemistry and BioScience – The Food: Environment Challenge.* London: Royal Society of Chemistry, pp. 55–63.

Hayes, T. B., Collins, A., Lee, M., Mendoza, M., Noriega, N., Ali Stuart, A. and Vonk, V. (2002). Hermaphroditic demasculinised frogs after exposure to the herbicide atrazine at low ecologically relevant doses. *Proceedings of the National Academy of Sciences*, **99**, 5476–5480.

Heaney, S. P., Hall, A. A., Davies, S. A. and Olaya, G. (2000). Resistance to fungicides in the QoI-STAR cross-resistance group – current perspectives. *Proceedings of the Brighton Crop Protection Conference, Pests & Diseases*, 755–762.

Heap, I. M. (1999). International survey of herbicide-resistant weeds: lessons and limitations. *Proceedings of the Brighton Crop Protection Conference, Weeds*, 769–776.

Hedgecott, S. (1996). *Proposed Environmental Quality Standards for Atrazine and Simazine in Water.* Final report for UK DoE. SAC. WRc report No. DoE 2197.

Hess, F. D. (2000). Light-dependent herbicides – a review. *Weed Science*, **48**, 160–170.

Higginbotham, S., Jones, R. L., Gatzweiler, E. and Mason, P. J. (1991). Point-source contamination: quantification and practical solutions. *Proceedings of the Brighton Crop Protection Conference, Weeds*, 681–686.

Hignett, R. R. (1991). Dietary intakes of herbicides and plant growth regulators. *Proceedings of the Brighton Crop Protection Conference, Weeds*, pp. 1285–1294.

Hillier, D. C. and White, S. L (2001). Pesticide trends in raw and drinking water. *Proceedings of the Brighton Crop Protection Conference 2001, Pesticide Behaviour in Soils and Water*, 307–312.

Hoar, S. K., Blair, A., Holmes, F. F., Boyson, C. D., Robel, R. J. and Hoover, R. (1986). Agricultural herbicide use and risk of lymphoma and soft-tissue sarcoma. *Journal of the American Medical Association*, **256**, 1141–1147.

Huckle, K. R. and Millburn, P. (1990). Metabolism, bioconcentration and toxicity of pesticides in fish. In: Hutson, D. H. and Roberts, T. R. (eds.) *Environmental Fate of Pesticides*, Volume 7. *Progress in Pesticide Biochemistry and Toxicology*. New York: John Wiley, pp. 175–244.

Keith, J. O. (1996). Residue analyses: how they were used to assess the hazards of contaminants to wildlife. In: Nelson Beyer, N., Heinz, G. H., Redmon-Norwood, A. W. (eds.) *Environmental Contaminants in Wildlife – Interpreting Tissue Concentrations*. Boca Raton, FL: Lewis Publishers.

Kirby, C. (1980). *The Hormone Herbicides*. Alton: BCPC Publications.

Kogevinas, M., Becher, H., Benn, T., Bertazzi, P. A., Boffetta, P., Bueno-de-Mesquita, H. B., Coggon, D., Colin, D., Flesch-Janys, D., Fingerhut, M., Green, L., Kauppinen, T., Littorin, M., Lynge, E., Mathews, J. D., Neuberger, M., Pearce, N. and Saracci, R. (1997). Cancer mortality in workers exposed to phenoxy herbicides, chlorophenols and dioxins. An expanded and updated international cohort study. *American Journal of Epidemiology*, **145**, 1061–1075.

Kolpin, D. W. and Martin, J. D. (2003). *Pesticides in Ground Water from Agricultural Land-use Wells. 1991–2001. Summary Statistics; Preliminary Results for Cycle 1 of the National Water Quality Assessment Program (NAWQA) 1992–2001.* At http://ca.water.usgs.gov/pnsp/pestgw/Pest-GW_2001_table1_ag.html.

Konradsen, F., van der Hoek, W., Cole, D. C., Hutchinson, D., Daisley, H., Singh, S. and Eddleston, M. (2003). Reducing acute poisoning in developing countries – options for restricting the use of pesticides. *Toxicology*, **192**, 249–261.

Krieger, N., Wolff, M. S., Hiatt, R. A., Rivera, M., Vogelman, J. and Orentreich, N. (1994). Breast cancer and serum organochlorines: a prospective study among white, black and asian women. *Journal of the National Cancer Institute*, **86**, 589–599.

Labrada, R. and Fornasari, L. (eds.) (2001). *Global Report on the Validated Alternatives to Methyl Bromide*. Report by FAO. At http://www.unepie.org/ozoneaction/library/tech/globrepo.pdf.

Lahdetie, J. (1995). Occupational and exposure related studies in human sperm. *Journal of Occupational and Environmental Medicine*, **37**, 922–930.

Lampkin, N., Measures, M. and Padel, S. (2003). *Organic Farm Management Handbook*. Aberystwyth: University of Wales.

Lasota, J. A. and Dybas, R. A. (1991). Avermectins; a novel class of compounds: implications for use in arthropod pest control. *Annual Review of Entomology*, **36**, 91–117.

Lomberg, B. (2001). *The Skeptical Environmentalist*. Cambridge: Cambridge University Press.

Lounibos, L. P. (2002). Invasions by insect vectors of human disease. *Annual Review of Entomology*, **47**, 233–266.

Lutman, P. J., Boatman, N. D., Brown, V. K. and Marshall, E. J. P. (2003). Weeds: their impact and value in arable ecosystems. *Proceedings of the BCPC International Congress, Crop Science and Technology*, 219–226.

Lyr, H., Russell, P. E., Dehne, H. W. and Sisler, H. D. (eds.) (1998). *Modern Fungicides and Antifungal Compounds*: II. *Intercept*.

Mammal Society (2002). *National Survey of Road Deaths*. At http://www.abdn.ac.uk/mammal/road_deaths.htm.

Marrs, T. C. (2004). Toxicology of herbicides. In: Marrs, T. C. and Ballantyne, B. (eds.) *Pesticide Toxicology and International Regulation*. New York: Wiley, pp. 305–348.

Marshall, E. J. P. (2001). Biodiversity, herbicides and non-target plants. *Proceedings of the Brighton Crop Protection Conference, Weeds*, 855–862.

Martin, H. (1964). *The Scientific Principles of Crop Protection*. London: Edward Arnold.

Mattheiessen, P. (1998). Background evidence for environmental effects of endocrine disruptors. *Proceedings of the Brighton Crop Protection Conference, Pests and Diseases*, 207–216.

Matthews, G. A. (1992). *Pesticide Application Methods*. London: Longman.

Matthiessen, P. and Gibbs, P. (1998). A critical appraisal of the evidence for tributyltin-mediated endocrine disruption in molluscs. *Environmental Toxicology and Chemistry*, **17**, 37–43.

McEwan, F. L. and Stephenson, G. R. (1979). *The Use and Significance of Pesticides in the Environment*. New York: John Wiley.

Mellanby, K. (1989). DDT in perspective. In: Macfarlane, N. R. (ed.) *Progress and Prospects for Insect Control*. BCPC Monograph No. 43. Alton: BCPC Publications.

Mellanby, K. (1992). *The DDT Story*. Alton: BCPC Publications.

Millburn, P. (1995). The fate of xenobiotics in mammals: biochemical processes. In: Hutson, D. H. and Paulson, G. D. (eds.) *The Mammalian Metabolism of Agrochemicals*. New York: John Wiley.

Miller, P. C. H. (1999). Factors influencing the risk of drift into field boundaries. *Proceedings of the Brighton Crop Protection Conference, Weeds*, pp. 439–446.

Miller, P. C. H. (2003). The current and future role of application in improving pesticide use. *Proceedings of the BCPC International Congress, Crop Science and Technology*, pp. 247–254.

Minh, T. B., Kunisue, T., Yen, N. T. H., Watanabe, M., Tanabe, S., Hue, N. D. and Qui, V. (2002). Persistent organochlorine residues and their bioaccumulation profiles in resident and migratory birds from North Vietnam. *Environmental Toxicology and Chemistry*, **21**, 2108–2118.

Minting, P. (1998). *Keeping Tabs of the Blue-Green Menace*. At http://www.edie.net/library/features/WTJ9859.html.

Morris, A. J., Bradbury, R. B. and Evans, A. D. (2003). Sustainable arable farming for an improved environment: the effects of novel winter wheat sward management on skylarks *(Alauda arvensis). Proceedings of the BCPC International Congress, Crop Science and Technology*, 227–232.

Morris, T. (2002). *Evidence of Indirect Effects of Pesticides on Birds*. Presented to the Pesticides Forum UK. At http://www.pesticides.gov.uk/pesticidesforum/papers/pdf/pf129.pdf.

Morrison, H. I., Wilkins, K., Semenciw, R., Mao, Y. and Wigle, D. (1992). Herbicides and cancer. *Journal of the National Cancer Institute*, **84**, 1866–1874.

Myllymaki, A. (1996). The prospects and conditions for the safe use of second generation anticoagulants in microtine pest control. *Proceedings of the Brighton Crop Protection Conference, Pests and Diseases*, 151–156.

Naumann, K. (1989). Acetylcholinesterase inhibitors. In: Macfarlane, N. R. (ed.) *Progress and Prospects for Insect Control*. BCPC Monograph No. 43. Alton: BCPC Publications, pp. 21–42.

Nickell, L. G. (1994). Plant growth regulators in agriculture and horticulture. In: Hedin, P. A. (ed.) *Bioregulators for Crop Protection and Pest Control*. Washington DC: American Chemical Society.

Nigg, H. N. (1998). Occupational monitoring. In: Echobichon, D. J. (ed.) *Occupational Hazards of Pesticide Exposure*. London: Taylor & Francis, pp. 51–80.

Nix, J. (2002). *Farm Management Pocketbook*. Wye: Imperial College at Wye.

O'Connor, R. J. (1992). Indirect effects of pesticides on birds. *Proceedings of the Brighton Crop Protection Conference, Pests and Diseases*, 1097–1104.

Oomen, P. A. (1998). Risk assessment and risk management of pesticide effects on non-target arthropods in Europe. *Proceedings of the Brighton Crop Protection Conference, Pests and Diseases*, 591–598.

Ouellet, M. (2000). Amphibian deformities: current state of knowledge. In: *Ecotoxicology of Amphibians and Reptiles*. Sparling, D. W., Linder, G. and Bishop, C. A. (eds.) Brussels: Society of Environmental Toxicology and Chemistry (SETAC), pp. 617–661.

Parsons, E. C. M. (2002). *The Impacts of Pollution on Humpback Dolphins*. Presented to the 54th meeting of the International Whaling Commission. At http://www.whales.gn.apc.org/downloads/IWC%20papers/SC54-SM5.pdf.

Peakall, D. B. (1996). Dieldrin and other cyclodiene pesticides in wildlife. In: Nelson Beyer, N., Heinz, G. H. and Redmon-Norwood, A. W. (eds.) *Environmental Contaminants in Wildlife – Interpreting Tissue Concentrations*. Boca Raton, FL: Lewis Publishers, pp. 73–97.

Perrin, R. M. (1995). Synthetic pyrethroids success story. In: Best, G. and Ruthven, D. R. (eds.) *Pesticides – Developments, Impacts, and Controls*. London: Royal Society of Chemistry, pp. 19–27.

Pesticides Safety Directorate (2003a). *The Plant Protection Products Directive (91/414/EEC)*. At http://www.pesticides.gov.ukec_process/EC_overview_general/91414background.htm.

Pesticides Safety Directorate (2003b). *Data Requirements Handbook*. At http://www.pesticides.gov.uk/applicant/registration_guides/data_reqs_handbook/contents.htm.

Pettersson, O. (1997). Pesticide use in Swedish Agriculture: the case of 75% reduction. In: Pimental, D. (ed.) *Techniques for Reducing Pesticide Use*. New York: John Wiley and Sons, pp. 79–102.

Petty, R. E. (2001). *Understanding Public Attitudes Towards Pesticides*. At http://www.ag.ohio-state.edu/~plantdoc/pubpest/6feature.html.

Phillips, M. (2002). *Suicide Blight Hits China's Women*. At http://news.bbc.co.uk/2/hi/asia-pacific/2526079.stm.

Phillips, M. and McDougall, J. (2003). Agrochemical product introduction and reregistration: the challenge to the generic industry. *Agrolook*, **4**, 23–28.

Portelli, M. J. and Bishop, C. A. (2000). Ecotoxicology of organic contaminants in reptiles: a review of the concentrations and effects of organic contaminants in reptiles. In: Sparling, D. W., Linder, G. and Bishop, C. A. (eds.) *Ecotoxicology of Amphibians and Reptiles*. Brussels: Society of Environmental Toxicology and Chemistry (SETAC).

Potts, G. R. (1986). *The Partridge: Pesticides, Predation and Conservation*. London: Collins.

Potts, G. R. (2003). Balancing biodiversity and agriculture. *Proceedings of the BCPC International Congress, Crop Science and Technology*, 35–44.

Prisbylla, M. P., Onisko, B. C., Shribbs, J. M., Adams, D. O., Liu, Y., Ellis, M. K., Hawkes, T. R. and Mutter, L. C. (1993). The novel mechanisms of action of the herbicidal triketones. *Proceedings of the Brighton Crop Protection Conference, Weeds*, pp. 731–738.

Rademacher, W. (2000). Growth retardants: effects on gibberellin biosynthesis and other metabolic pathways. *Annual Review of Plant Physiology*, **51**, 501–531.

Ratcliffe, D. A. (1970). Changes attributable to pesticides in egg breakage frequency and eggshell thickness in some British birds. *Journal of Applied Ecology*, **7**, 67–115.

Ratcliffe, D. A. (1993). *The Peregrine Falcon*, 2nd edn. New York: Poyser.

Reynolds, S. and Hill, A. (2002). Cocktail effects- stirred not shaken – yet. *Pesticide Outlook*, October 2000, 209–213.

Richards, R. P. and Baker, D. B. (1990). Estimates of human exposure to pesticides through drinking water: a preliminary risk assessment. In: Kurtz, D. A. (ed.) *Long Range Transport of Pesticides*. Boca Raton, FL: Lewis Publications, pp. 387–403.

Riley, D. (1990). Current testing in the sequence of development of a pesticide. In: Somerville, L. and Walker, C. H. (eds.) *Pesticide Effects on Terrestrial Wildlife*. London: Taylor and Francis, pp. 11–24.

Riley, D. (1991). Using soil residue data to assess the environmental safety of pesticides. *BCPC Monograph 47, Pesticides in Soil and Water*. Alton, Hampshire: BCPC Publications.

Riley, D. and Eagle, D. (1990). Herbicides in soil and water. *Weed Control Handbook*, 8th edn. Alton: BCPC Publications, pp. 243–259.

Roberts, D. R., Manguin, S. and Mouchet, J. (2000). DDT house spraying and re-emerging malaria. *Lancet*, **356**, 330–332.

Roberts, T. R. (1998). *Metabolic Pathways of Agrochemicals*. Part I. *Herbicides and Plant Growth Regulators*. London: Royal Society of Chemistry.

Roberts, T. R. (1999). *Metabolic Pathways of Agrochemicals*. Part II. *Insecticides and Fungicides*. London: Royal Society of Chemistry.

Royal Commission on Environmental Pollution (2005). *Crop Spraying and the Health of Residents and Bystanders*. At http://www.rcep.org.uk/cropspraying.htm.

Rudd, D. L. (1964). *Pesticides and the Living Landscape*. London: Faber and Faber.

Russell, D. A., Kranthi, K. R., Surulivelu, T., Jadhav, D. R., Regupathy, A. and Singh, J. (2000). Developing and implementing insecticide resistance management practices in cotton ICM programmes in India. *Proceedings of the Brighton Crop Protection Conference, Pests & Diseases*, 205–212.

Russell, P. (2003). Taking the path of least resistance. *Pesticide Outlook*, **14**, 57–61.

Saari, L. L. (1999). A prognosis for discovering new herbicide sites of action. In: Brooks, G. T. and Roberts, T. R. (eds.). *Pesticide Chemistry and Bioscience – the Food – Environment Challenge*. London: Royal Society of Chemistry, pp. 207–220.

Safe, H. (1995). Environmental and dietary oestrogens and human health: is there a problem? *Environmental Health Perspectives*, **103**, 346–351.

Sagenmuller, A. and Hewson, R. (2000). Global implementation of ICM in cotton. *Proceedings of the Brighton Crop Protection Conference, Pests & Diseases*, pp. 193–198.

Saito, S., Isayama, S., Sakamoto, N., Umeda, K. and Kasamatsu, K. (2002). Pyridalyl: a novel insecticidal agent for controlling lepidopterous pests. *Proceedings of the Brighton Crop Protection Conference, Pests & Diseases*, pp. 33–38.

Salgado, V. L. (1999). Resistant target sites and insecticide discovery. In: Brooks, G. E. and Roberts, T. R. (eds.) *Pesticide Chemistry and Bioscience: The Food and Environment Challenge*. London: Royal Society of Chemistry, pp. 236–246.

Sanborn, M., Cole, D., Kerr, K., Vahil, C., Sanin, L. H. and Bassil, K. (2004). *Systematic Review of Pesticides. Human Health Effects. Ontario College of Family Physicians*. At http://www.ocfp.on.ca/English/OCFP/Communications/CurrentIssues/Pesticides/default.asp?s=1.

Scholtz, K. and Spiteller, M. (1992). Influence of groundcover on the degradation of $^{14}$C-imidacloprid in soil. *Proceedings of the Brighton Crop Protection Conference, Pests and Diseases*, 883–888.

Shaner, D. L. (2003). Herbicide safety relative to common targets in plants and mammals. *Pest Management Science*, **60**, 17–24.

Sheail, T. (1985). *Pesticides and Nature Conservation: The British Experience 1950–1975*. Monographs in Science, Technology and Society, 4. Oxford Science Publications.

Shepherd, A. J. and Heather, A. I. J. (1999). Factors affecting the loss of six herbicides from hard surfaces. *Proceedings of the Brighton Crop Protection Conference, Weeds*, pp. 669–674.

Shore, R. F., Boyd, I. L., Leach, D. V., Stebbings, R. E. and Myhill, D. G. (1990). Organochlorine residues in roof timbers and possible implications for bats. *Environmental Pollution*, **64**, 179–188.

Shore, R. F., Malcolm, H. M., Weinburg, C. L., Turk, A., Horne, J. A., Dale, L., Wyllie, I. and Newton, I. (2002). *Wildlife and Pollution: 1999/2000 Annual Report*. JNCC Report No. 321.

Sieber, J. N. (1988). Principles governing environmental mobility and fate. In: Ragsdale, N. N., Kuhr, R. J. (eds.) *Pesticides: Minimising the Risks*. Washington, D.C.: American Chemical Society.

Smith, D. (1999). Worldwide trends in DDT levels in human breast milk. *International Journal of Epidemiology*, **28**, 179–188.

Sotherton, N. W. (1988). The cereals and gamebirds research project: overcoming the indirect effects of pesticides. In: Harding, D. L. J. (ed.) *Britain Since Silent Spring*. London: Institute of Biology.

Sparks, T. C., Crouse, G. D. and Durst, G. (2001). Natural products as insecticides: the biology, biochemistry and quantitative structure–activity relationships of spinosyns and spinosoids. *Pest Management Science*, **57**, 896–905.

Sparling, D. (2000). Ecotoxicology of organic contaminants to amphibians. In: Sparling, D. W., Linder, G. and Bishop, C. A. (eds.) *Ecotoxicology of Amphibians and Reptiles*. Brussels: Society of Environmental Toxicology and Chemistry (SETAC).

Spencer, W. F. and Cliath, M. M. (1990). Movement of pesticides from soil to the atmosphere. In: Kurtz, D. A. (ed.) *Long Range Transport of Pesticides*. Boca Raton, FL: Lewis Publications, pp. 1–16.

Stanley, P. I. and Bunyan, P. J. (1979). Hazards to wintering geese and other wildlife from the use of dieldrin, chlorvenfinphos and carbophenothion seed treatments. *Proceedings of the Royal Society, London B*, **205**, 31–45.

Stephenson, G. R. and Ritcey, G. M. (1998). Dislodgeable foliar residues of pesticides in agricultural, landscape and greenhouse environments. In: Echobichon, D. J. (ed.) *Occupational Hazards of Pesticide Exposure*. London: Taylor & Francis, pp. 51–80.

Stephenson, R. R. (1988). Testing insecticides for use in rice/fish cultivation. In: Greaves, M. P., Smith, B. D. and Greig-Smith, P. W. (eds.) *Environmental Effects of Pesticides*. BCPC Monograph 40. BCPC Publications.

Strong, F. M. (1973). *Toxicants Occurring Naturally in Foodstuffs*, 2nd edn. Washington, DC: National Academy of Sciences.

Tarrant, K. A., Field, S. A., Jones, A., McCoy, C., Langton, S. D. and Hart, A. D. M. (1994). Effects on earthworm populations of reducing pesticide use: part of the SCARAB project. *Proceedings of the Brighton Crop Protection Conference, Pests and Diseases*, 1289–1294.

Tatsukawa, R., Yamaguchi,Y., Kawano, M., Kannan, N. and Tanabe, S. (1990). Global monitoring of organochlorine insecticides – an 11-year case study (175–1985) of HCHs and DDTs in the open ocean atmosphere and hydrosphere. In: Kurtz, D. A. (ed.) *Long Range Transport of Pesticides*. Boca Raton, FL: Lewis Publications, pp. 127–141.

Taylor, W. A., Cooper, S. E. and Miller, P. C. H. (1999). An appraisal of nozzles and sprayers abilities to meet regulatory demands for reduced airborne drift and downwind fallout from arable crop spraying. *Proceedings of the Brighton Crop Protection Conference, Weeds*, pp. 447–452.

Thompson, C. M. and Richardson, R. J. (2004). Anticholineesterase insecticides. In: Marrs, T. C. and Ballantyne B. (eds.) *Pesticide Toxicology and International Regulation*. New York: Wiley, pp. 305–348.

Timchalk, C., Dryzga, M. D., Langvardt, P. W., Kastl, P. E. and Osborne, D. W. (1990). Determination of the effect of tridiphane on the pharmacokinetics of $[^{14}C]$atrazine following oral administration to male Fischer 344 rats. *Toxicology*, **61**, 27–40.

Timm, R. (1994). *Prevention and Control of Wildlife Damage. Cooperative Extension Division*. Institute of Agriculture and Natural resources, University of Nebraska, Lincoln USA. At http://wildlifedamage.unl.edu/handbook/handbook/allPDF/active.pdf.

Tomizawa, M. (2000). Neonicotinoid insecticide receptors. *Pesticide Outlook*, December 2000, 238–240.

UK Department of the Environment, Transport and the Regions (1999). *Design of a Tax or Charge Scheme for Pesticides*. Part A: *Rationale for an Economic Instrument Applied to Pesticides*. At http://www.environment.detr.gov.uk.

UK Department of the Environment, Transport and the Regions (2001). *Hormone (Endocrine) Disrupting Substances in the Environment*. At http://www.environemtn.detr.gov.uk/hormone/.

UK Drinking Water Inspectorate Information (2001). *Committee on Carcinogenicity of Chemicals in Food, Consumer Products and the Environment. Statement on Chlorinated Drinking Water and Cancer*. At http://www.defra.gov.uk/dwi/regs/infolett/1999/info1299.htm.

UK Marine SAC (Special Areas of Conservation) (2003). *Use and Impacts of TBT and Copper Anti-fouling Paints*. At http//www.ukmarinesac.org.uk/search/tbt.htm.

United States Environmental Protection Agency (1994). *An SAB Report: Assessment of Potential 2,4D Carcinogenicity: Review of the Epidemiological and Other Data on Potential Carcinogenicity of 2,4D by the SAB/SAP Joint Committee*. Washington DC, USA.

United States Environmental Protection Agency (2003). *Reducing Pesticide Risk*. At http://www.epa.gov/pesticides/health/reducing.htm.

Usha, S. (2000). Aerial spraying harms plantation workers in Kerala, India. *Pesticides News*, **47**, 6. At http://www.pan-uk.org/pestnews/pn47/pn47.p6.htm.

Vais, H., Williamson, M. S., Devonshire, A. L. and Usherwood, P. N. R. (2001). The molecular interactions of pyrethroid insecticides with insect and mammalian sodium channels. *Pest Management Science*, **57**, 877–888.

Van Maele-Fabry, G. and Willems J. L. (2003). Occupation related pesticide exposure and cancer of the prostate: a meta-analysis. *Occupational and Environmental Medicine*, **60**, 634–642.

Weber, J. B. (1994). Properties and behavior of pesticides in soil. In: Honeycutt, R. C. and Schabaker, D. J. (eds.) *Mechanisms of Pesticide Movement into Ground Water*. Boca Raton, FL: CRC Press, pp. 15–42.

Wells, J. W. (1994). The Food Safety Paradox – Perception vs Reality. *Proceedings of the Brighton Crop Protection Conference. Pests & Diseases*, 85–92.

White, S. L. and Pinkstone, D. C. (1995). The occurrence of pesticides in drinking water. In: Pesticide Movement to Water. BCPC Monograph 62, pp. 263–268. Farnham, UK: BCPC Publications.

White, S. L., Hillier, D. C., Evans, J. C., Hewson, R. T. and Higginbotham, S. (1997). A stewardship programme for isoproturon and water quality – a tale of two industries. *Proceedings of the Brighton Crop Protection Conference, Weeds*, pp. 1107–1106.

WHO (1995) *International Programme on Chemical Safety. Environmental Health Criteria: Anticoagulant Rodenticides*. At http://www.inchem.org/documents/ehc/ehc/ehc175.htm#SectionNumber: 1.3.

Wilkinson, C. F. and Barolo, D. M. (1999). The Food Quality Protection Act of 1996. *Proceedings of the Brighton Crop Protection Conference, Weeds*, 899–904.

Williams, R. J., Brook, D. N., Glendinning, P. J., Mattheisen, P., Mills, M. J. and Turnbull, A. (1991). Movement and modelling of pesticide residues at Rosemaund farm. *Proceedings of the Brighton Crop Protection Conference, Weeds*, 507–514.

Wong, C. K. C., Leung, K. M., Poon, B. H. T., Lan, C. Y. and Wong, M. H. (2002). Organochlorine hydrocarbons in human breast milk collected in Hong Kong and Guangzhov. *Archives of Environmental Contamination and Toxicology*, **43**, 364–372.

Wright, K. J., Seavers, G. P. and Wilson, B. J. (1997). Competitive effects of multiple weed species on wheat biomass and wheat yield. *Proceedings of the Brighton Crop Protection Conference, Weeds*, 497–502.

Zavatti, L. M. S. and Abakerli, R. (1996). Determination of pesticide residues in tomato fruits with a multiresidue method. In: *Uses and Environmental Safety in Latin America*. IUPAC/GARP Workshop on Pesticides. Sao Paulo, 1996, p. 78.

Zlotkin, E. (1999). The insect voltage-gated sodium channel as target of insecticides. *Annual Review of Entomology*, **44**, 429–455.

# Index